Analysis and Design of Bridges

NATO ASI Series

Advanced Science Institutes Series

A Series presenting the results of activities sponsored by the NATO Science Committee, which aims at the dissemination of advanced scientific and technological knowledge, with a view to strengthening links between scientific communities

The Series is published by an international board of publishers in conjunction with the NATO Scientific Affairs Division

A	Life Sciences	Plenum Publishing Corporation
B	Physics	London and New York
C	Mathematical and Physical Sciences	D Reidel Publishing Company Dordrecht and Boston
D	Behavioural and Social Sciences	Martinus Nijhoff Publishers The Hague/Boston/Lancaster
E	Applied Sciences	
F	Computer and Systems Sciences	Springer-Verlag Berlin/Heidelberg/New York
G	Ecological Sciences	

Series E: Applied Sciences – No. 74

Analysis and Design of Bridges

edited by

Çetin Yılmaz

Associate Professor
Civil Engineering Department
Middle East Technical University
Ankara, Turkey

S. Tanvir Wasti

Professor
Civil Engineering Department
Middle East Technical University
Ankara, Turkey

1984 **Martinus Nijhoff Publishers**
The Hague / Boston / Lancaster
Published in cooperation with NATO Scientific Affairs Division

Proceedings of the NATO Advanced Study Institute on Analysis and Design of Bridges, Çeşme, Izmir, Turkey, June 28-July 9, 1982

Library of Congress Cataloging in Publication Data

Main entry under title:

Analysis and design of bridges.

(NATO advanced science institutes series. Series E, Applied sciences ; 74)
Proceedings of the NATO Advanced Study Institute on Analysis and Design of Bridges held at Cesme and Izmir, Turkey, June 28-July 9, 1982.
"Published in cooperation with NATO Scientific Affairs Division."
1. Bridges--Design--Congresses. I. Yılmaz, Çetin. II. Wasti, Syed Tanvir, 1941- . III. NATO Advanced Study Institute on Analysis and Design of Bridges (1982 : Çeşme and İzmir, Turkey) IV. North Atlantic Treaty Organization. Scientific Affairs Division. V. Series.
TG300.A58 1984 624'.25 83-25461

ISBN-13: 978-94-009-6124-1 e-ISBN-13: 978-94-009-6122-7
DOI: 10.1007/978-94-009-6122-7

Distributors for the United States and Canada: Kluwer Boston, Inc., 190 Old Derby Street, Hingham, MA 02043, USA

Distributors for all other countries: Kluwer Academic Publishers Group, Distribution Center, P.O. Box 322, 3300 AH Dordrecht, The Netherlands

PREFACE

The Proceedings of the NATO Advanced Study Institute on Analysis and Design of Bridges held at Çeşme, İzmir, Turkey from 28 June 1982 to 9 July 1982 are contained in the present volume. The Advanced Study Institute was attended by 37 lecturers and participants from 10 different countries. The Organizing Committee consisted of Professors P. Gülkan, A. C. Scordelis, S.T. Wasti and Ç. Yılmaz.

The guidelines set by NATO for the Advanced Study Institute require it to serve not only as an efficient forum for the dissemination of available advanced knowledge to a selected group of qualified people but also as a platform for the exploration of future research possibilities in the scientific or engineering areas concerned. The main topics covered by the present Advanced Study Institute were the mathematical modelling of bridges for better analysis and the scientific assessment of bridge behaviour for the introduction of improved design procedures. It has been our observation that as a result of the range and depth of the lectures presented and the many informal discussions that took place, ideas became fissile, the stimulus never flagged and many gaps in the engineering knowledge of the participants were "bridged". Here we particularly wish to mention that valuable informal presentations of research work were made during the course of the Institute by Drs. Friedrich, Karaesmen, Lamas and Parker. Contributing tangibly to the success of the Advanced Study Institute were the several opportunities for social get-togethers in which the ladies also took part.

The editing of the proceedings has entailed a moderate amount of work in the case of a few manuscripts and only very slight interference with some others. In general the proceedings faithfully represent the lectures as transmuted from the spoken to the written word by the lecturers themselves. Unfortunately Professor B.T. Yen's contribution was not received in time for publication.

Inspite of the inevitable overlapping, it has been found suitable and convenient to group the lectures under the following five main headings :

I. Design Considerations
II. Analysis Methods for Steel Bridges
III. Concrete Slab and Box Girder Bridges
IV. Dynamics and Vibration
V. Earthquake and Wind Effects

The lectures in the present volume cover an impressive range of subjects and it is our hope that they will of use to a large number of engineers and researchers in civil, structural and bridge engineering.

While the manuscripts were being assembled for the publishers, we learnt with shock and dismay of the tragic death (in December 1982) of Professor Conrad P. Heins, Jr. in an air crash in China. Conrad "Connie" Heins was Professor at the University of Maryland, a well known researcher in Bridge Engineering and a dynamic personality. Characteristically, he was one of the first lecturers to send us his manuscripts. R.I.P.

We wish to take this opportunity to thank the many individuals and organizations whose assistance contributed to the success of the Advanced Study Institute.

1. The NATO Scientific Affairs Division, Brussels, Belgium for substantially funding the Institute.

2. Middle East Technical University, Ankara, Turkey for providing facilities and services.

3. AFA Prefabrike Beton Sanayi ve Ticaret A.Ş., İzmir, Turkey for generously hosting social and technical activities.

4. YÜKSEL PROJE, A.Ş., Ankara, Turkey for hosting social events and funding services.

5. SETAN Engineering Consultants, Ankara, Turkey for funding services.

6. BMB (United Engineering Consulting Group Co. Ltd.) Ankara, Turkey for funding services.

Last, but certainly not least, the book in your hands has been produced with care and skill by Martinus Nijhoff Publishers, The Hague. The Netherlands. We would like to thank them for their patience and understanding and, especially, for the prompt cooperation we received from Mrs. Henny Hoogervorst.

Ankara
April 1983

ÇETİN YILMAZ
S. TANVIR WASTI
Editors

TABLE OF CONTENTS

IV. DYNAMICS AND VIBRATION

I. DESIGN CONSIDERATIONS

PRESTRESSING OF CURVED BRIDGES

Christian Menn

Professor fur Baustatik und Konstruktion
Eidgenossische Technische Hochschule
Zurich

SUMMARY

A review of some structural concepts, prestressing schemes, analysis methods, and fundamental behavior of curved prestressed concrete bridges is presented.

The application of several structural schemes to curved bridges is presented. Equilibrium requirements, the effect of moment redistribution on equilibrium requirements, and the effects of external constraints are reviewed.

An approximate method for the analysis of curved girder bridges is developed. A comparison with an elasticity solution demonstrates the applicability of the method.

Based on the structural requirements and approximate analysis method presented, optimal prestressing requirements for the equalization of torsion in simple- and continuous-span curved prestressed box girder bridges are developed.

1. INTRODUCTION

Twenty years ago both the design and construction of curved bridges posed considerable problems. In comparison with straight bridges they were also much more expensive to erect. Today, however, the difference in cost between straight and curved bridges is much less significant. The alignment of highways is no longer determined by a desire for straight bridges. Thus in modern highway systems most bridges are at least partially curved.

1.1 Structural Systems for Curved Bridges

The customary structural systems for straight bridges are generally applicable for curved bridges, as long as the span and the curvature of the latter do not exceed certain values.

In the case of ordinary beam bridges, the girders are entirely adapted to the curvature of the roadway. The load capacity is thereby not significantly reduced.

In the case of arch bridges, the inner edge of the arch in plan is adapted to the curvature of the roadway and the outer edge of the arch is symmetric with respect to the straight axis of the arch (Fig. 1). The column faces on the inner side are vertical, while on the outer side they are oblique. This results in an aesthetically pleasant form. The span of the arch and the curvature of the roadway are, however, limited.

A cable stayed system can also be utilized in a curved bridge. If the pylons have the form of a V (Fig. 2), the stay cables are outside of the bridge deck throughout the bridge and do not encroach on the road space. The span and curvature of the roadway are, however, limited with the use of this scheme.

1.2 Load Carrying Behavior of Curved Girders

The load carrying behavior of a curved girder is characterized by the fact that torsion moments are required for equilibrium. This so called equilibrium torsion clearly differs from the constraint torsion which very often occurs.

In the case of equilibrium torsion, the torsion moments are in principle independent of the torsional stiffness of the girder (Fig. 3a). A large torsional stiffness yields small deformations; a vanishing torsional stiffness yields a mechanism.

In the case of constraint torsion, the torsion moments are directly dependent on the torsional stiffness (Fig. 3b). A large torsional stiffness results in large torsion moments. A vanishing torsional stiffness results in vanishing torsional moments; the

FIG. 1: CURVED ARCH BRIDGE

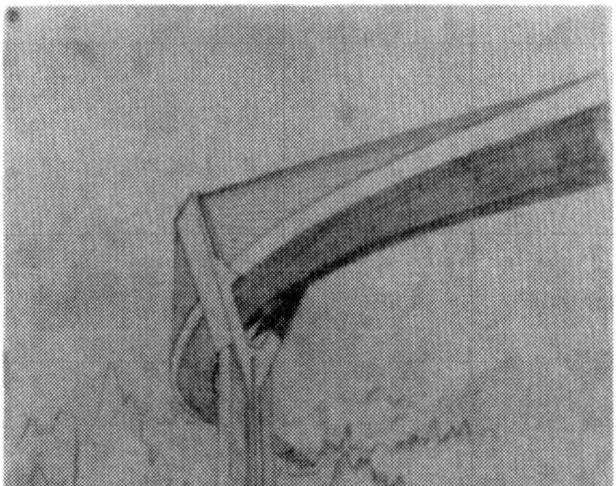

FIG. 2: CURVED CABLE STAYED SYSTEM

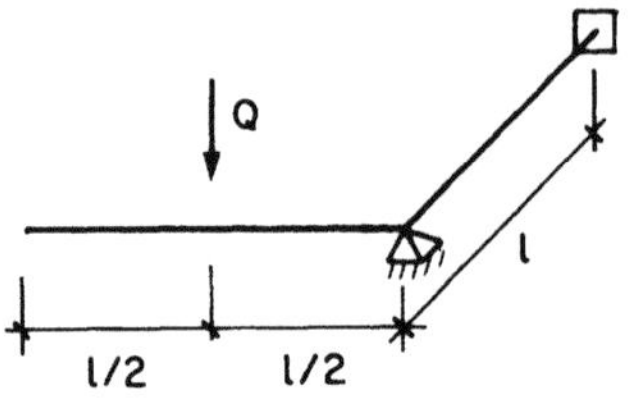

$$T = Q \cdot \frac{l}{2}$$

$$GK \rightarrow 0 \longrightarrow \text{collapse!}$$

Fig 3a

FIG. 3A: EQUILIBRIUM TORSION

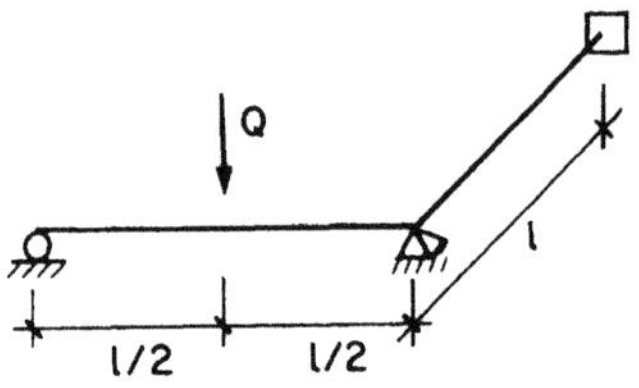

$$T = Q \frac{3 \cdot GK \cdot l}{16 \cdot (3 \cdot EI + GK)}$$

$$GK \rightarrow 0 \longrightarrow T \rightarrow 0$$

Fig 3b

FIG. 3B: CONSTRAINT TORSION

equilibrium of the system, however, remains.

Torsion moments and bending moments in a curved system are interrelated through equilibrium conditions. Therefore, a redistribution of the bending moments in a statically indeterminate system results in a corresponding redistribution of the torsion moments (Fig. 4). This effect must be considered in the case of moment redistribution resulting from a change of the statical system (e.g. cantilever construction bridges) or when checking the ultimate load capacity of a system in accordance with the theory of plasticity.

In most cases, the main beam section forces (N, V, M, T) in a curved bridge may be calculated from the basic requirements of static equilibrium. In T- and box- cross sections, however, transverse internal stresses which are dependent on the beam section forces are required in order to satisfy static equilibrium requirements. The investigation of these additional stresses is better based on a simple analysis than a finite element analysis, since the load carrying behavior can be clearly detected and thereby the reinforcement can be efficiently patterned.

In curved girders, prestressing can be used to modify not only the bending moments in the concrete section but also the torsion moments and the transverse stresses.

The cross section of a curved bridge is usually designed as a hollow box. In principle, T cross sections are also applicable. They are, however, weaker in torsion and exhibit larger transverse stresses. The subsequent deformations are therefore larger for T cross sections than they are for box cross sections.

1.3 Bearings and Constraints

Curved bridges should always be supported by torsion resistant bearing arrangements at the abutments. Point bearings are often used on the columns. Single columns with point bearings are suitable especially for overpasses, since they provide optimal transparency and roadway space. The rigidity of the system is greater, however, when torsion resistant bearings are used at the intermediate supports as well as at both the abutments (Fig. 5).

The displacements at the free end of a curved cantilever girder, due to temperature changes and shrinkage on one hand and due to prestressing and creep on the other, do not take the same direction (Fig. 6). Bearings and expansion joints are generally designed to allow free displacements only along the tangent to the bridge axis. Transverse restraint from the bearings results in redundant forces due to temperature changes and shrinkage. These forces are usually not significant. Bridges with a great degree of

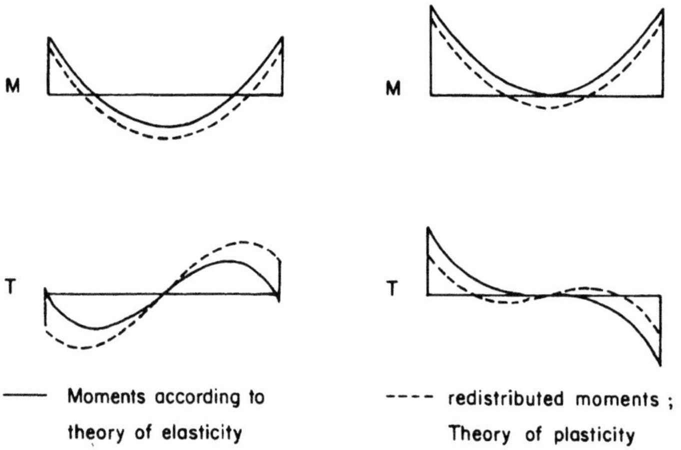

Fig 4 Continous girder Cantilever construction

FIG. 4: MOMENT REDISTRIBUTION

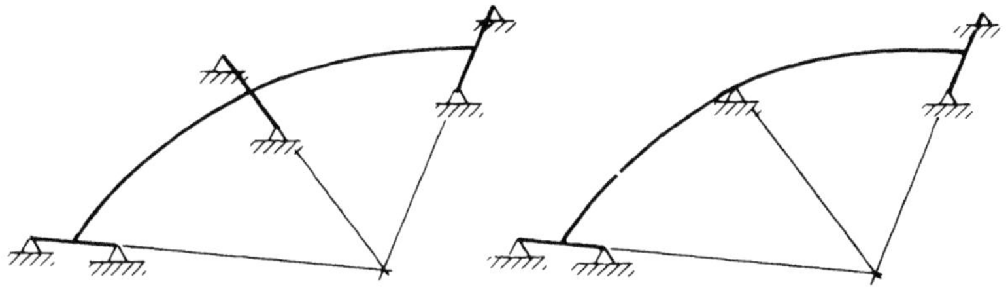

a) Torsion resistant bearings at all supports

b) Torsion resistant bearings at abutments; Point bearing at columne

Fig 5

FIG 5: ARRANGEMENT OF BEARINGS

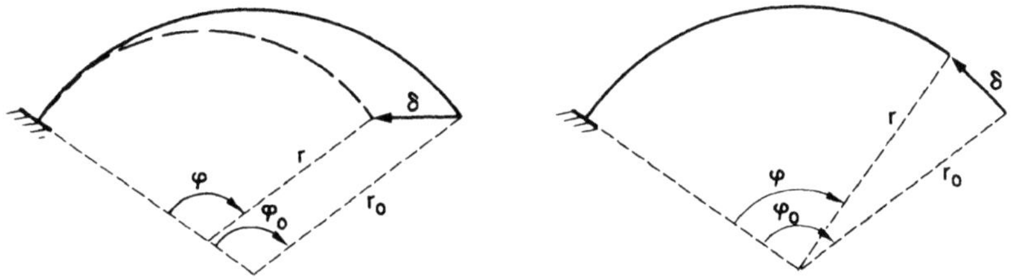

$\varphi = \varphi_0 \;;\; r = r_0 \cdot (1-\varepsilon)$

$\varepsilon = \omega \cdot \Delta T + \varepsilon_{cs}$

$\varphi = \varphi_0 \cdot (1-\varepsilon) \;;\; r = r_0$

$\varepsilon = \varepsilon(P)$

Fig 6

FIG 6: DISPLACEMENTS OF A CURVED GIRDER

curvature could conceivably be designed without any joints. Such a design, although certainly feasible, is seldom used.

2. CALCULATION OF SECTION FORCES

Calculation of beam section forces based on the theory of elasticity is relatively tedious. Therefore, the computation is usually performed with the assistance of a computer program. In the theory of elasticity, both equilibrium and compatibility conditions are satisfied. The determination of the deformations due to bending and torsion, however, is based on the assumed rigidity of the homogeneous concrete cross section and the assumed linear elastic behavior of the materials. At high stress levels these assumptions are of only limited applicability, so even the results of such an elastic computer analysis must be regarded as approximate.

A simple method for the calculation of the beam section forces can therefore be of significant value. Such a method is presented in the following.

This method always satisfies static equilibrium. Compatibility conditions are satisfied only for an infinite torsional stiffness GK. Such a calculation of the beam section forces is very simple and is hardly less reliable than the 'exact' theory of elasticity. Indeed, as long as static equilibrium is satisfied, such an approximation of the compatibility conditions is acceptable. This simplified method is therefore also suitable for the determination of optimal prestressing.

The equations of equilibrium for a curved girder element (Fig. 7) are as follows:

$$\begin{aligned} dV + q \cdot ds &= 0 \\ dT + M \cdot d\varphi + (e \cdot q + t) \cdot ds &= 0 \\ dM - T \cdot d\varphi - V \cdot ds &= 0 \end{aligned}$$

Thus,

$$\frac{d^2M}{ds^2} = -(q - \frac{1}{r} \cdot \frac{dT}{ds})$$

$$\frac{dT}{ds} = -(\frac{M}{r} + e \cdot q + t) = -m_t$$

The bending and torsion moments in the curved girder can in principle be calculated directly from these relations in the same manner as for a straight girder. An approximate method based on the requirements of static equilibrium may be applied here to some advantage.

For a curved continuous girder, the bending moments are

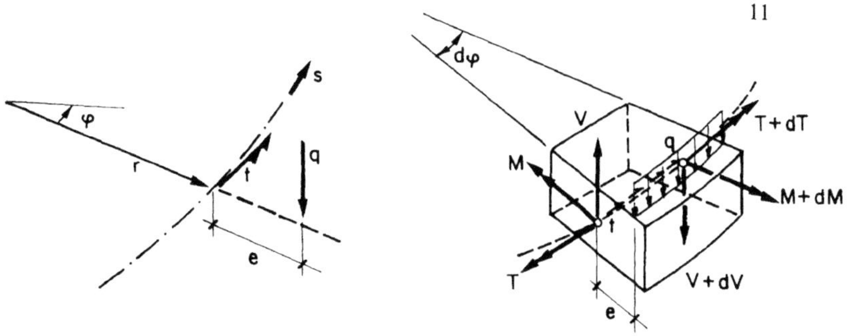

Fig 7

FIG. 7: GEOMETRY AND SECTION FORCES

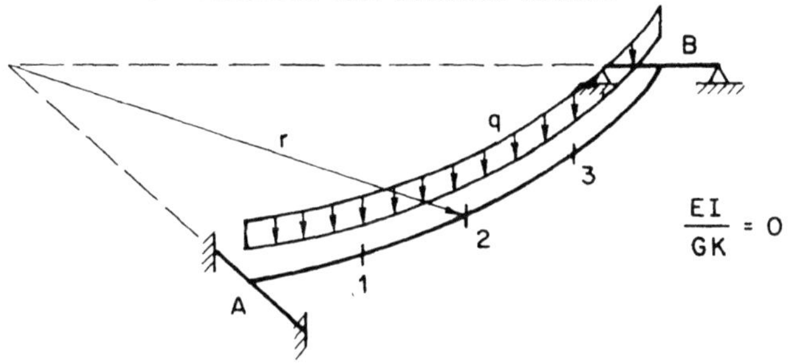

	simplified method		exact method	
	M*	T*	M	T
A	-125.00	0	-125.50	-0.023
1	0	2.865	-0.278	2.863
2	62.50	1.042	62.461	1.047
3	62.50	-2.344	62.549	-2.339
B	0	-4.167	0	-4.164

Fig 8

FIG. 8: EXAMPLE OF PROPOSED METHOD

calculated first as if it were a straight girder, satisfying the compatibility conditions of the straight girder. The radius of the straight girder is infinite, so the torsion term in the bending moment relation is neglected. The torsion moments are subsequently calculated from the bending moments resulting from the straight girder analysis, making use of the compatibility conditions for torsion. The method may be summarized as follows:

$$\frac{d^2M}{ds^2} = -q$$

$$\frac{dT}{ds} = -(\frac{M}{r} + e \cdot q + t)$$

This method guarantees satisfaction of static equilibrium in the curved girder, but the conditions of compatibility according to the theory of elasticity are not precisely met.

If necessary, the bending moments may be recalculated in a second step, taking into account the calculated torsion moments. The proposed method converges, however, very rapidly in the case of normal angles $\varphi_0 = 1/r$. The first step of the calculation is normally sufficient.

A comparison of an exact analysis with the one-step simplified analysis is shown in Fig. 8.

Calculation of the transverse internal stresses requires additional investigation. On a girder element, the deviation forces from the normal stresses are in equilibrium with the differential shear flow (Fig. 9).

The shear flow consists of two parts: the shear flow resulting from St. Venant Torsion and the shear flow resulting from the longitudinal bending of the cross-sectional panels. Exact calculation of these two parts is relatively difficult. It is based on assumptions concerning transverse and longitudinal stiffnesses and is therefore not of significant importance in the design of the reinforcement. In general an estimation of the two parts of the shear flow is sufficient: the deformations of the cross section due to transverse bending and longitudinal bending of the panels must be about equal.

3. PRESTRESSING

An arbitrarily acting prestressing force P (Fig. 10) yields the following section forces in the concrete cross section of a statically determinate curved girder:

$$N_c = -P_x$$

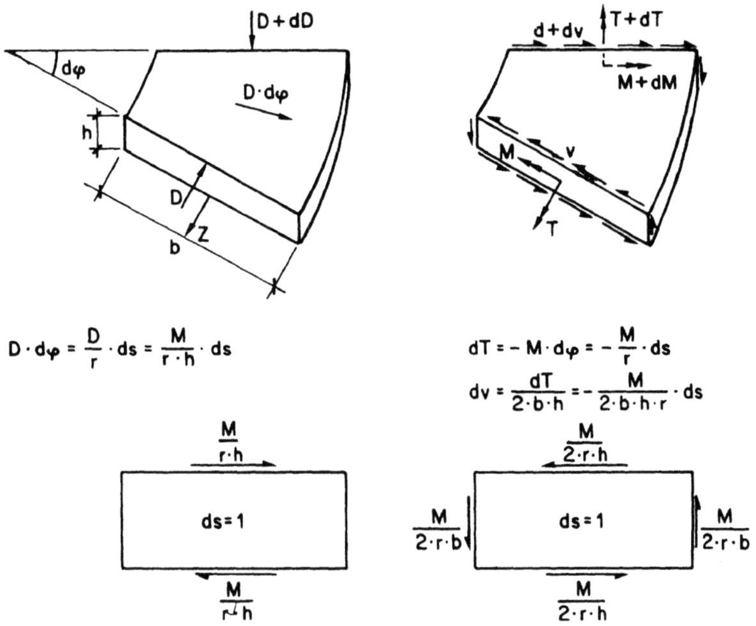

Fig 9

FIG. 9: DEVIATION FORCES AND DIFFERENTIAL SHEAR FLOW

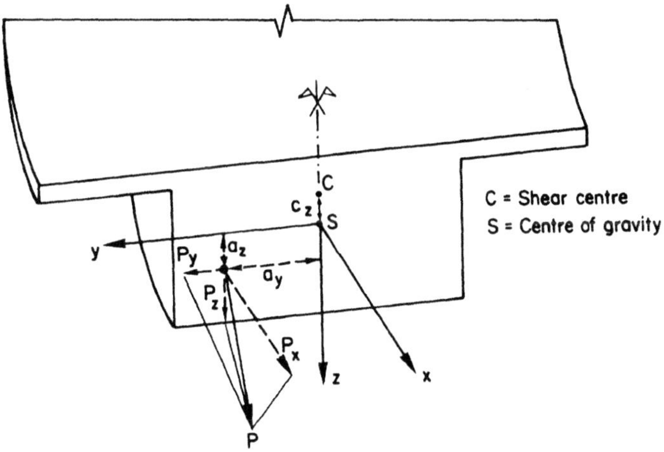

Fig 10

FIG. 10: COMPONENTS OF PRESTRESSING FORCE P

$$V_{cy} = -P_x \cdot \frac{da_y}{dx}$$

$$V_{cz} = -P_x \cdot \frac{da_z}{dx}$$

$$M_{cx} = -P_x \cdot (a_y \cdot \frac{da_z}{dx} - a_z \cdot \frac{da_y}{dx})$$

$$M_{cy} = -P_x \cdot a_z$$

$$M_{cz} = P_x \cdot a_y$$

$$T_c = -P_x \cdot [(a_y - c_y) \cdot \frac{da_z}{dx} - (a_z - c_z) \cdot \frac{da_y}{dx}]$$

As can be seen, torsion in the concrete cross section may be caused by tendons in the webs as well as by tendons in the flanges.

In the following, the required arrangement of the prestressing tendons in order to ensure a specific distribution of the torsion moments (e.g. $T_{cp} = - T_{cg}$) will be presented, presuming the girder cross section is a symmetrical one cell hollow box.

3.1 Tendons in the Webs

Tendons located only in the webs of the curved girder result in the following section forces in the concrete cross section:

$$P_x^l = P_x^r = (P_x)/2$$

$$T_{cp} \neq 0 \qquad V_{cy} = V_{cz} = 0$$

$$M_{cy} = M_{cz} = 0$$

$$a_y^r = -a_y^l = -b/2 \qquad \rightarrow M_{cz} = 0$$

$$\frac{da_y}{dx} = 0 \qquad \rightarrow V_{cy} = 0$$

$$a_z^r = -a_z^l \qquad \rightarrow M_{cy} = 0$$

$$\frac{da_z^r}{dx} = \frac{da_z^l}{dx} \qquad \rightarrow V_{cz} = 0$$

As a result the torsion moments become

$$T_{cp} = - \frac{P_x}{2} \cdot b \cdot \frac{da_z^l}{dx}$$

and the derivative provides the corresponding torque:

$$m_{tp} = - \frac{dT_{cp}}{dx} = \frac{P_x}{2} \cdot b \cdot \frac{d^2 a_z^l}{dx^2} = q_{pz}^l \cdot b$$

The distribution of torsion moments in a simply supported curved girder under dead load is shown in Fig. 11. The equalizing prestress torsion $T_{cp} = - T_{cg}$ requires the following geometry for the tendons in the web on the outer side of the curve:

$$\frac{da_z^l}{dx} = \frac{2 \cdot T_c}{P_x \cdot b}$$

$$a_z^l = \frac{2}{P_x \cdot b} \cdot \int T_c \cdot dx$$

and the reverse geometry in the web on the inner side of the curve (Fig. 12).

This tendon geometry for resistance of torsion may always be superimposed on the tendon geometry designed for the resistance of bending. Therefore, in simply supported girders, equalization of torsion and resistance to dead load bending may be provided with the same tendons (Fig. 12).

The distribution of torsion moments in a continuous curved girder under dead load is shown in Fig. 13. This distribution is clearly different from that in the simply supported girder under the same load (Fig. 11). Redundant moments due to prestressing designed to resist bending influence the torsional moments, generally resulting in the torsional moment distribution shown in Fig. 14.

In this case the equalizing prestress torsion $T_{cp} = - T_{cg}$ requires the following geometry for the tendons in the web on the outer side of the curve:

$$\frac{da_z^l}{dx} = \frac{2 \cdot T_g}{P_x \cdot b}$$

In contrast with the simply supported girder, this tendon geometry for equalization of torsion can not always be superimposed on the most effective tendon geometry for resistance of bending (Fig. 15).

3.2 Tendons in the Flanges

Tendons located only in the flanges of the curved girder result in the following section forces in the concrete cross section:

$$P_x^s + P_x^i = P_x \qquad a_z^s = -(h - a_z^i)$$

$$T_{cp} \neq 0 \qquad V_{cy} = V_{cz} = 0$$

$$M_{cy} = M_{cz} = 0$$

$$M_{cy} = 0 \quad : \quad -P_x^s \cdot a_z^s = P_x^i \cdot a_z^i \qquad \rightarrow P_x^s,\ P_x^i$$

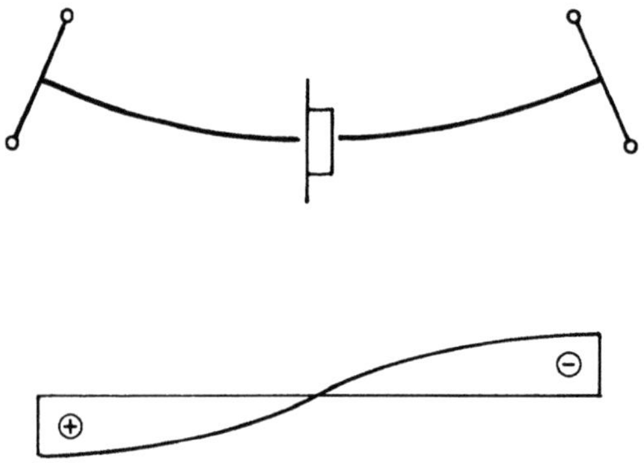

Fig 11

FIG 11: TORSION IN A SIMPLY SUPPORTED CURVED GIRDER UNDER DEAD LOAD

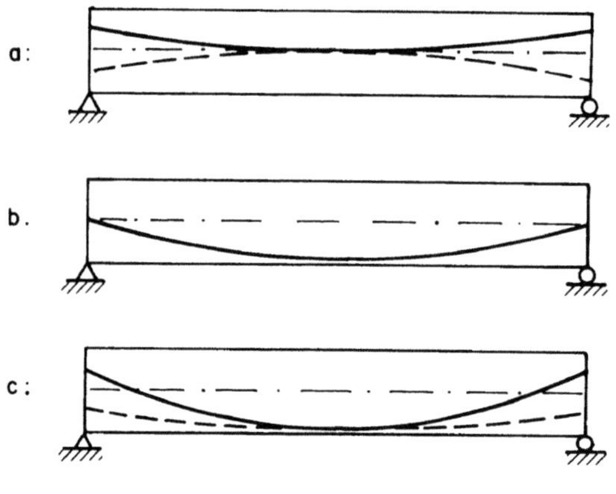

Fig 12

a: equalizing torsion moments
b: equalizing bending moments
c: superposition

FIG. 12: TENDON GEOMETRY FOR A SIMPLY SUPPORTED CURVED GIRDER

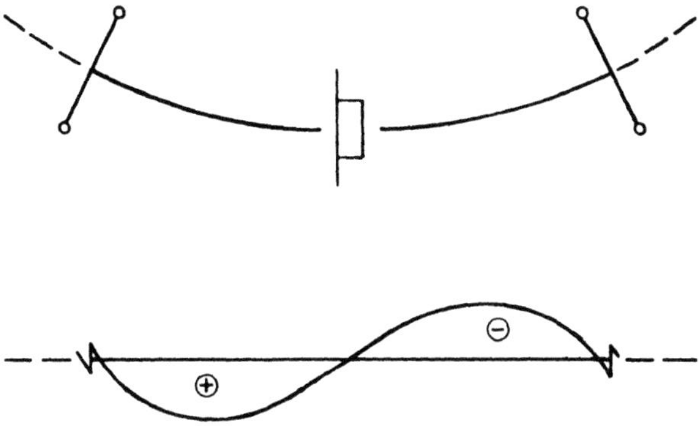

Fig 13

FIG 13: TORSION IN A CONTINUOUS CURVED GIRDER UNDER DEAD LOAD

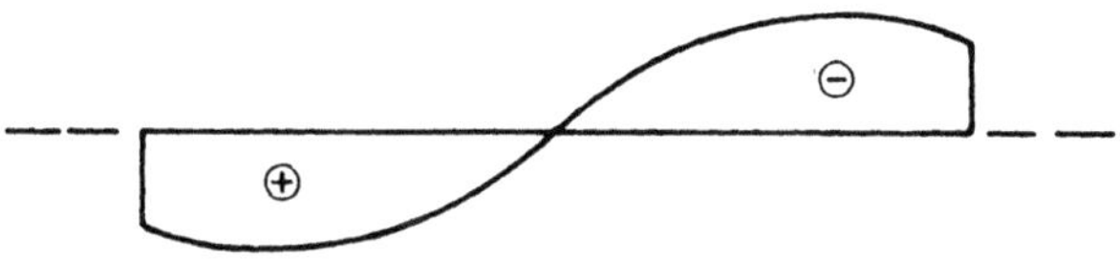

Fig 14

FIG. 14: TORSION MOMENT INCLUDING REDUNDANT MOMENT DUE TO PRESTRESSING

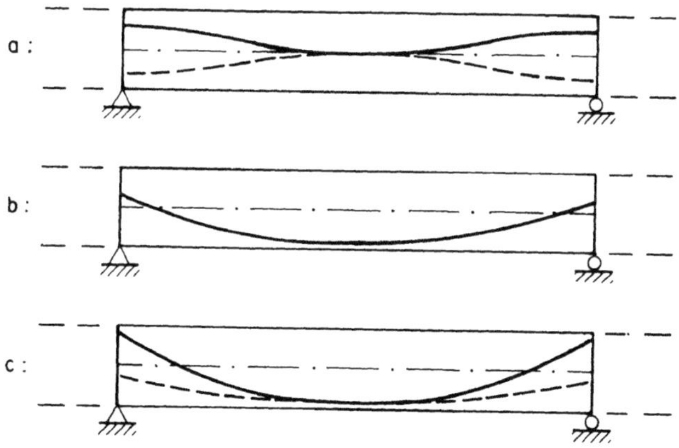

a: equalizing torsion moments
b: equalizing bending moments
c: superposition

Fig 15

FIG. 15: TENDON GEOMETRY FOR A CONTINUOUS CURVED GIRDER

$$V_{cz} = 0 \quad : \quad \frac{da_z}{dx} = 0$$

$$M_{cz} = 0 \quad : \quad -P_x^s \cdot {}_y^s = P_x^i \cdot a_y^i \qquad \rightarrow \quad a_y^s,\ a_y^i$$

$$V_{cy} = 0 \quad : \quad -P_x^s \cdot \frac{da_y^s}{dx} = P_x^i \cdot \frac{da_y^i}{dx} \qquad \rightarrow \quad \frac{da_y^s}{dx},\ \frac{da_y^i}{dx}$$

As a result the torsion moments become

$$T_{cp} = P_x^i \cdot h \cdot \frac{da_y^i}{dx} = -P_x \cdot a_z^s \cdot \frac{da_y^i}{dx}$$

$$T_{cp} = P_x^s \cdot h \cdot \frac{da_y^s}{dx} = -P_x \cdot a_z^i \cdot \frac{da_y^s}{dx}$$

and the derivative provides the corresponding torque:

$$m_{tp} = -\frac{dT_{cp}}{dx} = P_x \cdot {}_z^s \cdot \frac{d^2 a_y^i}{dx^2} = a_z^s \cdot q_{py}^i$$

$$m_{tp} = P_x \cdot a_z^i \cdot \frac{d^2 a_y^s}{dx^2} = a_z^i \cdot q_{py}^s$$

3.3 Transverse Stresses due to Prestressing

The deviation forces of the tendons consist in general of three components:

Deviation forces q_{py}^K due to the curvature of the girder in the x-y plane : $q_{py}^K = -\frac{P}{r}$

Deviation forces due to q_{py} : $q_{py} = P_x \cdot \frac{d^2 a_y}{dx^2}$

Deviation forces due to q_{pz} : $q_{pz} = P_x \cdot \frac{d^2 a_z}{dx^2}$

As a result the deviation forces q_{py} and q_{pz} cause the following torque about the shear center:

$$m_{tp}(q_{py}) = -q_{py} \cdot (a_z - c_z)$$

$$m_{tp}(q_{pz}) = q_{pz} \cdot (a_y - c_y)$$

The deviation forces q_{py}^K of the tendons are in equilibrium with the deviation forces q_{cy}^K of the normal stresses in the concrete.

$$q_{py}^K + q_{cy}^K = 0$$

$$q_{cy}^K = -\frac{1}{r} \cdot \int \sigma_{cx}(N_c,\ M_{cy},\ M_{cz}) \cdot dA_c$$

Based on equilibrium and the distribution of normal stresses

δ_{cx} the corresponding transverse stresses due to q^{K}_{py} can be readily determined. The deviation forces q_{py} and q_{pz} (Fig. 16) as well as the torque $m_{tp}(q_{py})$ and $m_{tp}(q_{pz})$ are in equilibrium with the differential shear flow.

$$q_{py} + \frac{dV_{cy}}{dx} = 0$$

$$\frac{dV_{cy}}{dx} = \int_{A_c} \frac{d\tau(V_{cy})}{dx} \cdot dA_c$$

$$q_{pz} + \frac{dV_{cz}}{dx} = 0$$

$$\frac{dV_{cz}}{dx} = \int_{A_c} \frac{d\tau(V_{cz})}{dx} \cdot dA_c$$

$$m_{tp}(q_{py}) + \frac{dT_c(q_{py})}{dx} = 0$$

$$m_{tp}(q_{pz}) + \frac{dT_c(q_{pz})}{dx} = 0$$

$$\frac{dT_c}{dx} = \int \frac{d\tau(T_c)}{dx} \cdot dA_c$$

The transverse stresses due to these deviation forces and the applied torque are calculated according to the distribution of the differential shear flow. In the case of a constant cross section the differential shear flow exhibits the same distribution as the shear flow due to the shear forces V_{cy} and V_{cz} and the torsion moment T_c (taking into account only St. Venant Torsion) (Fig. 17).

The deviation forces q_{py} and q_{pz} can be separated into two independent loading cases:

1: The resultant of the deviation forces passing through the shear center (no torque component).
2: The deviation forces causing only torques (the torque resultant).

For each loading case the transverse stresses can be calculated taking into consideration the equilibrium with the differential shear flow (Figs. 18, 19).

Superposition of the different transverse stresses shows that, supposing the geometry of the tendons such that $T_{cp} = - T_{cq}$, and respectively $V_{cy} = V_{cz} = 0$ (Section 3.1), for tendons in the webs the transverse stresses become twice as large, but for tendons in the flanges the transverse stresses can be compensated for.

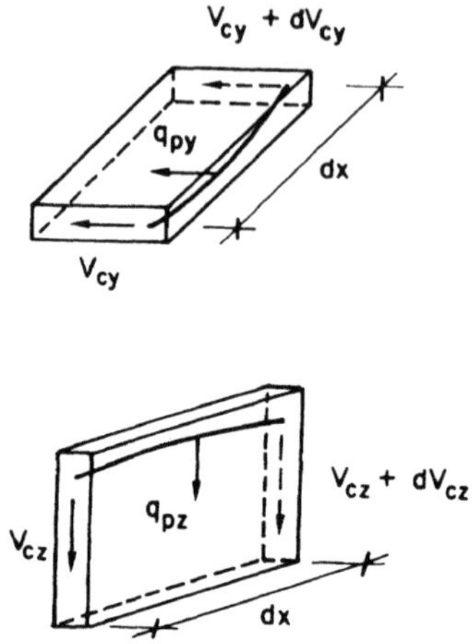

Fig 16

FIG. 16: DEVIATION FORCES q_{py}, q_{pz}

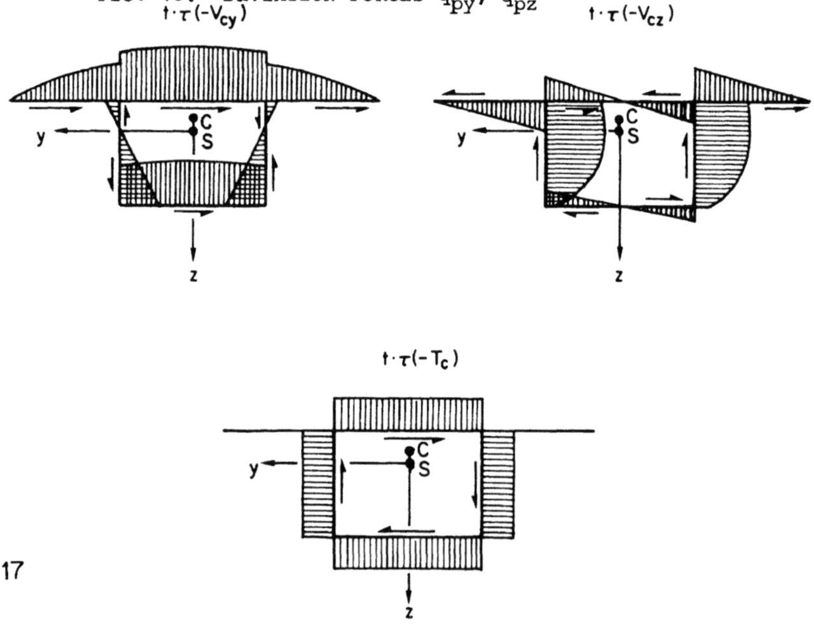

Fig 17

FIG. 17: SHEAR FLOWS DUE TO $-V_{cy}$, $-V_{cz}$, $-T_c$

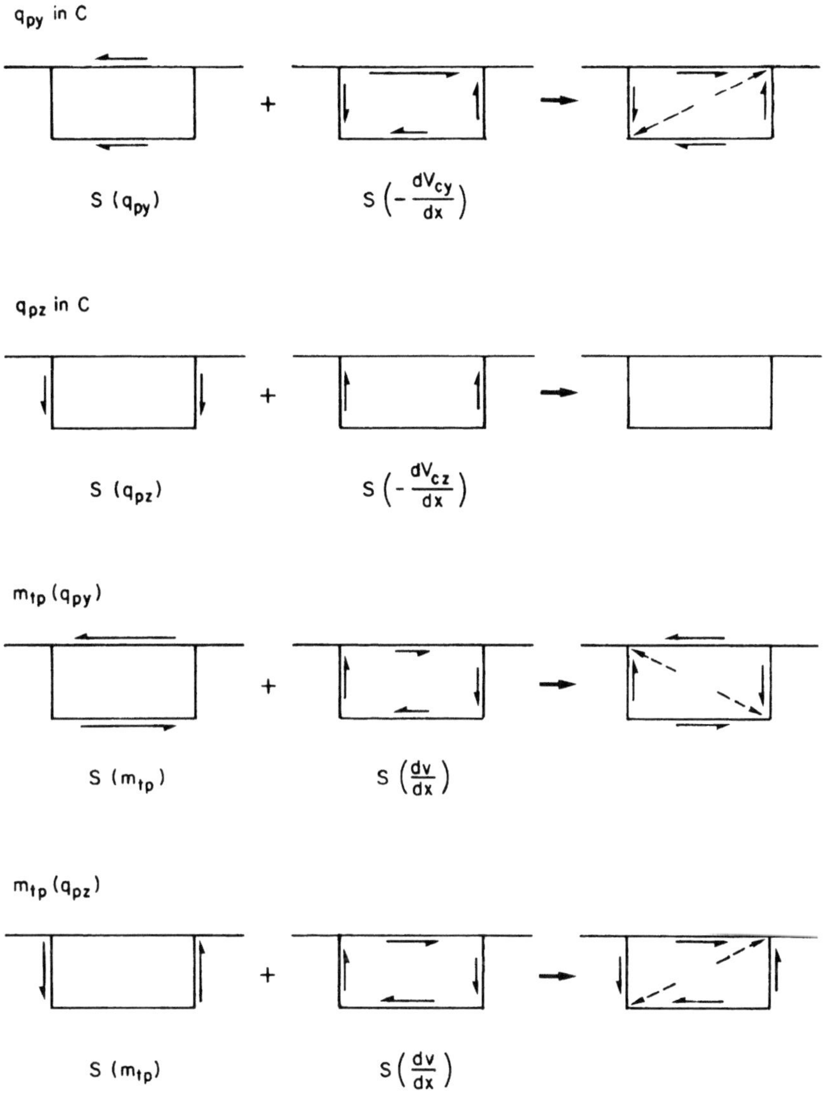

Fig 18

FIG. 18: TRANSVERSE STRESSES DUE TO q_{py}, q_{pz}, $m_{tp}(q_{py})$, $m_{tp}(q_{pz})$

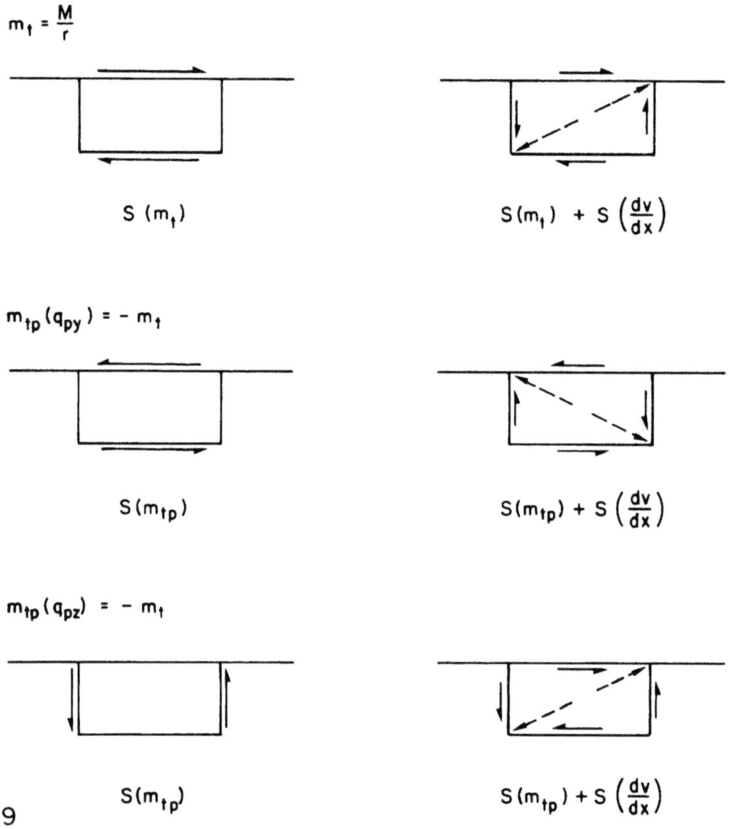

Fig 19

FIG. 19: TRANSVERSE STRESSES DUE TO TORSION AND PRESTRESSING

CERTAIN ASPECTS OF BRIDGE DESIGN

Prof.Dr. Fritz Leonhardt

Stuttgart, West Germany

ABSTRACT . During these lectures at the NATO Advanced Study Institute it is my intention to treat certain aspects of bridge design that stem from my experience of several decades as a researcher and bridge designer. From the viewpoint of general interest as well as applicability, the following topic has been chosen for the first part of the presentation : the importance of partial prestressing.

For the second session of lectures, as per the request of the organizers, it is my intention to treat the design and construction of cable-stayed bridges.

1. PARTIAL PRESTRESSING

The behaviour of fully prestressed concrete structures, especially bridges, during the last 30 years has not been satisfactory in all cases. Structures with I-beams for high ratios of live load to dead load often developed camber due to elastic and creep deformations of the precompressed tension chord member. The large number of tendons necessary for full prestressing reduce the concrete area in the critical section and do often not allow sufficient cover and spacing.

In quite a number of bridges, mainly continuous box girder bridges, cracks appeared in zones which should have had only compressive stresses according to design calculations. Such cracks had a rather large width, in one case 5 mm, proving that the small amount of reinforcing bars was stressed above yield at the moment of cracking . It was found that these cracks had mainly been caused by internal restraint forces due to temperature differences or differences of shrinkage and creep, especially if thin slabs are connected to thick portions of the structure. Some of these bridges almost failed due to these temperature cracks because the amount of reinforcing bars was too small. The possible range of these temperature differences was measured in several countries and was found to be as large as about 25°K. If the stresses caused by such temperature differences are calculated for the homogeneous concrete structure, then the engineer gets worried because these stresses can easily reach values far above the tensile strength of concrete or close to about half the amount of live load stresses. Therefore, if tensile stresses due to these temperature effects would have to be avoided by additional precompression, the amount of prestressing force would have to be increased considerably, causing difficulties in structural detailing, mainly for finding the necessary place for these additional tendons and their anchorages. In addition transverse vertical and horizontal prestressing would be necessary, which would make construction work far too complicated.

However, it would, indeed, be wrong to increase the degree of prestressing for avoiding such tensile stresses, because these restraint forces decrease and in many cases almost disappear as soon as a few cracks in the concrete relieve the degree of restraint by decreasing the bending or tensile rigidity (Fig. 1). Therefore, these restraint forces can practically be neglected as far as the safety against failure for the ultimate limit state is concerned. This ultimate limit state for full load actions governs the dimensioning of the necessary steel area (prestressing steel + rebars) for the tension chord members in girders under bending and no additional steel is needed against temperature restraint forces, if a sufficient portion of this steel is used for crack control and securing ductility.

It is well known and verified by many tests that fine hair cracks in concrete structures have no adverse influence on the durability or on safety against corrosion. The task is only to keep the crack width limited to harmless values.

Experience with testing of large scale prestressed concrete girders shows that the cracks, which must come if the load goes up above service load in order to find the ultimate capacity, will open quickly if the amount of reinforcing bars is very small. This unfavourable behaviour is due to the fact that the bond strength of grouted tendons, especially of large tendons, is only between 20 and 40% of the bond strength of normal reinforcing bars (1). Therefore, the bond gets destroyed over a considerable length if the rebars are stressed above yielding at cracking and are, therefore, not able to prevent this loss of bond.

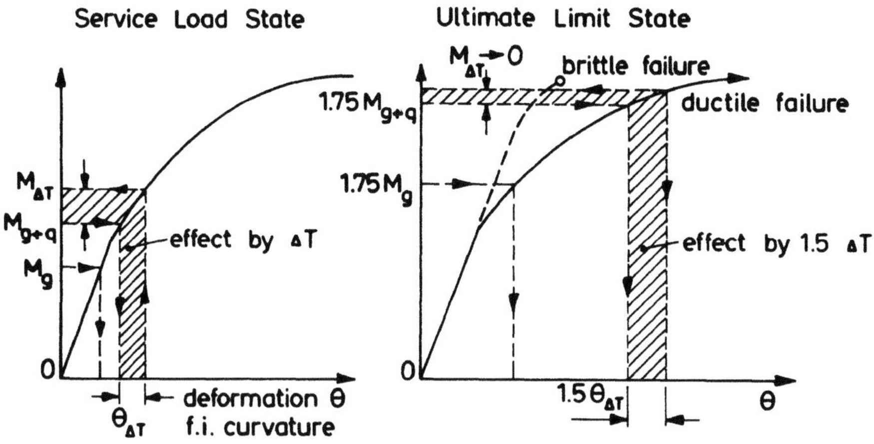

Fig. 1 Restrained forces, like $M_{\Delta T}$ = Moment due to ΔT caused by restrained deformation, decrease under load effects above service load towards ultimate load; high degree of prestressing leads to brittle failure.

We know from tests that, in such cases, we even do not obtain the full capacity which we calculate under the assumption of a perfect bond corresponding to a straight line strain diagram. This means that not only for crack control but also for ultimate safety a larger amount of rebars is needed if we use the big size grouted tendons for post tensioned structures.

Of course, this additional steel for rebars contributes to the ultimate load capacity and allows a reduction in the amount of prestressing steel. Simultaneously the post cracking behaviour of the structure becomes more ductile and less sensitive against unforeseen actions or against temperature stresses.

All these facts lead to the conclusion that we get a better behaviour of our prestressed concrete structures by reducing the degree of prestressing and increasing the amount of unstressed reinforcing bars, or in other words: partial prestressing improves the serviceability performance and simultaneously improves ultimate safety, if the structural details are well designed.

1.1 Performance requirements for serviceability

Besides the ultimate safety against load actions there are performance requirements for serviceability to be observed, which relate to admissible deformations and crack width, to durability and safety against corrosion (2).

For most structures, however, fine hair cracks can be allowed and the prescribed limit of the crack width can be secured with a low degree of prestressing or even with no prestressing, if the reinforcing bars are designed in order to fulfil the conditions for crack width limitation as described below. Even a small amount of prestressing helps to limit the crack width, because the jump of steel stress at cracking gets reduced considerably by a rather small amount of prestressing (Fig. 2). There is a wide range of freedom to choose the necessary degree of prestressing, if the area of prestressing steel plus the area of reinforcing bars satisfy the ultimate limit condition. There are cases in which the upward deflection due to creep of girders must be limited or kept zero in order to fulfil serviceability requirements; the degree of prestressing can then be governed by this condition, which cannot be fulfilled with a high degree of prestressing.

The quality of the structure and also the durability are practically independent of the degree of prestressing, if the serviceability performance requirements and those for the safety towards ultimate capacity are fulfilled.

1.2 Choice of the degree of prestressing

Practical experience suggests the following definition of the prestressing degree :

$$\varkappa = \frac{\text{moment of decompression after losses}}{\text{moment for maximum service load}} = \frac{M_D}{\max M_{DL+LL}}$$

This definition has the advantage that the engineer can directly see if the degree which he intends to choose will, for instance, cover the full dead load moment, M_{DL}, or will cover dead load plus 30% of live load or will cover only 80% of dead load.

Prof. H. Bachmannof Zürich has published charts giving the necessary amounts of prestressing and reinforcing steel for rectangular cross sections depending on the degree of prestressing (Fig. 3). He further gives the increase of steel stress in the prestressing steel and the steel stress in reinforcing bars due to full service load, again depending on the degree of prestressing (Fig. 4). For a prestressed concrete railroad bridge with I-beams such charts were calculated in my office (Fig. 5 and 6). In these

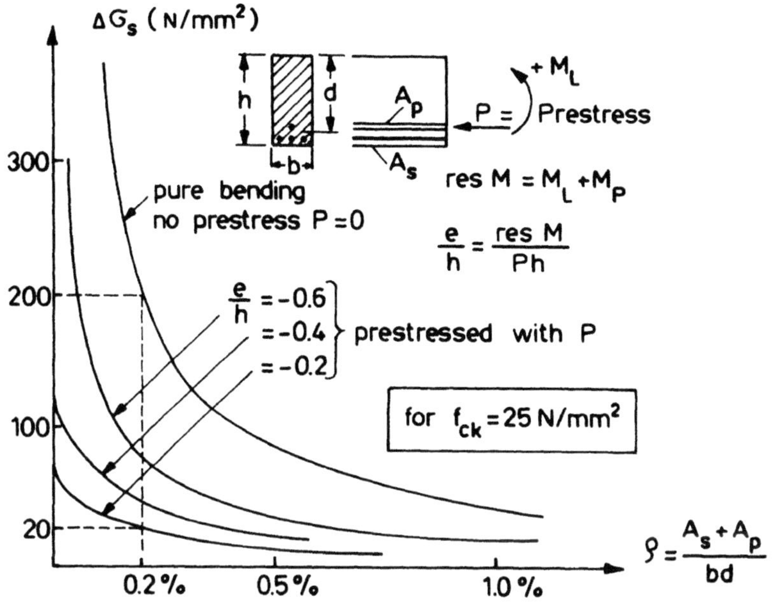

Fig. 2 Jump of steel stress at cracking
e = related eccentricity of resulting normal force
A_p can only be added to A_s for well bonded pre-tensioned steel.

charts a sufficiently large minimum percentage of reinforcement was assumed to prevent opening of cracks at the cracking load. The min $\rho_r = A_s/A_{c,ef}$, related to the affected concrete area, was assumed to be about 0.5%.

The charts give the minimum of total steel quantity $A_p + A_s$ at a prestressing degree of $\varkappa = 0.6$ for the rectangular cross section and of $\varkappa = 0.82$ for the I-beam. For higher prestressing degrees additional steel is needed to fulfil the condition of $\sigma_c = 0$ (full prestress) or $\sigma_c <$ admissible σ_{ct} (limited prestress). This additional steel is not needed for ultimate limit design. In the case of the rectangular cross section, the steel needed for full prestressing $\varkappa = 1.0$ gives a global factor of safety of $\gamma = 2.7$ where

γ = 1.8 (Swiss code) would be sufficient.

If the unit price for the prestressing steel compared to the unit price for reinforcing steel is assumed to be 3:1 (dotted lines in Fig. 3 and 5) one finds the minimum of cost for the necessary amount of steel to carry the bending moments. This minimum is given for a degree of prestressing between 0.45 and 0.6 depending on the type of cross section. One has to check yet if the necessary amount of reinforcing bars can easily be placed in the available concrete section. The largest space in between such bars depends on the allowed crack width. No crack width verification is needed, if these spacings in the tension chord zone do not exceed the following values :

Crack width limit mm	0.1	0.2	0.3
max.bar spacing mm	80	120	150

The charts of the steel stresses show that the stress amplitudes remain far below the fatigue limits if for rectangular cross sections the degree of prestressing is chosen above 0.4 and for I-beams above 0.5 and if the frequently repeated portion of the live load remains below about 70% of the dead load. This means that choosing partial prestressing in most practical cases does not cause any fatigue problems if the degree of prestressing and the ratio LL:DL are within these limits. Therefore, even for railroad bridges partial prestressing can be chosen.

1.3 Partial prestressing and shear or torsion

In some countries vertical or inclined prestressing bars in webs in box or I-girders can be found lately in order to avoid principal tensile stresses due to shear or torsion. In my experience this type of prestressing is in most cases not necessary, because inclined shear cracks have practically never been found in prestressed concrete bridges. Such cracks would appear only with loads causing bending moments well above the M_D (moment of decompression), normally above full service load. This is partially due to the fact that a part of the shear forces is carried inside the chord members and does not cause tension in the web. Of course, the spacing of the stirrups should be rather close, so that, if a shear crack would develop, the crack width would remain very small.

One of the biggest prestressed concrete bridges in Germany is the Kochertalbrücke, Geislingen, which has a box girder with 2 webs only 45 cm thick which have to carry a 6 lane freeway with a total width of 30 m over spans of 138 m. The maximum shear stress under service load in these webs runs up to 5.8 N/mm^2 and the principal tensile stresses were calculated as 3.8 N/mm^2. No web prestressing was chosen for shear or torsion in spite of these high shear and tensile stresses, but the stirrups for the shear reinforcement were spaced down to 8 cm near the intermediate supports. Even additional inclined stirrups were placed. But no shear crack could be found so far. It is therefore my hope that those authorities, who still stick to the request of full prestressing, may soon introduce codes allowing partial prestressing in order to get the economical and technical advantages of this design concept.

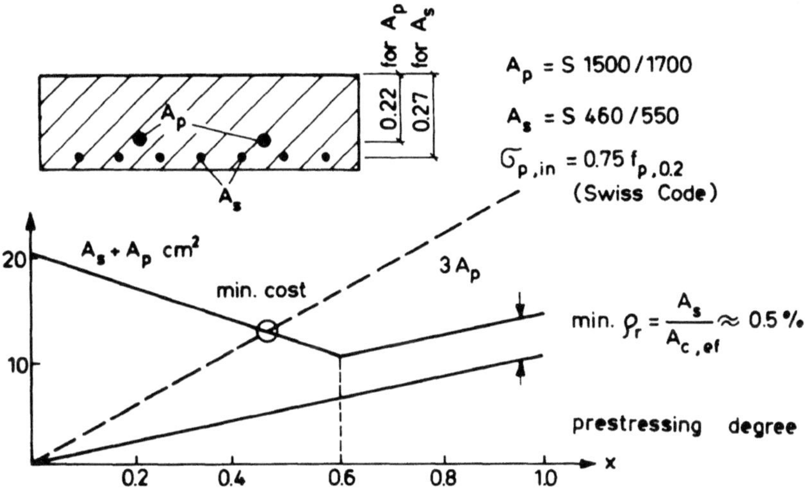

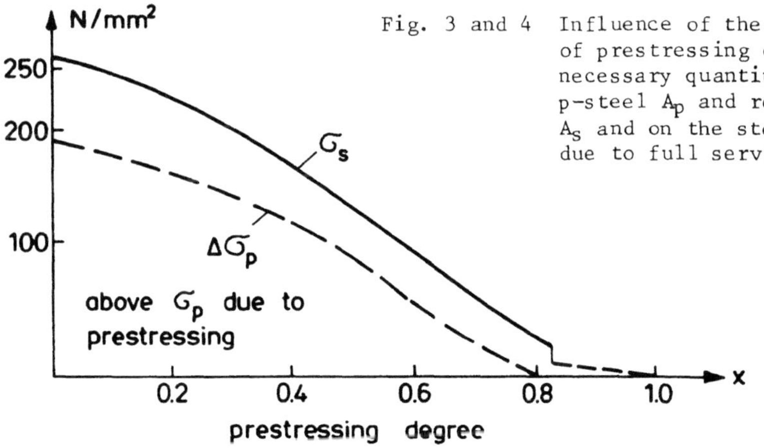

Fig. 3 and 4 Influence of the degree of prestressing on the necessary quantities of p-steel A_p and rebar-steel A_s and on the steel stress due to full service load.

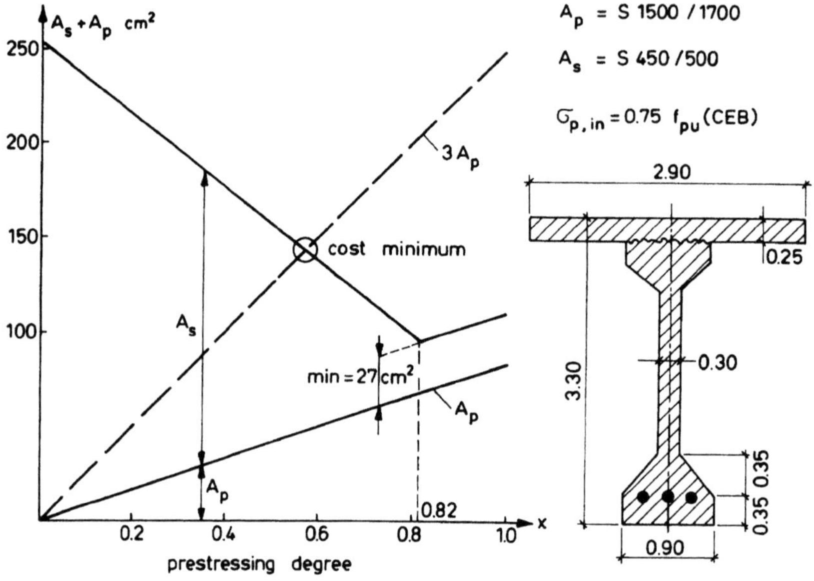

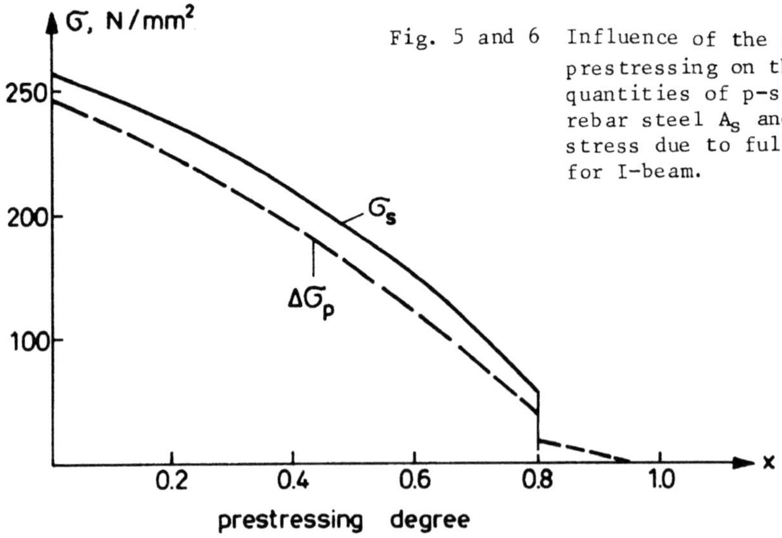

Fig. 5 and 6 Influence of the degree of prestressing on the necessary quantities of p-steel A_p and rebar steel A_s and on the steel stress due to full service load for I-beam.

2. CABLE - STAYED BRIDGES

2.1 The development to the multi-stay-cable system

The idea to support a beam by stays from a tower is very old. The rebirth of the system must be credited to F. Dischinger (3) who pointed out the advantages of high stresses in stays of high strength steel. The first modern cable-stayed bridges were beam bridges with only 2 to 6 stay supports in the main span. The spans between the stay supports were between 30 and 60 m and needed large bending stiffness of the beam with a depth of 3 to 4 m. The stay forces were large so that several ropes were needed to build up the cables. The anchorages of these cables were rather complicated. A considerable amount of auxiliary structures were needed to erect such bridges (Fig. 7).

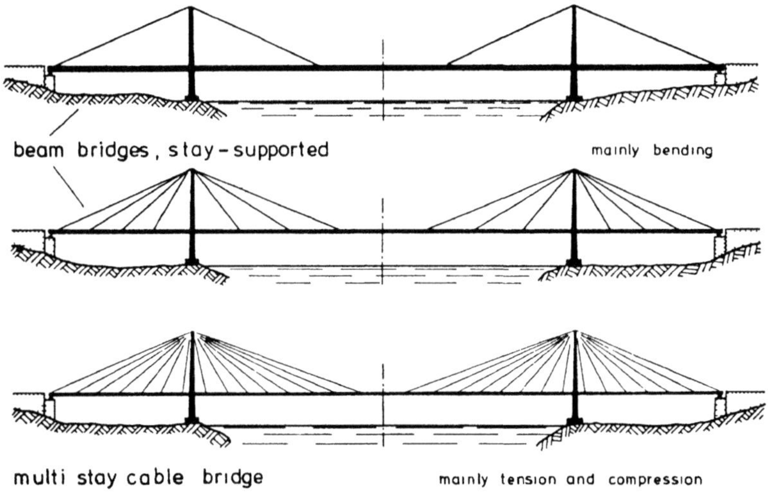

Fig. 7 Development from stayed beam to the proper multi-stay-cable system

In the further development it was found that most difficulties disappeared if a large number of stays is used with spacings at the deck anchorage of only 8 to 15 m, so that free cantilevering erection is possible without any auxiliary supports (Fig. 7). This development to the multi-stay-cable bridges practically led to a new type of main girder system which could no more be defined as a beam girder. It has its own proper qualities. What was formerly a beam girder, has become mainly the compressive chord member of a cantilever structure hung up to towers by inclined stay cables. Similar to a chord of a truss, this compressive chord member does not need much bending stiffness, because the triangle tower-stay-chord gives abundant stiffness for getting deflection lines with curvatures which fulfil the performance requirements for highway and railroad traffic. Therefore, the depth of the longitudinal girders or stringers in the deck structure is almost independent of the main span and should be chosen small, in order to avoid unnecessarily large longitudinal bending moments in the deck structure.

This multi-stay-cable system is in fact a new system, different from our classical systems of beam girders, arches, or suspension bridges with stiffening girders. Comparative calculations showed that this system deflects less under highway traffic loadings than slender continuous beams or suspension bridges. It has also superior qualities as far as the dynamic behavior is concerned, mainly due to its large amount of system damping.

These superior qualities are obtained if highly stressed cables of sufficient inclination are used so that vertical deflections will be small. The cable stiffness is the paramount parameter in this system,it is usually described by the formula

$$A_s E_{eff} = \frac{A_s E_o}{1 + \dfrac{\gamma^2 \ell_c^2 E_o}{12\sigma^3}} \qquad (1)$$

where :

A_s = area of cable steel

E_o = modulus of elasticity of straight vertical cable

γ = weight of cable (incl. corrosion protection) related to $A_s \cdot \ell_c$

ℓ_c = horizontal span of cable

σ = tensile stress of cable, influencing the sag of cable

The evaluation for E_{eff} is shown in Fig. 8. The stiffness of such cables increases with the third power of the steel stress and decreases with the second power of the horizontal span length due to the sag effect (change of sag by change of stress). The relation between live and dead load has influence on the dimensioning of the cables and therefore on the stiffness under dead load conditions. Depehding on this p:g ratio the cable stiffness might become insufficient for $\ell_c > 250$ m or for main span lengths of 500 m.

However, is is easy to reduce the sag effect for longer cables by introducing so-called stiffening ropes as shown in Fig. 9. In this way the high stiffness of the cable-stayed system can be maintained for very long spans, up to main spans of 1500 m for railroad and highway bridges.

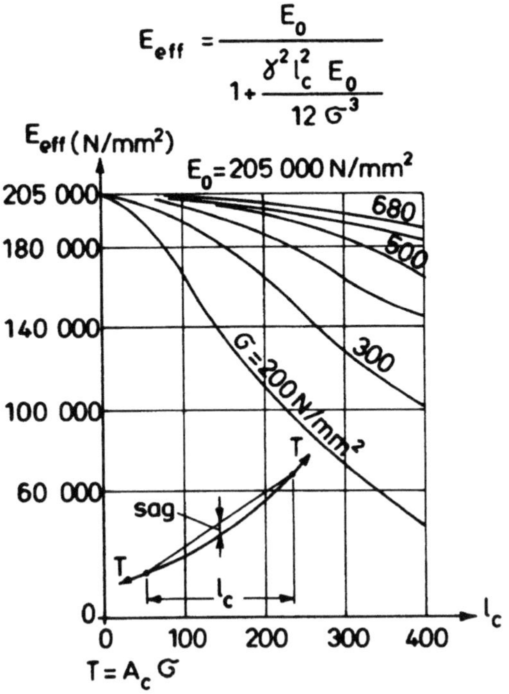

Fig. 8
Effective modulus of elasticity gives influence of the sag of the cable on its stiffness

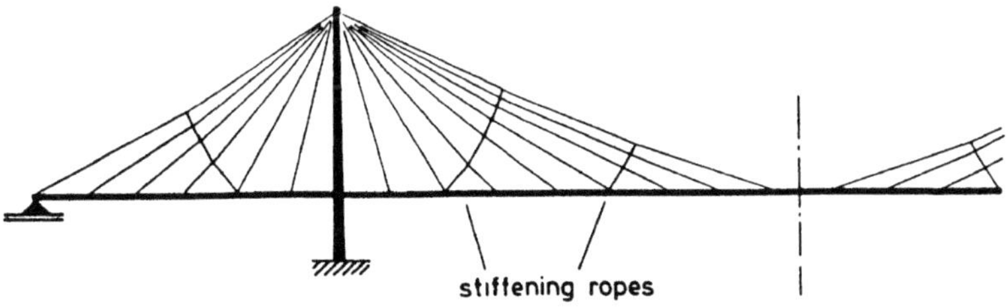

Fig. 9 **Stiffening ropes to reduce the sag effect on the stiffness of the cables**

2.2 The arrangement of the stay cables

There are many possibilities for the arrangement of the cables. There is the fan-shaped configuration (Fig. 10) in which all cables join at the head of the tower. In the harp-shaped arrangement (Fig. 11) all cables are parallel and their anchorages at the tower are distributed over the height of the tower. This arrangement needs more steel for the cables, results in higher compressive normal forces in the deck, and causes bending moments in the tower. From a technical and economical view it is, therefore, interior to the fan-shaped arrangement, but for the appearance of the bridge it might be superior, because all cables look parallel also in view under a skew angle. This aesthetic advantage was decisive for the choice of the harp-system in the famous Düsseldorf bridge family crossing the River Rhine (4).

If it is requested that cables should be easily replaceable in cases of accidents, then it is difficult to realize the pure fan arrangement in which all cables join theoretically at one point

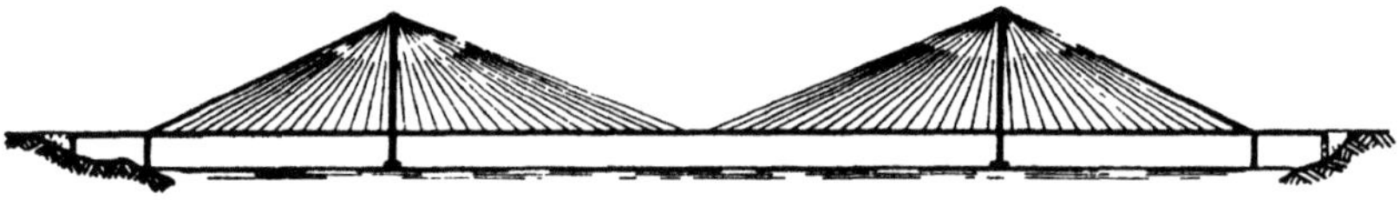

Fig. 10 Fan-shaped configuration of stay cables

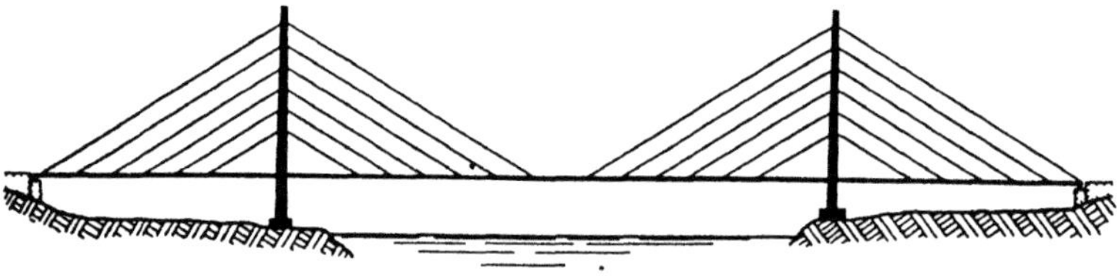

Fig. 11 Harp-shaped configuration of stay cables

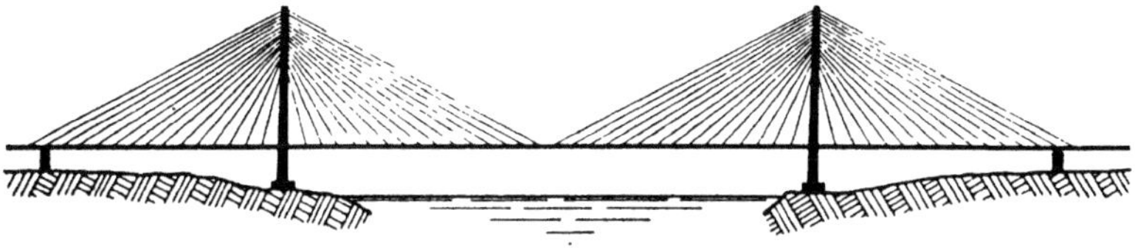

Fig. 12 **Fan-shaped arrangement, anchorages of** cables at tower distributed

above the tower top. For replaceable cables it proved to be simpler to have the anchorages of the cables distributed vertically over a certain length of the tower. This results in a cable configuration as shown in Fig. 12. The length of this distribution of the anchorages can also be larger but then bending moments in the tower leg are caused by differences in the horizontal components of the cable forces due to certain live load positions.

Of course, other configurations of the cables are possible mainly depending on local conditions for the ratios between main and side spans. A harmonic arrangement of the cables is important for the aesthetic quality of such bridges and, therefore, the choice should be made with care and diligence.

2.3 The ratio between main and side spans

The ratio between side span ℓ_1 and main span ℓ has influence on the stress changes mainly of the back stay cables, which hold the tower head back to the anchor pier. Live load in the main span increases these stresses, live load in the side span decreases them, and in long side spans with $\ell_1/\ell > 0.4$, the cables could become more or less slack. The back stay cables get the largest stress amplitudes of all cables and these amplitudes must be kept safely below the fatigue strength of the cables for that part of the live load which may occur a million times (for instance at highway bridges 40% of maximal live load).

The ratio ℓ_1/ℓ has further influence on the amount of vertical anchoring forces at the anchor pier. This anchor force decreases with increasing ℓ_1/ℓ. Towers on both sides of the main span are assumed here. If there is only one tower on one side (unsymmetrical case) then the main span acts similar as if the span length is about 1.8ℓ.

A good choice of the ratio between side and main span is important for a good and economical design. As an aid for this choice a chart has been calculated in which we find the parameter p:g plotted vertically, and horizontally the length of the main span in meters (5).

Of course, the choice between the length of side and main span depends also on local conditions of water depth, of foundation data or on the wish to have a tower on one side of the main span only.

2.4 The optimal height and stiffness of towers

The height of the towers has influence on the necessary amount of cable steel and on the longitudinal compressive forces in the bridge deck. The higher the tower, the smaller will be the quantities of the cable steel and the compressive forces. The curves in Fig.13 show that it is of no use to make the towers higher than about 0.2ℓ up to 0.25ℓ, because one has also to consider the quantities needed for the tower. For bridges with the tower on one side the h/ℓ must be related to 1.8ℓ.

In the longitudinal direction the towers should be slender and have a small bending stiffness, so that live loads in the main span do not cause large bending moments in the tower but get the back stay cables acting. Transmitting the unbalanced horizontal components from the top of the towers to the ground by back stay cables is much more economical than by bending resistance of the towers.

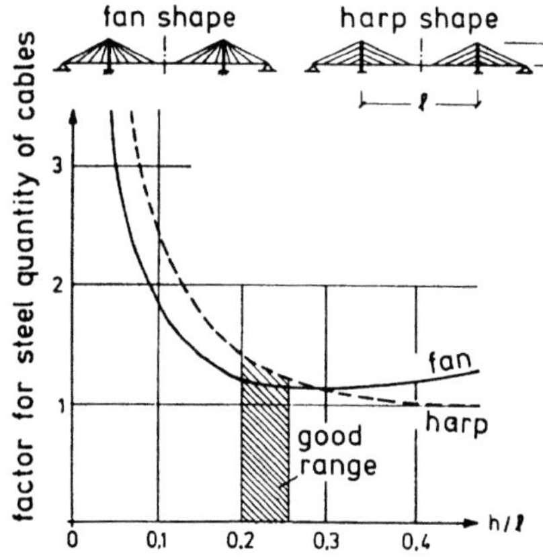

Fig. 13
Quantity of cable steel as a function of relative height of towers

Longitudinal bending stiffness of towers, which is characteristic for Morandi's early stayed beam bridges, get very large moments, which must be carried by the foundation and can easily double their cost. In order to avoid this, some towers of cable-stayed bridges have been built with foot hinges so that the foundations are concentrically loaded (Mannheim and Oberkassel bridges across the River Rhine) (4, 6).

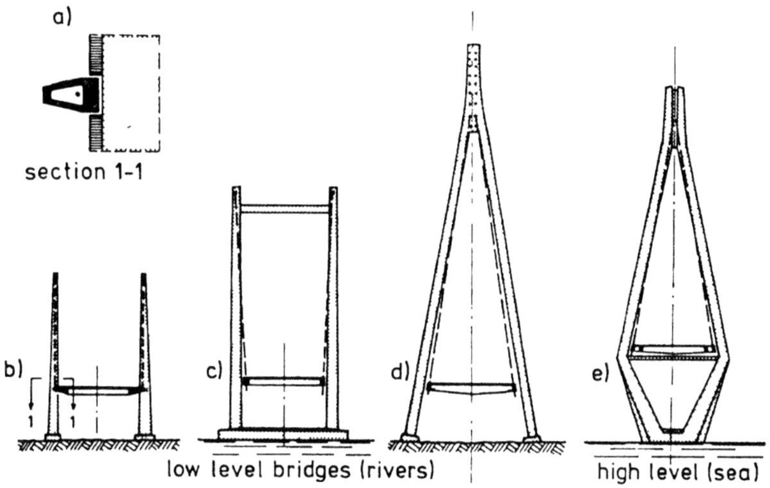

Fig. 14
Two cable planes along the edges of the deck with different solutions for the towers

2.5 The arrangement of the cables transversely and cross sections of the deck structures

In normal cases the bridge deck should be hung up along its edges, resulting in two planes of cables and two towers, standing just outside of the railing of the bridge deck. The towers should get an unsymmetrical cross section with most of the load carrying area and consequently the center of gravity close to the bridge deck, so that the cable planes can be vertical or must be only slightly inclined (Fig. 14a). The wind loads acting on the tower by the cables are small so that no horizontal bracing between the two towers is necessary, if the tower legs are tapered and fixed in the foundation.

If the height of the towers is considerably larger than the width of the bridge, then a horizontal beam connecting the towers at the top may be useful, allowing a small inclination of the cable plane or of the tower legs, so that the cable planes can be kept vertical (Fig. 14b).

For long spans, A-shaped towers add to the good appearance of such bridges, because all cables join at the one tower top (Fig.14c). A good example is the Pont de St.Nazaire with a main span of 404 m, crossing the River Loire near its mouth (7). Joining the cables of the two planes at tne top increases also the torsional stiffness of the bridge deck.

For high level bridges it is desirable to join the tower legs under the bridge deck in order to narrow the necessary width of the foundation, as was done in the design for the Faroe Bridge in Denmark (Fig. 14d).

The cross section of the deck structure can be very simple, if the bridge is hung up along its edges. No torsional rigidity is necessary because the cables give a stiff support along each edge and the deflection is small, so that unsymmetrical loading gives only a very small transverse inclination of the deck. Therefore, for a width of the bridge up to around 15 m, a simple massive or hollow concrete slab with ribs along the edges is sufficient (Fig. 15a). The edge rib allows us to anchor the cables at any point and ensures the buckling safety.

For wider bridges, cross girders are necessary which should be arranged with a spacing of only 3 to 5 m, so that the concrete slab or the orthotropic steel plate can easily span longitudinally. Here most of the reinforcing bars of the steel stiffening ribs run longitudinally and help to carry the compressive normal forces (Fig. 15b and c). The concrete deck slab is advantageous for main spans up to about 600 m, even if steel girders are used transversely for constructional reasons. Composite action should be provided. If the deck slab is of concrete, then all longitudinal girders should also be of concrete so that no creep problems arise due to the high longitudinal compressive stresses. The all steel bridge with an orthotropic plate deck becomes mandatory for the very big spans in order to reduce the dead loads. Even for very large highway bridges longitudinal edge girders with a depth between 1.0 and 2.5 m are sufficient and can be used for anchoring the cables.

2.6 Situation at the ends of the cable stayed bridges

Near the ends of the side spans or of a main span hung up to one tower only, there are the only regions with large bending moments. As a consequence the angular changes of the deflection line at such free ends of the longitudinal girder are large. This

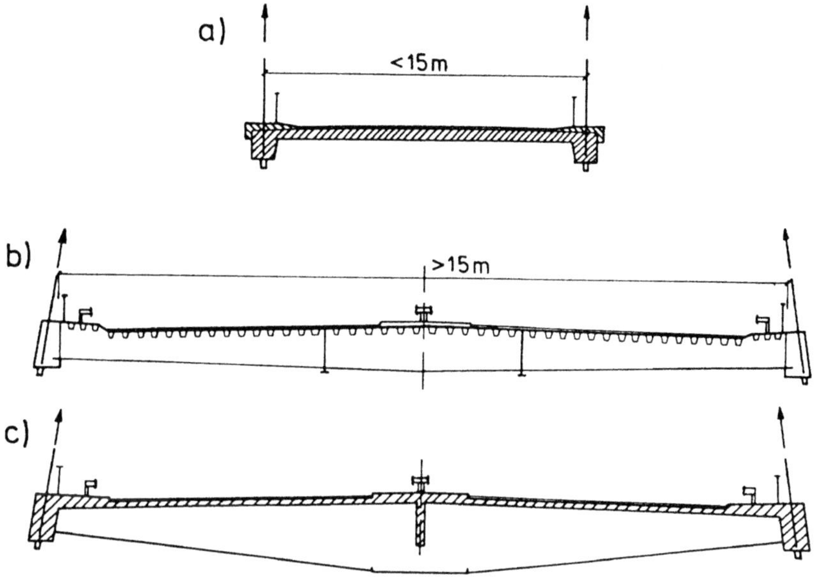

Fig. 15 a Cross section of a concrete bridge with a width b < 15 m

Fig. 15 b c Cross section of concrete or steel bridge with b > 12 m

is acceptable for normal highway bridges but is causes difficulties for railroad bridges which can easily be avoided, if tne girder continues with an increased depth into a small approach span. This solution can generally be recommended, if approach bridges follow behind the side or main spans. By such continuity the uplift forces of the back stay cables can be counteracted by the weight of this adjoining span and by ballast concrete within the depth of this girder extending to both sides of the anchor pier (Fig. 16). It might be sufficient to have only a cantilever, which carries a hinged bearing of the approach bridge. The continuity allows us also to distribute the anchorages for the back stay cables over a certain length behind the axis of the anchor pier.

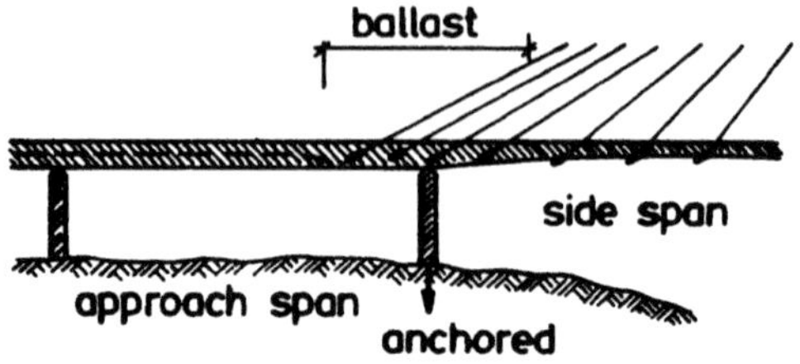

Fig. 16 Continuity to the approach bridge allows lengthwise distribution of the back stay cables

2.7 Arrangement of bearing and joints

Vertical bearings should only be arranged at the end of the side spans but not at the tower, where it is better to continue the elastically deformable support condition which is given by the stay cables. If stiff vertical bearings at the towers would be chosen, then large longitudinal bending moments would cause trouble and require more bending stiffness than necessary for the rest of the bridge.

The bearings for horizontal transverse loads, like wind loads, must be arranged in a way to allow angular changes of the wind girder in a horizontal plane. At the towers these wind bearings can be simple rubber pads which act directly against the tower legs and which should have an open gap of about 3 mm for the unloaded condition to allow vertical and horizontal movements of the deck easily. At the end of the side span the wind bearing can be placed in the center-line of the bridge. It must allow angular changes in horizontal and vertical planes and longitudinal movements of the bridge.

For horizontal loads in the longitudinal direction, which are mainly caused by braking forces, different arrangements are possible; for symmetrical bridges no fixed bearing is needed, if at the end of each side span speciay fluid bearings are used, which allow for the slow temperature elongations without much resistance but which are sufficiently stiff to react to breaking forces without much deformation. In this case two expansion joints are needed.

In seismic regions it is easily possible to design the bearings with bolted steel angles limiting the movement for normal service conditions, which, however, break away if extreme seismic amplitudes occur. Shock absorbers have to damp further excessive movement, so that the big mass of the deck structure would not cause damaging forces to the towers and piers. In this way cable-stayed bridges can be made very safe against earthquake attacks.

2.8 Maximum span length

The experience gained by designing and constructing a large number of cable-stayed bridges during the last 25 years allow us to state that multi-stay-cable bridges as described can be built for highway traffic with spans up to 700 m, for railroads up to about 500 m with prestressed concrete. Designs for steel bridges have been made with main spans of about 1 300 and 1 500 m (6 lanes highway and 2 tracks railroad) for the Messina Straits Crossing. No structural difficulties were found (5).

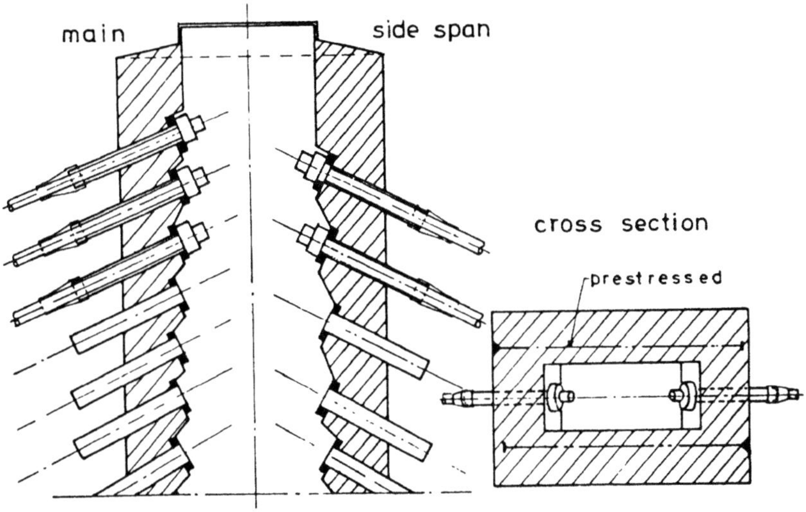

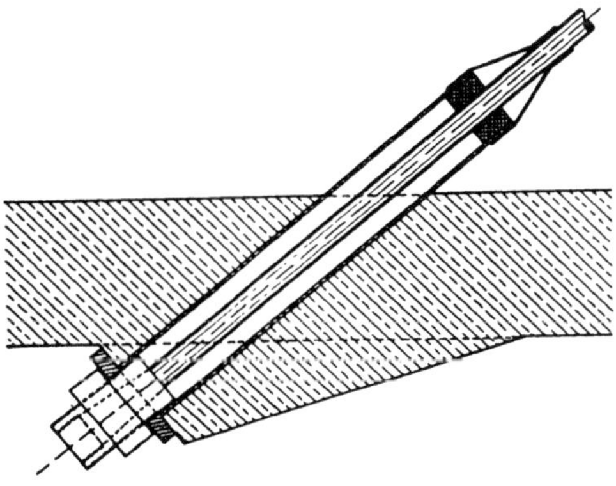

Fig. 17
Anchorage of HiAm-cables in the tower and in the concrete edge girder of the deck

2.9 The towers

The shape of the towers is sometimes a subject for special architectural treatment. As engineers we should try to keep the forms as simple as possible, making use of tapering, good proportions and suitable profiles of the cross sections to obtain a pleasing appearance.

Experience proves that concrete towers are cheaper than steel towers, and the cost difference increases with the size of the bridge. Box sections are preferable to solid sections, so that access to the cable anchorages at the top can be provided inside the tower shafts.

Over the years, many different types of saddles or anchorages for the cables have been designed. However, there is now wide agreement that the cables should be anchored in a way which allows easy replacing of each cable. This led to a rather simple solution as shown in Fig. 17. The cable socket is pulled through a steel pipe embedded in the concrete and a ring nut is turned on the thread, to hold the socket in place. The anchor chamber in the tower head must be sufficiently large to handle the equipment which is needed for this work. The horizontal tensile force, which results from two opposite anchor forces, is counteracted by horizontal prestressing tendons just outside the anchor chamber, which keep the concrete of the tower head under horizontal compression.

2.10 The cables and their anchorage

The cables are the most important members of this system, they must, therefore, be safe against fatigue and corrosion. The best quality cables must be chosen. A large amount of testing results and practical experience is available for the judgement, which type of cable would be the best (8, 9, 10, 11). In this lecture there is not sufficient opportunity to present this knowledge in detail. The following recommendations can be given, based on this knowledge :

1. Steel ropes, protected only by painting should not be used for large cable-stayed bridges as such ropes have failed at several large bridges and have had to be replaced.

2. Parallel wire or parallel strand cables are superior to ropes by their fully elastic behaviour and well defined modulus of elasticity.

3. The corrosion protection should be absolutely secured by placing the steel of the cables inside a tube, which is tightly connected to the anchorages. The tube can be of black polyethylene (PE), which has been proven to keep its qualities for at least 20 years and will probably need no maintenance for 40 or 50 years, if

correctly handled during the transport, erection and injection.

4. The fatigue strength of the cables depends mainly on the fatigue strength of the anchorage. The normal zinc filled sockets of ropes give only a rather low fatigue strength, because the high strength of the wires gets damaged by the high temperature, which must be used for pouring the metal. Fatigue tests proved that the stress changes, which can safely be carried by such anchorages, are as 100 to 120 N/mm^2 for large ropes with a diameter bigger than 80 mm.

   Thick walled steel pipes around the cables can also be used to help carrying the live load. As a result the stress amplitudes of the wires are reduced. This was done at the Main Bridge in Höchst (12) and at the Brotonne Bridge (13).

5. Special anchorages have been developed, for instance the BBR HiAm-anchorage (9) with a cold filling material in the conical cone or other types like the one of Freysinnet International, which was used for the Vigo Bridge in Spain (14).

2.11 Dynamic Behaviour

If stiff cables are chosen with E_{eff} > 180 000 N/mm^2 then very favourable dynamic behaviour of the cable-stayed bridges can be expected. This is especially true for the aerodynamic behaviour and can be traced to the following phenomena :

1. The strain behaviour of the cables is non-linear due to the influence of the sag effect.

2. Each of the many inclined cables with the mass of the deck belonging to it has a different frequency. Whenever forces act, which excite oscillations of the bridge in a certain mode, then the development of the amplitudes is immediately interrupted by the interference of cables with different frequencies. The multi-stay-cable system develops hereby a system damping of normal concrete or steel structures like beams , arches or suspension bridges. Due to this system damping resonance oscillations with large amplitudes are impossible and only such reasonance oscilla- tions can become dangerous for long span bridges.

   This favourable dynamic behaviour was first proved by a few tests at the 90 m long cable-stayed pedestrian bridge across the Schillerstrasse in Stuttgart, a light weight steel bridge (only 150 kg/m^2 dead weight), which can easily be excited but develops amplitudes of not more that about 5 mm. The second proof was given by dynamic model tests at the ISMES Institute of Bergamo, which were conducted for the railroad bridges over the Parana (15). Fig. 18 shows a typical oscillation diagram from these tests.

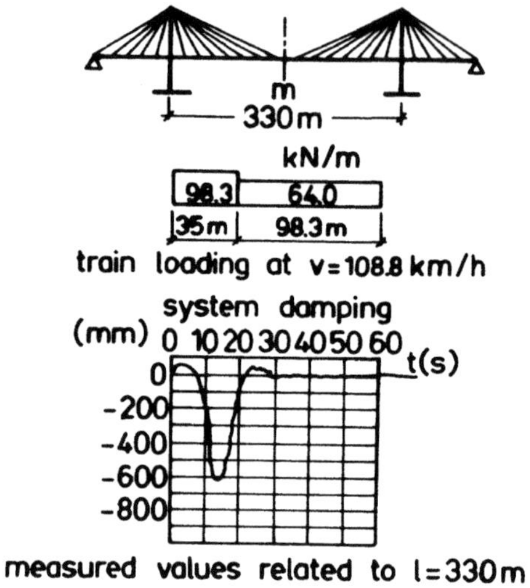

Fig. 18
Oscillation diagram of a cable-stayed bridge

The good behaviour was confirmed by the experience during the free cantilevering construction of these and other bridges and also by the behaviour under railroad traffic.

This system damping prevents any resonance oscillations and secures hereby the aerodynamic safety even for very long spans, if a large number of cables and sufficiently stiff cables are involved and if the span/width ratio is not larger than about 40. No aerodynamic shaping is necessary. Theories, which have been developed to verify aerodynamic safety of suspension bridges have, therefore, only limited validity for these multi-stay-cable bridges; the same is true for sectional model tests in wind tunnels, as long as the system damping cannot be imitated with sufficient similarity in such tests.

The system damping cannot prevent torsional oscillations of bridges which are hung up by cables in one plane along the center-line; furthermore, the frequencies of bending and torsional oscillations are independent of each other. There may be a danger of getting torsional resonance if the box girder has a shape which would give a large c_M factor (shape factor related to pitching moments), so that the wind forces could start torsional deformations.

Our knowledge in this respect is still limited and, therefore, a diligent investigation must be recommended. Another dynamic problem is the oscillation of the cables themselves, as was observed at several cable-stayed bridges, especially at the Brotonne Bridge, where car dampers had been installed at the roadway level. The authors use neoprene dampers at the end of the steel pipe, in which the end of the cable is enclosed over a certain length near the anchorages (Fig. 17). This neoprene damper has to prevent also angular changes of the cable at the anchorage which would cause additional bending stresses. So far these dampers have actually prevented amplitudes larger than about 20 mm in cables with lengths up to about 150 m, if the cables have cement injection inside the PE-tubes, which give a high damping to the cables themselves.

2.12 Cable-stayed bridges for railroad

The strong damping of multi-stay-cable bridges makes this system very suitable for railroad traffic, especially for modern high speed railroads. Of course, the dead load masses should be large so that dead load stresses in the cables are sufficiently high to make them very stiff. Sufficient dead load is easily obtained, if the ballast for the tracks continues over the bridge as is requested now by most railroad companies.

Further, the deck should be a prestressed concrete structure for main spans $\ell<400$ m. The depth of the longitudinal girders must be slightly larger than for highway bridges in order to keep the curvature and the change of the gradient of the deflection line sufficiently small. Short side spans and a combination with highway traffic help to reduce the stress amplitudes. Several long span cable-stayed railroad bridges have been built or are under construction.

REFERENCES

1. Leonhardt, F. : "Vorlesungen über Massivbau-Spannbeton", Volume 5, Stuttgart, 19.

2. CEB-FIP Model Code, Table 15.1, Paris, 1978.

3. Dischinger, F. : Hängebrücken für schwerste Verkehrslasten Der Bauingenieur, 3/1949, p.65 and following and p.107 and following.

4. Tamms, F., Beyer, E. : Kniebrücke Düsseldorf Book edited by Beton Verlag Düsseldorf 1969.

5. Leonhardt, F. and Zellner, W. : "Cable Stayed Bridges", IABSE Surveys S-13/80, Zürich, May 1980.

6. Volke, E. ; Rademacher, C.-H. : Nordbrücke Mannheim - Ludwigshafen (Kurt-Schumacher-Brücke) Der Stahlbau 1973, Heft 4, p. 97, Heft 5, p. 138, Heft 6, p. 161.

7. Vanbourdolle, M, Ciolina, J., Bacarrere, J.: Le pont de Saint Nazaire-Saint Brévin , Annales de l'Institute Technique du Bâtiment et des Travaux Publics 347 (1977), pp. 13-43.

8. Andrä, W., Zellner, W. ; Zugglieder aus Paralleldrahtbündeln und ihre Verankerung bei hoher Dauerschwellbelastung, Die Bautechnik, 1969, Hefte 8 and 9, pp. 263-268, 309-315.

9. Andrä, W., Saul, R. : Versuche mit Bündeln auz parallelen Drahten und Litzen für die Nordbrücke Mannheim-Ludwigshafen und das Zeltdach in München, Die Bautechnik, 1974, Hefte 9, 10, 11, pp. 289-298, 332-340, 371-373.

10. Andrä, W., Saul, R. : Die Festigkeit, insbesondere Dauerfestigkeit langer Paralleldrahtbündel, Die Bautechnik, 1979, Heft 4, pp. 128-130.

11. Freyssinet Cable Stays, Technical Outline, Brochure by Freyssinet International, Ref. Fl. 246A/08.79.

12. Schambeck, H. : Bau der zweiten Mainbrücke der Farbwerke Hoechst AG - Konstruktion und Ausführung Vortrage Betontag 1973, pp. 359-172, Wiesbaden, Deutscher Beton-Verein e.V.

13. Mathivat, J. : The Brotonne Bridge, Proceedings of the Eight Congress of the Féderation Internationale de la Précontrainte, May 1978, Part 2, pp. 164.

14. De Miranda F., Leone, A., Passaro, A. : Il ponte strallato sullo stretto di Rande presso Vigo (Spagna) delle Autopistas del Atlantico - Costruzioni Metalliche 2/1979.

15. Leonhardt, F., Zellner, W., Saul, R. : Zwei Schragkabelbrücken fün Eisenbahn - und Stra enverkehr üser den Rio Parana (Argentinien) - Der Stahlbau, 1979, Hefte 8 and 9.

II. ANALYSIS METHODS FOR STEEL BRIDGES

OVERLOADING AND INELASTIC RESPONSE OF STEEL MULTIGIRDER BRIDGES

Celal N. Köstem, Professor of Civil Engineering

Fritz Engineering Laboratory, 13, Lehigh University
Bethlehem, Pennsylvania 18015 USA

1 INTRODUCTION

Highway bridges may be designed in accordance with the allowable stress, load factor, or load and resistance factor design approach. When these bridges are loaded with the regular design vehicles the response of the structure will be linearly elastic, or very close to it. The collapse load levels of these bridges can be computed by any one of the accepted methods available in the literature. However, the overloading of the bridge may propel its structural response to a level that is above the proportional limit but far below the collapse load. In the bridge engineering literature there exists a major gap in the prediction of the structural response of the bridge at post-proportional limit, and also the determination of the extent of damage to the superstructure. This paper provides an overview of the inelastic response of highway bridges composed of steel girders, rolled, built-up or plate girders, and reinforced concrete deck slab.

The various research programs that are summarized herein have all employed the finite element displacement method, Derivations and the development of the mathematical models have been extremely laborious. Presentation of any formulation would not have increased the engineering contributions of this paper. It is recommended that readers interested in the mathematical derivations of the employed models, and the verification of these models, should refer to the references cited. The paper is aimed at bridge engineers.

1.1 The Overloading of Bridges

The extensive research carried out on the overloading behavior of simple span prestressed concrete I-beam bridges, which were designed in accordance with the AASHTO Standard Specifications for Highway Bridges (9), indicated that prior to the occurrence of any distress in the beams the reinforced concrete deck slab will undergo substantial damage in the form of cracking of the concrete for increased overload levels (5,7). These cracks occur approximately at the mid-spacing between the beams at the bottom of the slab, that is positive bending of the deck. The cracks are essentially parallel to the girders. Regardless of the extent of the damage to the deck the prestressed concrete I-beams did not have any distress up until extremely high load levels, at which level the deck slab was essentially "destroyed."

Bridge engineers speculated that since both steel girder and prestressed concrete I-beam bridges are designed under the same specifications (9), which reflects the same structural response expectations, similar structural response should be expected from steel multigirder highway bridges. The research have showed that there exists substantial differences between the response characteristics of steel and prestressed concrete bridges at load levels that are above the proportional limit (2,3,7).

2 THE METHOD OF ANALYSIS

The "routine" methods of analysis of beam-slab highway bridges-the structures which inherently have a very high degree of structural indeterminancy-employ the oversimplified "distribution factor" approach (9,10). The studies have indicated that even for the elastic regime of the structure this approach is far from being satisfactory (2,7). Thus, a more detailed scheme needs to be employed which will account for the structural indeterminancy, as well as the gradual initiation and spread of material nonlinearities. The finite element modeling with plate bending elements, with membrane stiffness properties, and the eccentric beam elements would provide a realistic approach to the elastic simulation of the superstructure. To account for the material nonlinearities, it is essential that both the beam and the plate bending elements be subdivided into layers, through the depth of the structural member (Fig. 1). Throughout the analysis with the gradually increased load levels, using the tangent stiffness formulation, stress and strain, as well as the damage, history of each layer of each element need to be monitored. A complicated analysis scheme as such requires the development of a detailed software package. Program BOVAS (Bridge Overload Analysis-Steel) has been developed to analyze simple or continuous steel multigirder highway bridges (4). A number of benchmark studies have been conducted to verify the accuracy of the program (2,3,4). It has been found that the program can predict the full load versus deformation (or damage) history of a given bridge for a given loading condition, up to the collapse of the bridge superstructure, if an acceptable finite element mesh is employed. On the average the maximum discrepancy noted between the analytical simulation via program BOVAS versus the available full scale or the model studies was found to be about 5%. Thus, the observations and conclusions presented herein have a high degree of reliability as compared to any other method of "approximation" of the overload and inelastic response of bridge superstructures.

3 NONLINEARITIES ASSOCIATED WITH OVERLOADING

In the simulation of the post-linear elastic response of steel multigirder bridges the following possible sources of nonlinearities need to be initially considered: nonlinearity of stress-strain curve for concrete and steel, cracking and crushing of the concrete, yielding of steel, local and global stability of the steel girders and their components, post-buckling behavior of steel components, and slip between the deck and girders. In addition, the question regarding the small versus large deflection formulation needs to be looked into. The research have illustrated that formulation of the problem on undeformed geometry does not cause any inherent inaccuracies in the analysis scheme (3).

3.1 Interface Slip

The slip between the slab and the girders could theoretically be a major problem, especially at increased overload levels. This problem could be further compounded due to the recent trend in the

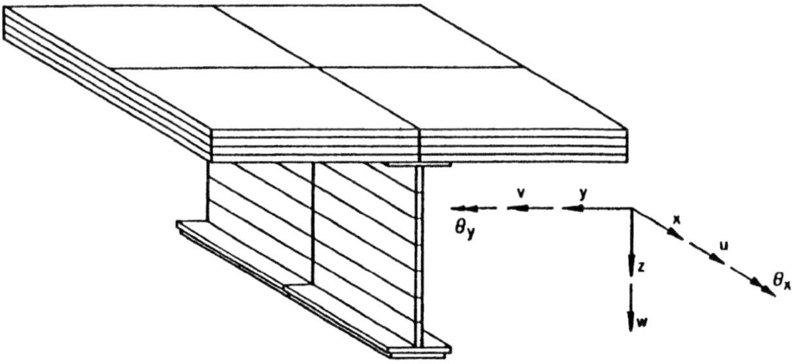

Fig. 1 Slab and Girder Layering

increased use of "partial shear connectors."

Detailed analytical research, and its correlation with the available test data, have demonstrated that for fully composite construction the slip is not a progressive phenomenon for increased overload levels (8). In the case of the use of partial shear connectors slip does not occur under service loads, not even at load levels in excess of service loads. For higher load levels, a slip of a finite amount can take place, which causes a slight "jump" in the load-deformation curve, corresponding to a slight incremental loss of stiffness (8). The load deformation curve continues to climb, with the previous slope, without any further disturbance, as far as the slip contribution is concerned, until the spread of extensive damage to the superstructure. Additional finite slip can also take place. It has been noted that for fully composite or partial shear connection cases the neglect of the slip will not result in any major inaccuracy in the prediction of the post-linear elastic response of the bridge superstructure. The result of this simplification will be the slight overestimation of the stiffness of the bridge.

In the case of fully noncomposite construction it is noted that there still exists a substantial initial composite interaction between the deck and the girders at design load levels. For increased overload levels there will be intermittent slip of undefinable magnitude. The slip, in this case, again is not progressive. For increased live load levels the frictional forces between the deck and the girders increase, thus arresting the slip in part. In any engineering analysis and research the contribution of the slip should preferably not be overemphasized. The complications arising from its inclusion do not warrant the increase in accuracy.

3.2 Material Nonlinearities

The research have demonstrated that the inclusion of material nonlinearities is essential in the correct assessment of the inelastic response of steel multigirder highway bridges (2,3,7). This is especially the case if the study requires the simulation of the full-range response, i.e. from zero load level up to the collapse of the bridge.

If an investigation is aimed at the determination of the rated capacity of a given bridge, and if serviceability criteria will have to be considered, as has been the case with the common practice in the USA (10), then the allowable stress levels for the girder steel and for the reinforcing bars are far below the levels at which material nonlinearities will be pronounced. In these types of analysis it is noted that serious considerations that might be given to the representation of the nonlinearity of the steel stress-strain curve may not be critical (3). However, the nonlinearity of the concrete stress-strain curve can not be neglected, especially if the rating provisions do not specify the stress levels in concrete (10).

The cracking of the concrete must be simulated as realistically as possible (7). Even though the serviceability considerations require the determination of the crack width in reinforced concrete structures, the experience of the researchers with the use of any given crack width formula for any given structure-carefully monitored throughout the loading-has not been encouraging (7). Due to various factors, which are beyond the control of the bridge engineer, the crack width formulae do not give a reliable estimate for an existing structure, for which very little may be known as far as the exact placement of the reinforcing bars. Instead, it was found that the use of the crack depth, which can be reliably predicted, is a better measure of the serviceability limit in the definition of the allowable load levels in the overload analysis (5). The critical parameter, especially in the negative moment region of continuous bridges, where the deck cracking will initiate at the top of the slab, is the determination of the crack depth for infrequent overloadings. Due to this infrequency, the cracks will be "working cracks," such that after the traverse of the vehicle the cracks will close due to the rebound of the bridge superstructure (6).

3.3 Stability Considerations

The research have indicated that for rolled or built up sections the web buckling, with the exception of at the supports, could easily be retarded by vertical stiffeners. In the case of plate girder construction, especially those with deep and slender webs, the web buckling takes place almost at the same load level that some relatively deep cracks, for example, one-fourth the slab thickness, form in the deck. Regardless of the negative connotation of the web buckling, this is a local constrained phenomenon. After the web panel buckling, the "truss action" takes over (3). The post buckling strength of the web is usually more than sufficient to permit the traverse of infrequent overloads (2,3).

In the case of continuous construction lateral torsional buckling is a possibility in the negative moment region. However, even though this buckling mode is theoretically possible, due to the contribution of the wind bracing, and X-bracing or diaphragms, the lateral torsional buckling becomes an unlikely phenomenon (2).

3.4 Serviceability Limits

The main thrust of the research summarized herein is the ability to predict the post-linear elastic behavior of the bridge. Such a response is not uncommon for many special permit vehicles.

The key issue for the bridge engineer is to define the extent of permissible "recoverable damage" to the bridge. In the US bridge engineering practice, i.e. AASHTO Bridge Maintenance Manual, the allowable stresses for the steel girders have been set low enough to prevent the damage to the steel superstructure. It should be noted that these low stresses are set forth to account for the adverse effects of the residual stresses and the stress concentrtion at the connection details. The current serviceability limits require extensive in depth studies in order to relate the possible post-linear elastic behavior of the bridge components to the bridge life expectancy as well as the permissible load levels.

4 EXAMPLE BRIDGE

To illustrate the capabilities of the research summarized herein one example is presented. A four span continuous bridge with span lengths 21.34 m + 27.43 m + 27.43 m + 21.34 m (70 ft. + 90 ft. + 90 ft. + 70 ft.) was field tested by the researchers at the University of Tennessee. The cross section of the bridge is shown in Fig. 2. Figure 3 shows the load-deflection curve for the field test and the finite element analysis. It can be seen that there exists an excellent agreement between the actual behavior and the analytical prediction at elastic as well as at the transition into highly nonlinear "knee," and at the ultimate load level. The analysis indicated that slab cracking initiates at 1154 kN (259.5 kip) load, which is far in excess of the service loading. At 2475 kN (556.4 kips) first yielding begins in the bottom flange of the girder immediately under the loading. Careful inspection of the load-deflection curve indicates that for these load levels the bridge response is almost linear. Thus the initiation and spread of the damage have no perceptible effects on the overall structural response of the bridge. Additional observations on the behavior of

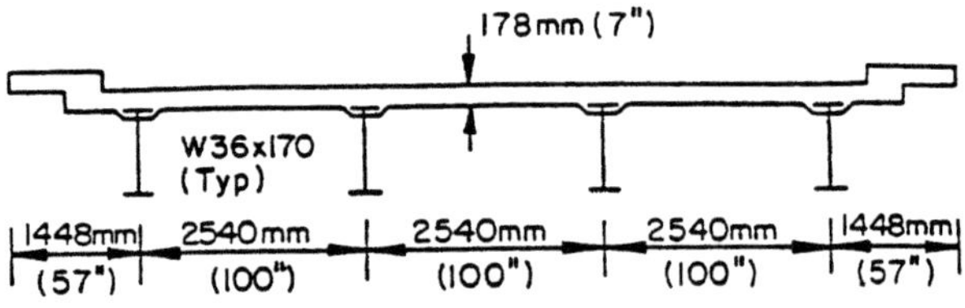

Fig. 2 Cross Section

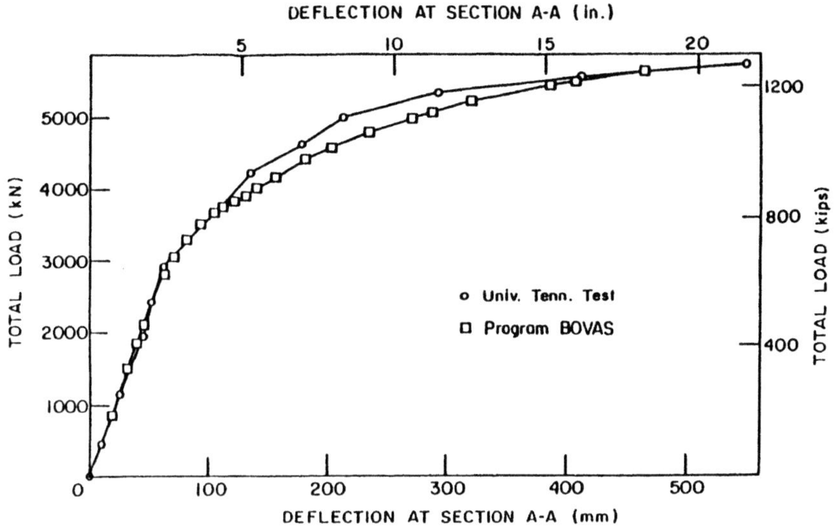

Fig. 3 Load versus Deflection Curve

this bridge testing and analysis can be found in reference 2.

The load levels given above are above the practical limits. If the analysis includes the residual stresses (results of which will not be presented herein), there may be a 20-30% decrease in the load levels that will cause yielding in the girders. It should be recalled that there are substantial stress concentrations in the steel connection details, causing grave concern for the initiation of the fatigue crack growth.

5 GENERAL OBSERVATIONS

Extensive observations can be listed with applicability to certain types of construction, and also dependent on the type of loading (6). However, it is possible to arrive at general observations and conclusions that are applicable to all steel multi-girder bridges, regardless of loading and the construction practice:

(1) In continuous beam-slab bridge superstructures the first failure is the cracking of the slab concrete in the negative moment region.

(2) Regardless of the type of loading, the negative moment regions suffer the greatest amount of damage.

(3) The effects of cross-bracing in more uniform lateral live load distribution are negligible.

(4) In deep plate girders the web panel buckling frequently occurs at load levels higher than the design loads. This buckling does not lead to any progressive damage. Exclusion of the web panel buckling in any analytical simulation can lead to the incorrect assessment of the stresses in the superstructure.

(5) In cover-plated girders, the plastification of the flange initiates near the tip of the cover-plate, a most undesirable location in view of fatigue considerations.

1. Fisher, T.A., J.W. Fisher, C.N. Kostem and D.R. Mertz. Design and Retrofit for Fatigue Damage in Web Gap, Fatigue of Steel and Concrete Structures, IABSE Colloquium Proceedings (Lausanne, Switzerland, 1982) 535-543.
2. Hall, J.C. and C.N. Kostem. Inelastic Analysis of Steel Multigirder Highway Bridges, Fritz Engineering Laboratory Report No. 435.1 (Lehigh University, Bethlehem, Pennsylvania, USA, 1980).
3. Hall, J.C. and C.N. Kostem. Inelastic Overload Analysis of Continuous Steel Multigirder Highway Bridges by the Finite Element Method," Fritz Engineering Laboratory Report 432.6 (Lehigh University, Bethlehem, Pennsylvania, USA, 1981).
4. Heishman, C.A. and C.N. Kostem. Users Manual for Program BOVAS - Detailed Version, Fritz Engineering Laboratory Report No. 435.2 (Lehigh University, Bethlehem, Pennsylvania, USA, 1982).
5. Kostem, C.N. Further Parametric Studies on the Overloading of Highway Bridges, Fritz Engineering Laboratory Report No. 434.2 (Lehigh University, Bethlehem, Pennsylvania, USA, 1980).
6. Kostem, C.N. A Parametric Study on the Overloading of Steel Multigirder Bridges, Fritz Engineering Laboratory Report No. 435.3 (Lehigh University, Bethlehem, Pennsylvania, USA, 1983).
7. Peterson, W.S. and C.N. Kostem. The Inelastic Analysis of Beam-Slab Highway Bridge Superstructures, Fritz Engineering Laboratory Report No. 378B.5 (Lehigh University, Bethlehem, Pennsylvania, USA, 1975).
8. Tumminelli, S.C. and C.N. Kostem. Finite Elements for the Elastic Analysis of Composite Beams and Bridges, Fritz Engineering Laboratory Report No. 432.3 (Lehigh University, Bethlehem, Pennsylvania, USA, 1978).
9. Standard Specifications for Highway Bridges. The American Association of State Highway and Transportation Officials, Washington, D.C., USA, 1977.
10. Manual for Maintenance Inspection of Bridges. The American Association of State Highway and Transportation Officials, Washington, D.C., USA, 1978.

STEEL CURVED I AND BOX GIRDER DESIGN AND ANALYSIS

C. P. Heins, Professor
Institute for Physical Science and Technology and
Civil Engineering Department
University of Maryland, College Park, MD 20742

INTRODUCTION

During the past ten years extensive research studies have been undertaken in the area of the design and analysis of curved bridge structures. Such studies were necessitated by increased demands for curved highway alignment and aesthetic considerations. It is therefore the purpose of this paper to present the results of some of these studies and the resulting design specifications.

ANALYSIS TECHNIQUES

The determination of the induced forces in a horizontally curved bridge system requires consideration of the interaction of the entire bridge structure. The influences of torsion (St. Venant and warping) must be included in such analysis.

As indicated in the AASHTO code (2), the effects of curvature may be neglected in evaluating the bending moments in the longitudinal members, according to the following;

Number of Girders	Angle for 1 span	Angle for 2 or more spa
2	2°	3°
3,4	3°	4°
5 or more	4°	5°

However, the influence of such curvature on the induced torsional forces should be examined. As a guide for such examination the following has been found (26);

- For $\psi \leq 0.4$ evaluation of stresses due to pure torsion may be neglected, but warping stresses must be included.
- For $\psi \geq 10.$ evaluation of stresses due to warping may be neglected.

where: $\psi = L[GK_T/EI_w]^{1/2}$ and is plotted in Fig. 1

Analysis techniques. - There are various analytical methods available for determining the response of curved bridges (23,24,28∿33). In general these methods can be categorized as follows;

1. Static Method-U.S. Steel (29)
2. Computer Matrix Grid Method (23,24,28)
   a) Three DOF
   b) Five DOF
3. Space Frame Grid Method (23)
4. Finite Element Method (23)

5. Finite Differences Method (8,9,10,15)

The first method "Static Method-U.S. Steel" (27,29) allows for direct dtermination of the forces in the longitudinal girders and diaphragms by making certain assumptions relative to load or force distribution, and generally does not require a computer. The method can be explained by examining a free body of a curved flange, Fig. 2 and 3. Noting that the resultant forces (F) in the flange are not in equilibrium in the vertical direction, therefore an artifical force q is developed, and equals;

$$q = \frac{M}{R} \cdot \frac{1}{h} \qquad (1)$$

where; q = horizontal load, along each flange, required to balance the flange forces

The effective concentrated torque T at a given diaphragm, can now be computed using this q force; which gives;

$$H = q \cdot \ell$$

however; $T = H \cdot h$

therefore; $T = q \cdot \ell \cdot h$ and

$$T \quad \frac{M}{R} \cdot \frac{1}{h} \cdot \ell \cdot h$$

therefore; $T = \frac{M}{R} \cdot \ell$ (2)

where: ℓ = distance between diaphragms

The total torque applied to the entire bridge, shown in Fig. 4, is therefore;

$$T_T = \Sigma \frac{M}{R} \cdot \ell \qquad (3)$$

The resistance to this total torque is in the form of shears V, which will be assumed linear with respect to the CG of the girders; therefore;

$$V_1/\frac{3D}{6} = V_2/\frac{D}{6}$$

$$V_2 = V_1 \cdot \frac{6}{3D} \cdot \frac{D}{6}$$

$$V_2 = V_1 \cdot \frac{1}{3} \text{ and } V_1 = V_4,\ V_2 = V_3$$

The shear on the outside girders can now be determined from equilibrium;

$$V_1 \cdot \frac{3D}{6} \times 2 + V_2 \cdot \frac{D}{6} \cdot \Sigma = 2 \frac{M}{R} \ell$$

$$V_1 \cdot D + V_1 \cdot \frac{D}{9} = \Sigma \frac{M}{R} \ell$$

$$V_1 \cdot \frac{10}{9} D = \Sigma \frac{M}{R} \ell$$

$$V_1 = \Sigma \frac{M}{R} \ell \cdot \frac{1}{1 \cdot 11\ D} \qquad (4)$$

$$V_2 = V_{1/3} \qquad (5)$$

In general the V forces can be written as;

$$V = \Sigma \frac{M\ell}{RD} \cdot \frac{1}{K} \qquad (6)$$

where k is given as follows for the various number of girders;

Girders	2	3	4	5	6	7	8	9	10
K	1.0	1.0	1.11	1.25	1.40	1.56	1.72	1.88	2.04

The shear forces V can now be applied to the individual girders as additional loads. The diaphragms are designed for the shear forces and end moments, shown in Fig. 4.

The second method to be utilized consists of a computer oriented matrix method, which may consist of three degrees of freedom (28) or five degrees of freedom (24). The 3 DOF grid, accounts for torsional effects, but excluded warping. The 5DOF grid, contains the warping in the top and bottom flanges, thus increasing the degreees of freedom by two.

The space frame grid method, consists of 6 DOF, and considers the girder flanges modeled as beams and the webs as a series of cross and vertical elements. The modeling of the girders in this manner include warping influences by summing the effects of axial loads and vertical bending moments on the flanges.

The finite element method, idealizes the bridge as a series of intersecting plates with specified boundary conditions. Although, the method, in general gives an excellent representation of the bridge, excessive data and computer time is required.

The finite difference procedure permits direct solution of the different curved girder equation. This equation, as developed by Vlasov (5), represent the load-deflection response of curved beams, including warping influences. The details of this technique will now be given. The load deformation response of a single curved girder is given by the following expressions:

$$\left(\frac{EI_w}{R^2} - EI_x\right)\eta'''' + \frac{GK_T}{R^2}\eta'' - \frac{EI_w}{R}\phi'''' + \frac{EI_w}{R}\phi'' + q_y = 0 \qquad (7)$$

$$-\frac{EI_w}{R}\eta'''' + \left(\frac{EI_x+GK_T}{R}\right)\eta'' - EI_w\phi'''' + GK_T\phi'' - \frac{EI_x}{R^2}\phi + m_z = D \qquad (8)$$

where:

η = Vertical Deflection

ϕ = Transverse Rotations

EI_x = Bending Stiffeners

GK_T = Torsional Stiffeners

EI_w = Warping Stiffness

R = Radius

Equations (7) and (8) can be written in central difference form, resulting in the following Equations (9) and (10);

$$\left[\begin{array}{ccccc} \frac{EI_w}{R^2}+EI_x & -\left[4\left(\frac{EI_w}{R^2}+EI_x\right)+\frac{\Delta^2 GK_T}{R^2}\right] & 6\left(\frac{EI_w}{R^2}+EI_x\right)+\frac{2\Delta^2 GK_T}{R^2} & -\left[4\left(\frac{EI_w}{R^2}+EI_x\right)+\frac{\Delta^2 GK_T}{R^2}\right] & \frac{EI_w}{R^2}+EI_x \\ \frac{EI_w}{R} & -\left[4\left(\frac{EI_w}{R}\right)+\frac{\Delta^2(EI_x+GK_T)}{R}\right] & 6\left(\frac{EI_w}{R}\right)+\frac{2\Delta^2(EI_2+GK_T)}{R} & -\left[4\left(\frac{EI_w}{R}\right)+\frac{\Delta^2(EI_x+GK_T)}{R}\right] & \frac{EI_w}{R} \\ n-2 & n-1 & n & n+1 & n+2 \end{array}\right]\begin{array}{c}\eta \\ \phi\end{array} = q_y\Delta^4 \qquad (9a)$$

$$\left[\begin{array}{ccccc} \dfrac{EI_w}{R} & -\left[4\left(\dfrac{EI_w}{R}\right)+\dfrac{\Delta^2(EI_x+GK_T)}{R}\right] & 6\left(\dfrac{EI_w}{R}\right)+\dfrac{2\Delta^2(EI_w+GK_T)}{R} & -\left[4\left(\dfrac{EI_w}{R}\right)+\dfrac{\Delta^2(EI_x+GK_T}{R} & \dfrac{EI_w}{R} \\ EI_w & -\left[4(EI_w)+\Delta^2 GK_T\right] & 6(EI_w)+2\Delta^2 GK_T+\dfrac{\Delta^4 EI_x}{R^2} & -\left[4(EI_w)+\Delta^2 GK_T\right] & EI_w \\ n-2 & n-1 & n & n+1 & n+2 \end{array}\right] = m_z\Delta^4 \qquad (10a)$$

or

$$[K_1]\{\eta\} + [K_2]\{\phi\} = q_y\Delta^4 \qquad (9b)$$

$$[K_3]\{\eta\} + [K_4]\{\phi\} = m_z\Delta^4 \qquad (10b)$$

where the node points are prescribed as shown in Fig. 5.

If the bridge now consists of a series of such curved girders, attached by a series of diaphragms, as shown in Fig. 6, then Eqs. (9) and (10) can be appropriately modified. Consider first the response of a single diaghragm #1, where the end actions developed are given by;

$$\begin{bmatrix} M_{Da} \\ Q_{Da} \\ M_{Dn} \\ Q_{Dn} \end{bmatrix} = [K_5]\cdot\begin{bmatrix} \phi_a \\ \eta_a \\ \phi_n \\ \eta_n \end{bmatrix} \qquad (11)$$

and [K] is a 4x4 matrix and can represent either beam or truss type diaphragms.

Similarly the end actions developed in diaphragm #2, are given by:

$$\begin{bmatrix} M_{Dn} \\ Q_{Dn} \\ M_{Db} \\ Q_{Db} \end{bmatrix} = [K_6] \cdot \begin{bmatrix} \phi_n \\ \eta_n \\ \phi_b \\ \phi_b \end{bmatrix} \tag{12}$$

The total diaphragm induced forces acting at point "n" can be obtained by adding Eqs. (11) and (12), giving;

$$\begin{bmatrix} M_{Dn} \\ Q_{Dn} \end{bmatrix} = \begin{bmatrix} K_5 \\ K_6 \end{bmatrix} \cdot \begin{bmatrix} \phi_a \\ \eta_a \\ \phi_n \\ \eta_n \\ \phi_b \\ \eta_b \end{bmatrix} \tag{13}$$

where:

$$[K_5] = [K_{31}, K_{32}, K_{33} + K_{11}, K_{34} + K_{12}, K_{13}, K_{14}] \tag{14}$$

$$[K_6] = [K_{41}, K_{42}, K_{43} + K_{21}, K_{44} + K_{22}, K_{23}, K_{24}] \tag{15}$$

and K_{ij} = force at points due to a deflection at point j, or points 2 and 4 represent vertical deflections and shears and points 1 and 3 represent rotations and moments.

The interaction of the curved girders and diaphragms can be obtained by adding Eq. (13) to Eqs. (9b) and (10b). It should be noted that Eqs. (9b) and (10b) represent the general response of curved girder #1, subjected to a uniform loads q_y and m_z. At node "n" an additional load per length will be imposed at the diaphragm and is equal to $M_{Dn/\Delta}$ and $Q_{Dn/\Delta}$. Therefore the final equations are;

$$[K_7]\{\eta\} + [K_8]\{\phi\} + [K_9]\{\phi\}'\Delta^3 + [K_{10}]\{\eta\}'\Delta^3 + m_z\Delta^4 = 0 \quad (16)$$

$$[K_{11}]\{\eta\} + [K_{12}]\{\phi\} + [K_{13}]\{\phi\}'\Delta^3 + [K_{14}]\{\eta\}'\Delta^3 + q_y\Delta^4 = 0 \quad (17)$$

where:

$$[K_7] = [K_3] + [0, 0, K_{33} + K_{11}, 0, 0]$$

$$[K_8] = [K_4] + [0. 0. K_{34} + K_{12}, 0, 0]$$

$$[K_9] = [K_{31}, K_{13}]$$

$$[K_{10}] = [K_{32}, K_{14}]$$

$$[K_{11}] = [K_1] + [0, 0, K_{43} + K_{21}, 0, 0]$$

$$[K_{12}] = [K_2] + [0, 0, K_{44} + K_{22}, 0, 0]$$

$$[K_{13}] = [K_{41}, K_{23}]$$

$$[K_{14}] = [K_{42}, K_{24}]$$

$$\{\phi\}' = \begin{bmatrix} \phi_a \\ \phi_a \end{bmatrix}$$

$$\{\eta\}' = \begin{bmatrix} \eta_a \\ \eta_a \end{bmatrix}$$

If the node at which Eqs. (16) and (17) are being applied does not correspond to a diaphragm connection, then the K_{ij} terms are set equal to zero.

The above referenced equations have been incorporated into a computer program (31) and utilized in developing design aids, as will now be described.

DESIGN AIDS

As mentioned previously, the analysis of curved girder systems requires the girder section properties I_x, K_T, and I_w . The designer, therefore, is required to estimate the size of the girders in order to compute these stiffnesses. This size estimation can be achieved if the primary internal girder forces, bending moment M_x and bimoment B_i, which induce normal stresses, were known with some degree of confidence. Then the basic bending equation,

$$\sigma_b = \frac{M_x C}{I_x} = \frac{M_x}{(I_x/C)} \qquad (18)$$

and warping equation,

$$\sigma_w = \frac{BiW_n}{I_w} = \frac{Bi}{(I_w/W_n)} \qquad (19)$$

could be applied by assuming a proportion of the design stress for σ_b and σ_w and computing the required (I_x/C) and (I_w/W_n) properties. Estimation of these stresses and/or forces has been obtained by performing a thorough system analysis of single and multispan curved bridges (31,32). The resulting force equations and graphs, in addition to impact factors and girder property equations, for I-girder bridges, will now be presented.

Impact Factors. - The dynamic response of single span curved girder bridges subjected to a sprung mass vehicle has been predicted by a Fourier Series and Lump Mass Techniques (35). These techniques were then applied in determining the response of typical highway bridges, and thus, the evaluation of impact factors. The impact factor for the bending, Fig. 7, and Fig. 8 lists the impact factor for the torsional effects. The curves are listed for various girder span lengths and central angles $\theta = (L/R)$. The curves are for a vehicle speed of V = 60 mph; other curves are available for velocities of 20 and 40 mph (35).

The impact factors are applied as in conventional pactice;

$$M = M_{STATIC}(1 + I_F) \tag{20}$$

$$Bi = Bi_{STATIC}(1 + I_T) \tag{21}$$

where

M = Induced Bending Moment

Bi = Induced Bimoment

Live Load Design. - The design of any bridge element requires the establishment of forces in that element. If the bridge has a straight alignment, the forces can be determined readily through use of distribution factors and simple beam and torsional forces creates an indeterminate situation. The following equations were, therefore, developed (34), utilizing relationship between single curved responses (33) and the system (30,32), to permit evaluation of the live load forces developed in a curved composite bridge I-girder system. These forces can then be used to establish the induced stresses and thus, proper girder section

Amplification Factor - K_1

All the internal forces and deformations for a single curved and straight girder have been evaluated using the various computer programs (33,36). The ratio of the reactions for these two girders gives the following,

$$K_1 = \frac{f(S.C.G.)}{f(S.S.G.)} \qquad (22)$$

This factor describes the immediate effect of curvature relative to a straight member. The graphical representation of these data and its analysis gives the following general equations (23) and (24),

$$K_{moment} = \frac{0.15}{R}(L/R) + 1 \qquad (23)$$

$$K_{bimoment} = [(35n)(L/R)^2 - 15(L/R)] \times 10^3 \qquad (24)$$

where

$n = R/100, \quad R > 100'$

R = Radius - Ft.

L = Span Length - Ft.

Distribution Factor - K_2

The evaluation of the true distribution of load to each girder, and this realistic values of internal forces, can be considered by analyzing the curved girder as a system. The number of tracks used in the analysis would be dependent on the number of lanes. The ratio of these resulting maximum forces to those in a single curved girder gives

$$K_2 = \frac{f(Sy.C.G.)}{f(S.C.G.)} \qquad (25)$$

A plot of this ratio vs. R and L will yield the following general equations (26) and (27). In all instances, the parameters R and L refer to the midlength and corresponding radius of the system. These equations are valid for girder systems which contain four, six, and eight girders spacings of 7', 8', 9', and 10'.

$$K_{moment} = (n + 3)\left(\frac{0.4L}{R}\right) + 0.6 \qquad (26)$$

$$K_{bimoment} = \frac{.11}{M}(R/L) \quad L < 70' \qquad (27a)$$

$$K_{bimoment} = \frac{(M - 1)}{6}(R/L) \quad L > 70' \qquad (27b)$$

where

$n = R/100, \quad R > 100'$

$M = L/50, \quad L > 50'$

R = Radius - Ft.

L = Span Length - Ft.

Reduction Factor K_3

Because many bridge structures are continuous, it is desirable to obtain some factors which can be applied to the simple span data to give preliminary forces in continuous spans. This factor can be written as

$$K_3 = \frac{f(Sy.C.G.)N}{f(Sy.C.G.)} \qquad (28)$$

where

$$N = \text{Number of spans (2 or 3)}$$

Utilizing a computer program (36), the maximum forces in a two or three span curved bridge system of four, six, and eight girders were evaluated under various critical loadings. A study of all the data and the resulting K_3 values give the following values listed in Table 1. The data is described relative to number of spans and is independent of number of girders. It should be emphasized that the two and three span girder systems must all contain equal span lengths with a maximum given span length of 100.0'. For example, for a three span system, the total maximum bridge length would then be 3 x L = 300.0'.

Evaluation of Girder Forces and Deformation

With the various factor equations available, it is now possible to evaluate preliminary forces in a curved girder bridge, relative to the forces in a striaght girder. The procedure is as follows:

1. Evaluate maximum function "F" for a single straight girder of length L subjected to a line of AASHO wheel loads. This function would be $F_{bending.}$ The function for $F_{bimoment}$ is assumed to be equal to one.
2. Evaluate amplification factors K_1 equations (23) and (24) for the midspan length L and radius R of the bridge system.

3. Evaluate distribution factors K_2 equations (26) and (27) for the given midspan length L, number of girders in system, and radius R.

4. Select reduction factor K_3 from Table 1 if system is continuous span.

5. Determine maximum function F of curved girder system, i.e.

$$\text{Max. Moment Static} = \text{M straight beam} \times (K_1 \times K_2 \times K_3) \quad (2$$

and

$$\text{Max. Bimoment Static} = 1.0(K_1 \times K_2 \times K_3) \quad (30)$$

6. Multiply Max. Moment and Max. Bimoment by (1 + I) Equations (20) and (21), respectively.

In order to account for girder spacing S, the resulting action of single straight girser subjected to a line of truck wheels should be multiplied by a Factor $\overline{S}$, where

$$\overline{S} = 1.29 \quad \text{for} \quad S = 7' \text{ or } 8'$$

$$\overline{S} = 1.57 \quad \text{for} \quad S = 9' \text{ or } 10'$$

Also, all of the parameters R and L are referenced to the centerline of the bridge system.

Dead Load Design. - The dead load response of curved bridge systems has been predicted by Murphy (31), which more realistically represents the response of the bare steel frame system. This technique has been computerized and then applied in the development of design curves. The curves represent the maximum bending stress (σ_b) and warping stress (σ_w) as a function of R, L, I_x, and diaphragm spacing S. The curves were based on typical girder properties obtained from a survey of bridge design as a function of girder length, as given in Figure 15. If the actual design properties are different from those referenced values given in Figure 15, a correction factor is required to modify the chart values.

It has been found that for bending stress,

$$\sigma_b = \bar{\sigma}_b C_1 \tag{31}$$

where

σ_b = Actual or Corrected Bending Stress

$\bar{\sigma}_b$ = Bending Stress form Referenced Properties

C_1 = $\bar{S}/S$

$\bar{S}$ = Referenced Section Modulus = $f_1(L)$

S = Design Section Modulus

The warping stress is related by the following,

$$\sigma_w = \bar{\sigma}\, C_2 \tag{32}$$

σ_w = Actual or Corrected Warping Stress

σ_w = Warping Stress from Referenced Properties

C_2 = $(W_n/I_w)(W_n/I_w0$

(W_n/I_w) = Referenced Value = $I_x f_x(L)$

(W_n/I_w) = Design Value

The design charts, Figures 9 through 14, list the induced stresses $\bar{\sigma}_b$ and $\bar{\sigma}_w$ normalized relative to the applied dead load (W) per length (K/in.). Thus, any variation in dead load may be considered in an actual design. The normalized stress vs. the (R/L) values are plotted for span lengths of L = 100', 125', and 150'. These plots are dependent on stiffness (I_x), number of girders, and diaphragm spacing.

The plots of $\frac{\bar{\sigma}_b}{w}$ vs. (R/L) for L = 100', 125', and 150' (Figures 4,6,8) are independent of diaphragm spacing and (I_x), as determined in the development of the curves. (4).

The plots of $\frac{\bar{\sigma}_w}{w}$ vs. (R/L) are dependent on diaphragm spacing and stiffness (I_x), as given in Figures 10, 12, and 14. These curves are limited to a four-girder system and are for specified stiffnesses, however, for other stiffnessed, the following equation may be applied,

$$\left(\frac{\bar{\sigma}_w}{w}\right)_e = \left(\frac{\bar{\sigma}_w}{w}\right)_{chart} \times \frac{\bar{I}_{rx}}{I_x} \tag{33}$$

where

$\left(\frac{\sigma_w}{w}\right)_e$ = Referenced Modified Factor

I_x = Actual Stiffness

$\bar{I}_{rx}$ = Reference Chart Value given in Figures 10, 12, and 1

Warping Stress Adjustment

The curves that are presented herein are the results relative to the four-girder system. In order to account for the behavior of six and eight-girder systems, the $\left(\frac{\bar{\sigma}_w}{w}\right)$ chart value must be modified. These chart values are modified by the following equation,

$$\left(\frac{\bar{\sigma}_w}{w}\right)_{exact} = \left(\frac{\bar{\sigma}_w}{w}\right)_{chart} - \Delta\left(\frac{\bar{\sigma}_w}{w}\right) \qquad (34)$$

where

$$\Delta\left(\frac{\bar{\sigma}_w}{w}\right) = \frac{I_x}{I_x} \frac{S^2}{(25^2)} \left(\frac{\bar{\sigma}_w}{w}\right)^1 \qquad (35)$$

$\left(\frac{\bar{\sigma}_w}{w}\right)^1$ = Change in the value of the chart due to an increase in the number of girders (six or eight) beyond the referenced four girders, as given in Figure 10

$\bar{I}_w$ = Reference Property given in Figure 16

I_w = Design Property

S = Design Diaphragm Spacing

Girder Spacing

The curves have been developed for girder spacing of 8.0', however spacings up to 10' are acceptable.

Torsional Properties. - In order to determine design stresses or distortions, the girder section property is required. If the girder is subjected to torsion and bending, the torsional properties, in addition to bending properties, will be required. The exact solution of the torsional properties of composite sections has been demonstrated (37). However, by idealizing the composite section, a series of simplifi equations can be developed and used for design (11).

Figure 17 describes a typical composite girder and pertinent dimensions. Neglecting the top girder flange and modifying the concrete slab thickness, as shown in Figure 12, with the dimensions defined as follows: $t_1 = nt_s$, $d_1 = bs$, $t_2 = w$, $d_2 = d_g + t_s/2 - t_f/2$, $d_3 = b_f$, $t_3 = t_f$, $n = E_c/E_s$, $m = G_c/G_s$. Using these dimensions, the resulting torsional properties are determined as follows,

<u>Shear Center</u>

$$\alpha = \frac{d_3^3 t_3}{(d_1^3 t_1 + d_3^3 t_3)^2} d_2 \tag{36}$$

<u>Normalized Warping Functions</u>

$$\text{Slab:} \quad W_{n_c} = \frac{ad_1}{2} \tag{37}$$

$$\text{Beam:} \quad W_{n_s} = \frac{(d_2-\alpha)}{2} d_3 \tag{38}$$

<u>Warping Stiffness</u>

$$I_w = \frac{\alpha^2}{12} t_1 d_1^3 + (d_2-\alpha)^2 \frac{t_3 d_3^3}{12} \tag{39}$$

<u>Torsional Constant</u>

$$K_T = \frac{1}{3} [\, d_3 t_3^3 + d_2 t_2^3 + md_1 t_s^3] \tag{40}$$

where

$$m = G_c/G_s$$

With the evaluation of these torsional parameters, the resulting normal stresses in the composite section due ti bimoment can be evaluated as follows:

Warping Nominal Stress

$$\text{Slab:}\quad \sigma_{w_c} = \frac{BiW_{n_c}}{I_w} \tag{41}$$

$$\text{Steel:}\quad \sigma_{w_s} = \frac{BiW_{n_s}}{I_w} \tag{42}$$

DESIGN SPECIFICATIONS

The AASHTO design criteria (1,2), as applied to steel I and Box girder bridges, is given in Table 2. These criteria are listed for both straight and curved elements.

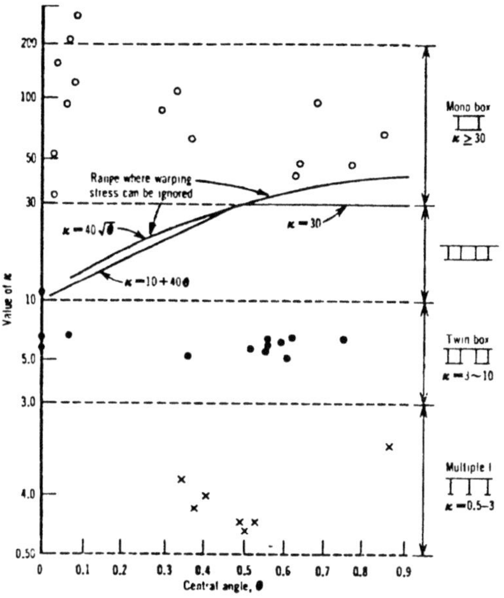

200
100
50
30
10
5.0
3.0
4.0
0.50
Value of κ
0
0.1
0.2
0.3
0.4
0.5
0.6
0.7
0.8
0.9
Central angle, θ
Range where warping stress can be ignored
κ = 40 √θ
κ = 30
κ = 10 + 40θ
Mono box
κ ≥ 30
Twin box
κ = 3 ~ 10
Multiple I
κ = 0.5–3

Figure 1

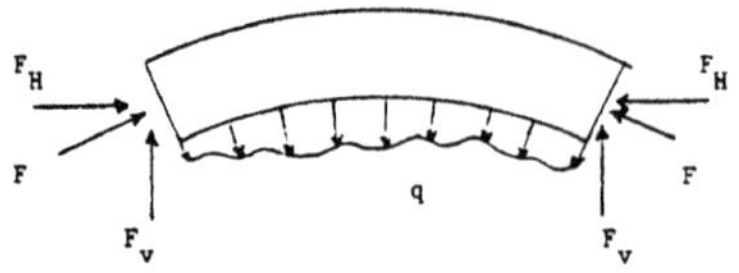

Fig. 2

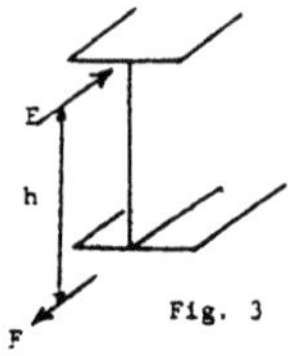

Fig. 3

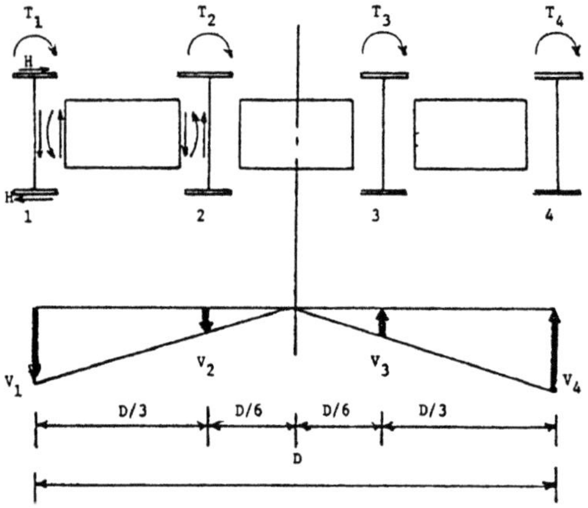

Fig. 4

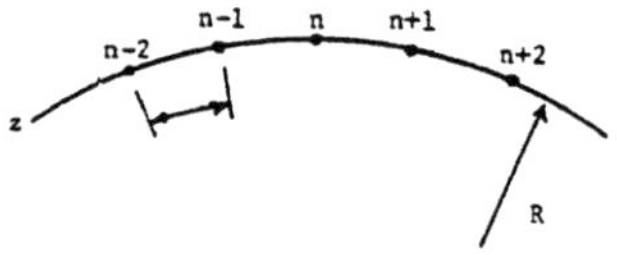

Fig. 5

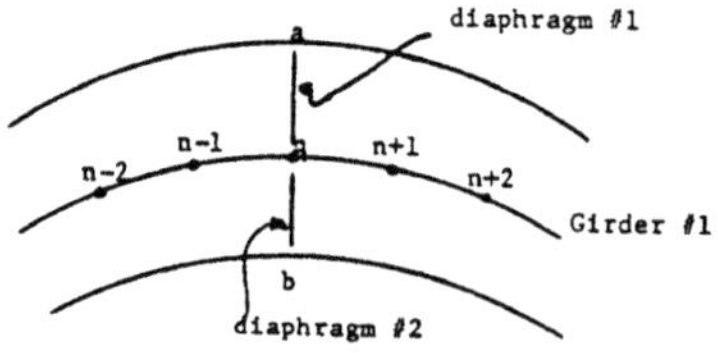

Fig. 6

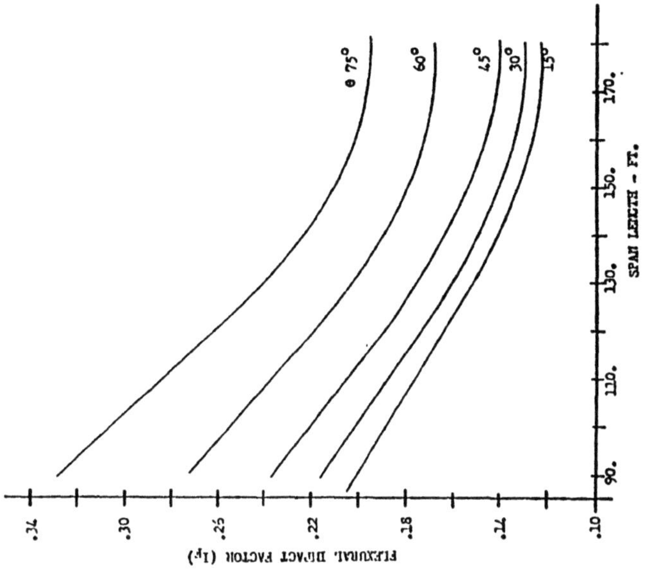

Figure 7

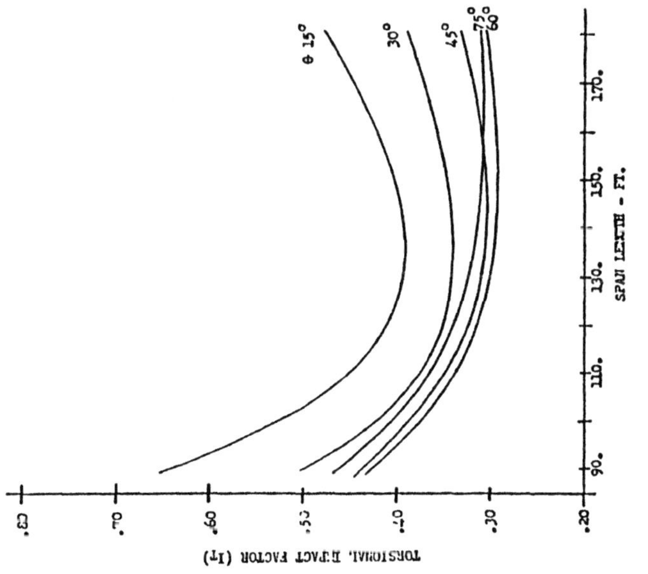

Figure 8

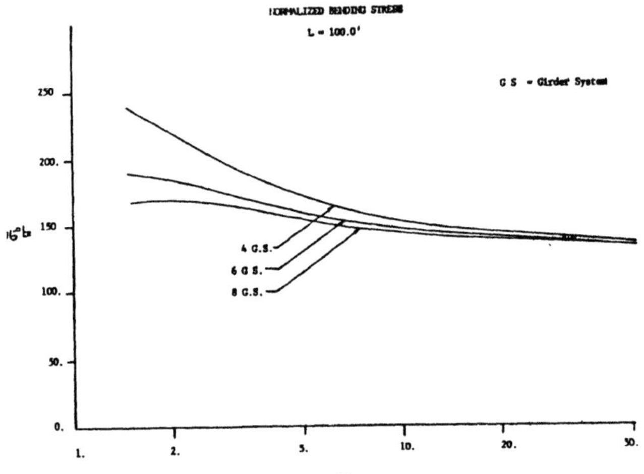

Figure 9

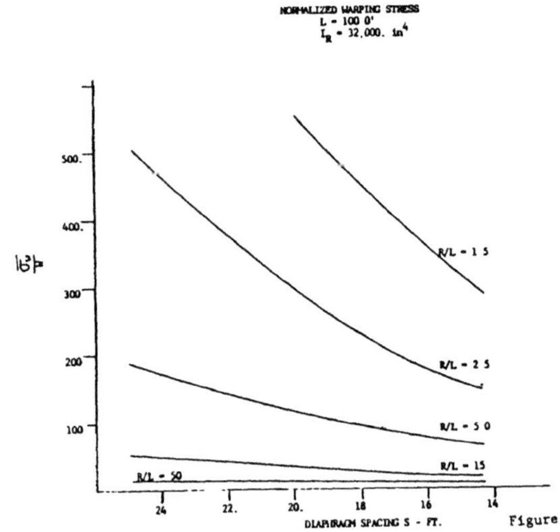

Figure 10

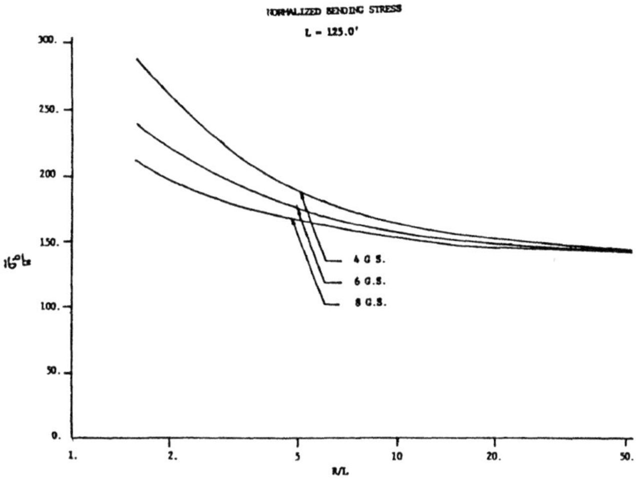

Figure 11

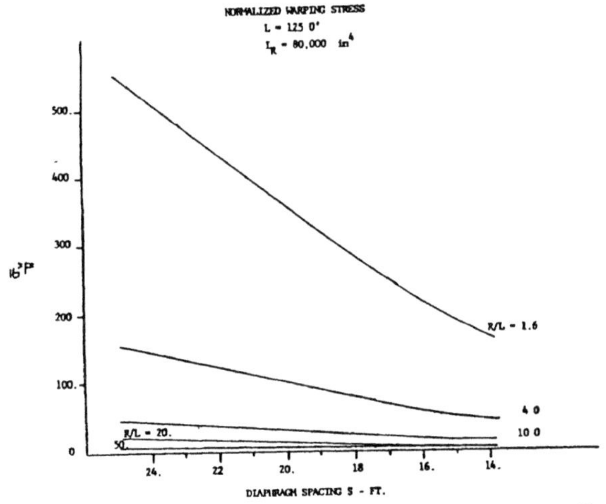

Figure 12

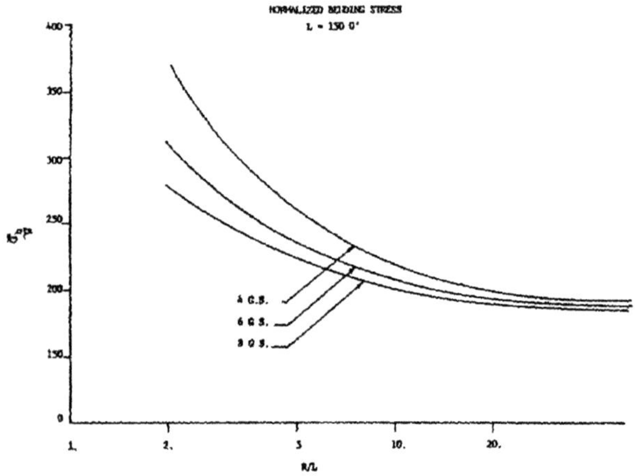

Figure 13

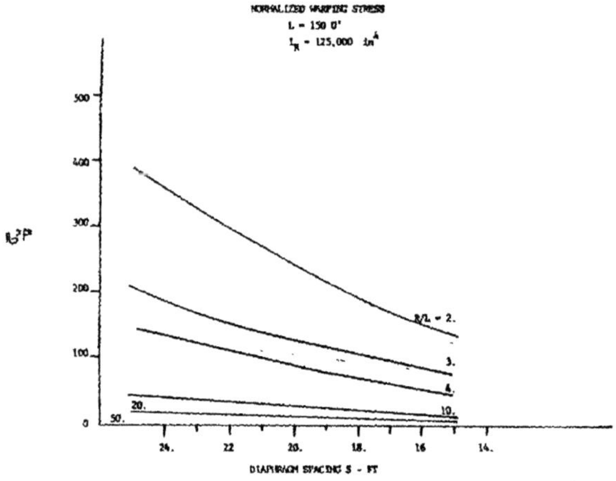

Figure 14

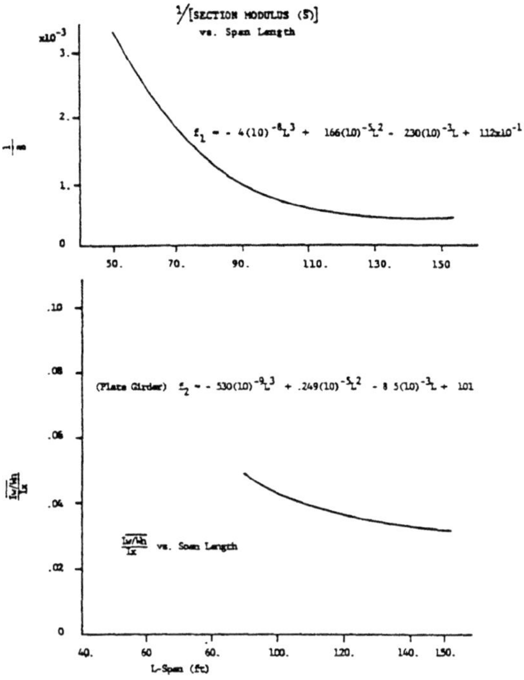

Figure 15

NORMALIZED WARPING STRESS VARIATION

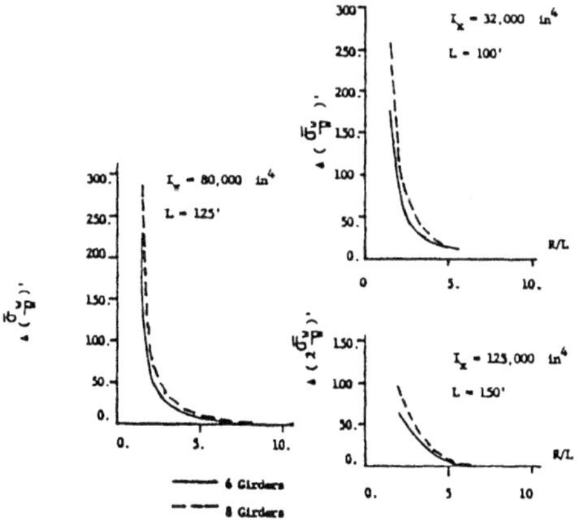

Figure 16

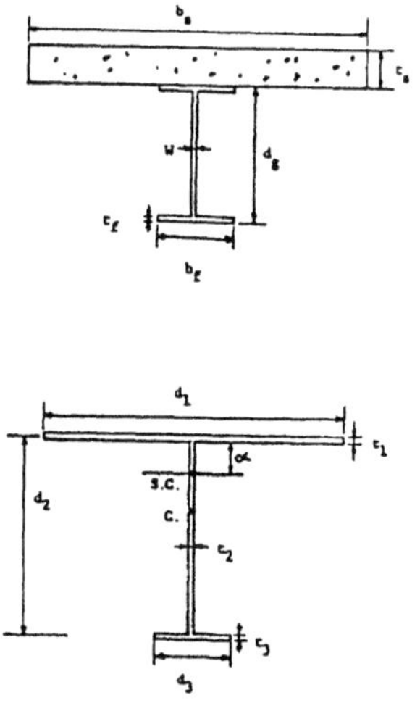

Figure 17

TABLE 1

K_3 - REDUCTION FACTOR FOR MAXIMUM FUNCTION IN TWO AND THREE SPAN BRIDGES

Number of Spans	$K_{Bending\ Moment}$	$K_{Deflection}$	$K_{Rotation}$	$K_{St.\ Venant}$	$K_{Warping\ Torsion}$	$K_{Bimoment}$	K_{Shear}
Two Span	.75	.70	.70	.75	.45	.35	1.00
Three Span	.65	.60	.70	.75	.40	.35	.90

TABLE 2 **Working Stress Design**

Item	Straight	Curved
Compression Flange (positive moment) 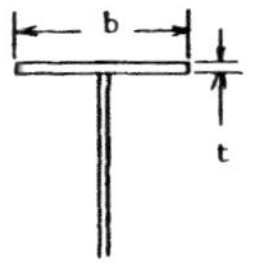 	$\frac{b}{t} \leq \frac{3250}{\sqrt{f_b}} \leq 24$	$\frac{b}{t} \leq \frac{4400}{\sqrt{F_y}}$ and $F_b = 0.55\ F_y[1 - \frac{(\frac{\ell}{r})^2 F_y}{4\pi^2 E}]\rho_B \rho_w$ where $\rho_B = \frac{1}{1 + (\frac{\ell}{R})(\frac{\ell}{b})}$ and $\rho_w = \rho_{w1}$ or ρ_{w2}, where; $\rho_{w1} = \frac{1}{1 - (\frac{f_w}{f_b})[1 - \frac{(\frac{\ell}{b})}{75}]}$ or $\rho_{w2} = \frac{0.95 + \frac{\frac{\ell}{b}}{[30+8000(0.1-\frac{\ell}{R})^2]}}{1 + 0.6\ (\frac{f_w}{f_b})}$ if $\frac{f_w}{f_b}$ (+) use smaller ρ_{w1} or ρ_{w2} $\frac{f_w}{f_b}$ (-) use ρ_{w1}

TABLE 2 **Working Stress Design** (Continued)

Item	Straight	Curved
Compression Flange (negative moment)	$\frac{b}{t} \leqq \frac{6140}{\sqrt{F_y}}$	$\frac{b}{t} \leqq \frac{6140}{\sqrt{F_y}} \cdot X$
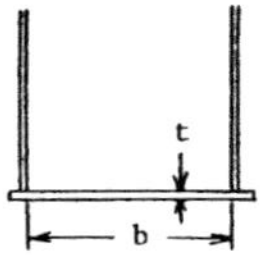	$f_b \leqq 0.55\ F_y$	$F_b = 0.55\ F_y \sqrt{1-9.2[\frac{f_v}{F_y}]^2}$
	$\frac{6140}{\sqrt{F_y}} \leqq \frac{b}{t} \leqq 60$ or $\frac{13,300}{\sqrt{F_y}}$	where $X = 1 + 4/3(\frac{f_v}{\sqrt{F_y}} - 0.15) \geqq 1$
	$f_b \leqq 0.55F_y - 0.224F_y[1-\sin\frac{\pi}{2}\ (\frac{13300 - {}^b/_t\sqrt{F_y}}{7160})]$	$\frac{6140}{\sqrt{F_y}} < \frac{b}{t} \leqq \frac{13,300}{\sqrt{F_y}}$ or 60
	$\frac{13300}{\sqrt{F_y}} < \frac{b}{t} \leqq 60$	$F_b = \left[0.326F_y + 0.224F_y\left\{\sin\frac{\pi}{2}\ (\frac{13,300 - \frac{b\sqrt{F_y}}{t}}{13,300-6140X})\right\}\right]\Delta$
	$f_b \leqq 57.6 \times 10^6\ (\frac{t}{b})^2$	where $\Delta = \sqrt{1-9.0(\frac{f_v}{F_y})^2}$
		if $\frac{b}{t} \geqq \frac{13300}{\sqrt{F_y}}$
		F_b is smaller of the following;
		$F_b = 57.6(\frac{t}{b})^2 \cdot 10^6\Delta$
		$F_b = 57.6(\frac{t}{b})^2 \times 10^6 - \frac{f_v^{\ 2}}{113.4(\frac{t}{b})^2} \times 10^6$

TABLE 2 **Working Stress Design** (Continued)

Item	Straight	Curved
Compression Flange (negative moment)	$\frac{w}{t} \leqq \frac{3070\sqrt{K}}{\sqrt{F_y}}$	$\frac{w}{t} \leqq \frac{3070\sqrt{K}}{\sqrt{F_y}} \cdot X_1$
	$f_b \leqq 0.55F_y$	$F_b = 0.55F_y \sqrt{1-912(\frac{f_v}{F_y})^2}$
with Stiffener	$\frac{3070\sqrt{K}}{\sqrt{F_y}} < \frac{w}{t} \leqq 60$ or $\frac{6650\sqrt{K}}{\sqrt{F_y}}$	where
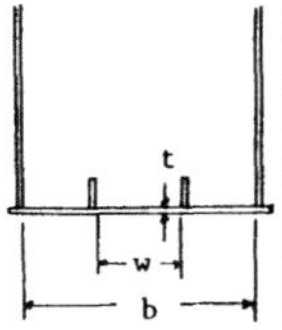	$f_b \leqq 0.55F_y - 0.224F_y\left[1 - \sin\frac{\pi}{2}\frac{6650\sqrt{K} - \frac{w}{t}\sqrt{F_y}}{3580\sqrt{K}}\right]$	$X_1 = 1$ (n>1)
	$\frac{6650\sqrt{K}}{\sqrt{F_y}} < \frac{w}{t} \leqq 60$	$X_1 = 0.93 + (1.6 - \frac{K}{K_s})(\frac{f_v}{F_y}) \geqq 1$ (n=1)
	$f_b \leqq 14.4 \times 10^6 K(\frac{t}{w})^2$	$2 \leqq K \leqq 4$
	Stiffener requirement	$K_s = \frac{5.34 + 2.84\left(I_s/wt^3\right)^{1/3}}{(n+1)^2} \leqq 5.34$
	with longitudinal Stiffener	
	$I_s \geqq \phi t^3 w$	
	where $\phi = \begin{cases} 0.07K^3n^4 & \text{for } n>1 \\ 0.125K^3 & \text{for } n=1 \end{cases}$	

TABLE 2 **Working Stress Design** (Continued)

Item	Straight	Curved
Compression Flange (negative moment) with stiffener 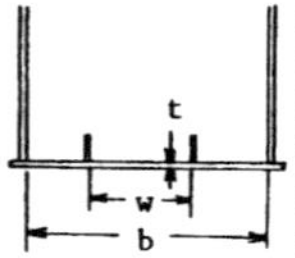		$\frac{3070\sqrt{K}}{\sqrt{F_y}} X_1 < \frac{w}{t} \leqq \frac{6650\sqrt{K}}{\sqrt{F_y}} X_2$ or 60 $F_b = \left[0.326F_y + 0.224F_y\left\{\sin \frac{\pi}{2} \frac{6650\sqrt{K}\,X_2 - \frac{w\sqrt{F_y}}{t}}{6650\sqrt{K}\,X_2 - 3070\sqrt{K}\,X_1}\right\}\right]\Delta$ Where $\Delta = \sqrt{1-9.0(\frac{f_v}{F_y})^2}$ $X_2 = 1 - 2.13(\frac{f_v}{F_y})$ $+ 0.1\left\{(\frac{K}{K_s}) - 5.34\right\}^2 (\frac{f_v}{F_y})$ $\frac{6650\sqrt{K}\,X_2}{\sqrt{F_y}} < \frac{w}{t} \leqq 60$ F_b is smaller value of $F_b = 14.4K(\frac{t}{w})^2 \Delta\ 10^6$ or $F_b = 14.4K(\frac{t}{w})^2 \times 10^6 - \frac{f_v^2 K}{14.4(K_s)^2(\frac{t}{w})^2} \times 10^6$

TABLE 2 **Working Stress Design** (Continued)

Item	Straight	Curved
With Longitudinal and Transverse Stiffener	Use same formula, but use K_1 instead of K $K_1 = \frac{[1+(\frac{a}{b})^2]^2 + 87.3}{(n+1)^2(\frac{a}{b})^2[1+0.1(n+1)]}$ $I_s \geq 8t^3w$ $I_t \geq 0.10(n+1)^3 w^3 \frac{f_s}{E} \frac{A_f}{a}$ where a: spacing of transverse stiffener f_s: maximum longitudinal bending stress A_f: Area of flange including longitudinal stiffener	

TABLE 2 **Working Stress Design** (Continued)

Item	Straight	Curved
Web Without Stiffener 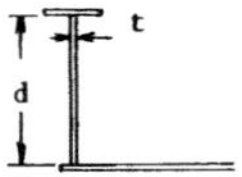	$\frac{d}{t} \leq 150$	Same
	$f_v \leq \frac{5.625 \times 10^7}{(^d/_t)^2} \leq \frac{F_y}{3}$	
	$\frac{d}{t} \leq \frac{23,000.}{\sqrt{F_b}} \leq 170$	
With Transverse Stiffener	$d_o \leq 1.5d$	If $^{d_o}/_R \leq 0.02$ use straight Girder criteria
	$f_v \leq \frac{F_y}{3} [C + \frac{0.87(1-C)}{\sqrt{1 + (^{d_o}/_d)^2}}]$	If $^{d_o}/_R \geq 0.02$
	$C = \frac{2.2 \times 10^8 [1 + (^d/_{d_o})^2]}{F_y(^d/t)^2} \leq 1.0$	$\frac{d}{t} \leq \frac{23000}{\sqrt{F_b}} \left\{1.19 - 10(\frac{d_o}{R}) + 34 (\frac{d_o}{R})^2\right\} \leq 170$
	d_o = stiffener spacing	$d_o \leq 1.5d$
		$f_v = \frac{F_y}{3} [C + \frac{0.87(1-C)}{\sqrt{1 + (^{d_o}/_d)^2}}]$
		$C = \frac{2.2 \times 10^8 [1 + (^d/d_o)^2]}{F_y(^d/t)^2} \leq 1.0$

TABLE 2 **Working Stress Design** (Continued)

Item	Straight	Curved
Web with Transverse Stiffener	Stiffener Criteria $I \geq \frac{d_o t^3}{10.92} J$ $J = 25 (\frac{d_o}{d})^2 - 20 \geq 5.0$	Stiffener Criteria $I \geq d_o t^3 J$ $J = [25(\frac{d_o}{d})^2 - 20] X \geq 5.0$ $X = 1.0$ for $\frac{d_o}{d} \leq 0.78$ $X = 1.0 + \left(\frac{(\frac{d_o}{d} - 0.78)}{1775}\right) Z^4$; $0.78 \leq \frac{d_o}{d} \leq 1.0$ $Z = 0.95 \frac{d^2}{Rt}$ $\frac{b}{t} \leq \frac{2600}{\sqrt{F_y}}$

REFERENCES

1. "Standard Specifications for Highway Bridges," 12th Edition, AASHTO, Washington, D.C., 1977.

2. "Guide Specifications for Horizontally Curved Highway Bridges," AASHTO, Washington, D.C., 1980.

3. T.V. Galambos, "Tentative Load Factor Design Criteria for Curved Steel Bridges," Rept. No. 50, Washington, Univ., St. Louis, MO, May 1978.

4. R. Wolchuck, R. Mayrbaurl, :Proposed Design Specifications for Steel Box Girder Bridges," FHWA Rept. TS-80-205, January 1980.

5. C.P. Heins, "Box Girder Design," AISC Engineering Journal, Vol. 15, No. 4, December 1978.

6. C.P. Heins, L.J. Hua, "Proportioning Box Girder Bridges," ASCE St. Div. Jour., Vol. 106, No. ST11, November 1980.

7. C.P. Heins, D.H. Hall, "Designers Guide to Steel Box Girder Bridges," Bethlehem Steel Corp., Bethlehem, PA, 1981.

8. C.P. Heins, J.C. Olenick, "Curved Box Beam Bridges Analysis," Computers and Structures Journal, Vol. 6, pp. 65-73, Pergamon Press, London, 1976.

9. C.P. Heins, F.H. Sheu, "Computer Analysis of Steel Box Girder Bridges," Civil Engineering Dept. Report, Univ. Of Maryland, June 1981.

10. C.P. Heins, J.C. Olenick, "Diaphragms for Curved Box Beam Bridges," ASCE St. Div. Jour., Vol. 101, No. ST10, October 1975.

11. C.P. Heins, A, Sahin, "Natural Frequency of Curved Box Girder Bridges," ASCE St. Dvi. Jour., Vol. 105, No. ST12, December 1979.

12. C.P. Heins, R. Humphrey, "Bending and Torsion Interaction of Box Girders, " ASCE St. Div. Jour., Vol. 105, No. ST5, May 1979.

13. C.P. Heins, et al., "Curved Steel Box Girder Bridges: A Survey," ASCE St. Div. Jour., Vol. 104, No. ST11, November 1978.

14. C.P. Heins,et al., "Curved Steel Box Girder Bridges: State of Art," ASCE St. Div. Jour., Vol. 104, No. ST11, November 1978.

15. C.P. Heins, W.H. Lee, "Curved Box Girder Field Test, " ASCE St. Div. Jour., Vol. 107, No. ST2, February 1981.

16. C.H. Yoo, J. Buchanan, C.P. Heins, W.L. Armstrong, "Loading Response of a Continuous Box Girder Bridge," Proceeedings ASCE Specialty Conference of Metal Bridges, St. Louis, MO, November 1974.

17. C.P. Heins, J.Y. Shyu, "Moment Capacity of Box Girders," Institute for Physical Science and Technology Report, University of Maryland, June 1981.

18. Heins, C.P., Huo, C., "Truck and Wind Load Effects on Girder Slab Bridge With/Without Wind Bracing," AISI Project 308, Effects of Bracing on I Girder Bridges, Inst. for Phy. Sci. and Tech., Tech. Note BN-988-A, University of Maryland, College Park, MD, April 1980.

19. Heins, C.P., Wind Load Effects on Tangent and Skew I Girder Bridges With/Without Lateral Wind Bracing," Addendum, AISI Project 308, Inst. for Phys. Sci. and Tech., University of Maryland, College Park, MD, July 1980.

20. Heins, C.P., Jin, J., "Load Distribution of Braced Curved I Girder Bridges," AISI Project 308, Effects of Bracing I Girder Bridges, Inst. for Phys. Sci. and Tech., University of Maryland, College Park, MD, June 1982.

21. Heins, C.P., Kato, H., "Bottom Lateral Bracing Effects of Curved I Girder Bridges," AISI Project 308, Effects on I Girder Bridges, Inst. for Phys. Sci. and Tech., University of Maryland, College Park, MD, February 1981.

22. Heins, C.P., Kuo, T.C., "Live Load Distribution of Simple Span Steel I-Beam Composite Highway Bridge at Ultimate Load," Department of Civil Engineering, University of Maryland, April 1973.

23. Wilson, L., Bathe, K.J., Peterson, F.E., Dovey, H., "User's Guide to Program SAP," University of California, Department of Civil Engineering, November 1972.

24. Schelling, D.R., Heins, C.P., Sikes, G.H., "State of the Art of Curved Girder Bridge Programs," Computer and Structures, Vol. 9, pp. 27-37, Pergamon Press, London, 1978.

25. Heins, C.P., Sheu, F.H., "Design/Analysis of Curved Box Girder Bridges," Computers and Structures, Vol. 15, pp. 241-258, Pergamon Press. London, 1978.

26. Nakai, H., Heins, C.P., "Analysis Criteria for Curved Bridges," ASCE St. Div. Jour., Vol. 103, No. ST7, July 1977.

27. Merritt, F.S., "Structural Steel Designers Handbook," McGraw Hill Book Co., New York, N.Y. , 1972.

28. Lavelle, F.H.m Boick, L.S., "A Program to Analyze Curved Girder Bridges", Engr. Bullentin No. 8, Univ. of Rhode Island, Kingston, Rhode Island, 1965.

29. United States Steel Corporation, Highway Structures Design Handbook, Vol. 1, 1967.

30. Bell, L.C., Heins, C.P., "The Solution of Curved Bridge Systems Using the Slope-Deflection Fourier Series Method", Civil Engineering Rept. No. 30, Univ. of Maryland, College Park, MD, June 1968.

31. Murphy, E.L., Heins, C.P., "Dead Load Analysis of Single Span Curved Bridges", Civil Engineering Rept. No. 52, Univ. of Maryland, College Park, MD, June 1973.

32. Heins, C.P., "The Presentation of the Slope-Deflection Method for the Analysis of Curved Orthotropic Highway Bridges", Civil Engineering Rept. No. 15, June 1967.

33. Spates, K.R., Heins, C.P., "The Analysis of Single Curved Girders with Various Loadings and Boundary Condition", Civil Engineering Rept. No. 20, June 1968.

34. Siminou, J., Heins, C.P., "Proposed Equations for Preliminary Design of Curved Girder Bridge Systems", Civil Engineering Rept. No. 25, June 1969.

35. Vashi, K.M., Heins, C.P., "Impact Factors for Curved Highway Bridges", Civil Engineering Rept. No. 32, Spetember 1969.

36. Bell, L.C., Heins, C.P., "Curved Girder Computer Manual", Civil Engineering Rept. No. 30, September 1968.

37. Kuo, J.T.C., Heins, C.P., "Torsional Properties of Composite Steel Bridge Members", Civil Engineering Rept. No. 37, June 1970.

38. Kuo, J.T.C., Heins, C.P., "Behavior of Composite Beams Subjected to Torsion", Civil Engineering Rept. No. 39, February 1971.

39. Heins, C.P., "Curved I Girder Design", Highway Research Record No. 547, Highway Research Board, January 1975.

SIMPLIFIED METHODS FOR THE ANALYSIS AND DESIGN OF BRIDGES OF CELLULAR CROSS-SECTION

H.R. Evans

Department of Civil and Structural Engineering, University College, Cardiff.

SYNOPSIS

The paper describes a number of simplified methods that have been developed for the analysis of cellular bridge decks; the behaviour in the linear elastic and in the non-linear post-buckling range is considered. Each method assumes an idealized structural behaviour to provide a simplified solution that may be carried out on a microcomputer. The methods have been developed specifically for use during the preliminary stages of design when repeated analyses may be required to establish optimum proportions. The paper includes some typical comparisons of the results given by the proposed methods with the results of other established methods.

1 INTRODUCTION

In recent years there has been a great increase in the computing power that is available to the bridge designer and this has made it possible for powerful techniques, requiring extensive computations, to be applied to the analysis of bridge structures. Of major importance in this context is the finite element method (1) which, by virtue of its adaptability, enables a wide range of bridge loading, geometry and support conditions to be taken into account in analysis.

Despite the increase in available computing power, the finite element method must still be regarded as a rather expensive method of analysis. Certainly, at the initial design stage, when the designer is endeavouring to optimize the proportions of his structure by repeated analysis, the costs involved in using the method are unacceptable. Furthermore, since most modern design codes are based on limit state considerations, it is essential to be able to determine the ultimate loads of bridge components. In such analyses, since due account must be taken of both geometric and material non-linearity, an incremental procedure must be adopted; the consequent demands upon computing time made by a finite element solution are exceptionally high.

There is, therefore, a real need for simplified alternatives to the finite element method for use at the preliminary design stage. Ideally, once the designer has established a reasonably efficient structure by repeated analysis employing these approximate techniques, a final check analysis by the finite element method should be carried out.

The advent of more powerful computers in no way lessens the value of such simplified alternative methods. In parallel with the development in main frame computing power in recent years there has been an even more significant development in microcomputers and programmable calculators. The simplified techniques to be discussed in this paper may be programmed for solution on micro-computers of the types that are becoming increasingly available to the design engineer.

The development and application of several simplified, approximate methods will be outlined in this paper. The first part of the paper will concentrate upon the elastic, global analysis of cellular girders, whereas the second part will consider the calculation of the collapse load of complete multi-cellular girders. Where possible, comparisons will be made between the approximate results calculated by the simplified methods and those obtained, either from a full finite element analysis, or from experiments.

2 GLOBAL ELASTIC ANALYSIS OF CELLULAR STRUCTURES

Three techniques associated with the global analysis of cellular girders in the elastic range will be discussed in some detail. The methods are :

(a) the nodal section method
(b) the bar simulation method
(c) the grillage method

2.1 The Nodal Section Method (2,3)

The nodal section method has been developed to provide a relatively simple method of predicting the moments, stresses and deflections in single-span and continuous box girders. It may be applied to girders where the geometry of cross-section varies along the span and to girders that are curved in plan. In the case of a straight girder of uniform cross-section the solution becomes extremely simple and can be carried out conveniently on a micro-computer.

The method is based on the adaptation of the "ordinary" method of analysis for folded plate roofs to the analysis of closed box sections.

In general, any component plate of a cellular girder bends both in its plane and perpendicular to its plane. It is assumed in the nodal section method that the bending action of each plate perpendicular to its plane can be represented by considering a series of transverse one-way slab strips and that the in-plane bending action of each plate can be idealized as that of a beam spanning in the longitudinal direction between the supports.

On the basis of these assumptions, the simplified structural behaviour shown in Fig.1, may be considered. The structural action is assumed to consist of the action of a series of one-way, transverse frames, elastically supported at the joints by a series of inter-connected longitudinal plate beams. The frames are assumed to transmit shears and moments in the transverse direction only, this action being termed the "transverse frame action". The plate beams, on the other hand, only transmit those force components within their planes and this action is termed the "longitudinal plate action".

The diagram gives an outline of the procedure followed in the nodal section solution. The structure is first idealized by taking a number of transverse nodal sections and a transverse frame is considered to be positioned at each nodal section; in the case of a girder of varying cross-sectional geometry, the dimensions of each frame will be different. The transverse frame action under the external loading is analysed by assuming that each frame is rigidly supported at the joints and the transverse bending moments and joint reactions are then determined simply for each frame in turn.

Since the frame and plate systems are considered to be inter-connected at the joints, the reactions calculated for the frame system are applied as joint loads on to the plate system.

The longitudinal plate system is then analysed by considering each individual plate to act as a beam spanning between the supports. It is ensured that equilibrium of longitudinal stresses and compatibility of longitudinal strains is maintained at the edges between plates and the longitudinal membrane stresses in the plates and the joint deflections are calculated. Since the transverse frames were analysed assuming the joints to be rigidly supported and since these joints were allowed to deflect in the longitudinal plate analysis, the displacements of the plate and frame systems are not compatible at this stage of the solution. These incompatibilities can be conveniently expressed in terms of the sway displacement of each component plate, see Fig. 2, and removed by an iterative technique that is termed the "sway correction procedure".

In this procedure, the sway displacements are applied to the transverse frame and the resulting moments evaluated by carrying out another transverse frame analysis. The associated reaction forces are then applied to the plate system and new joint

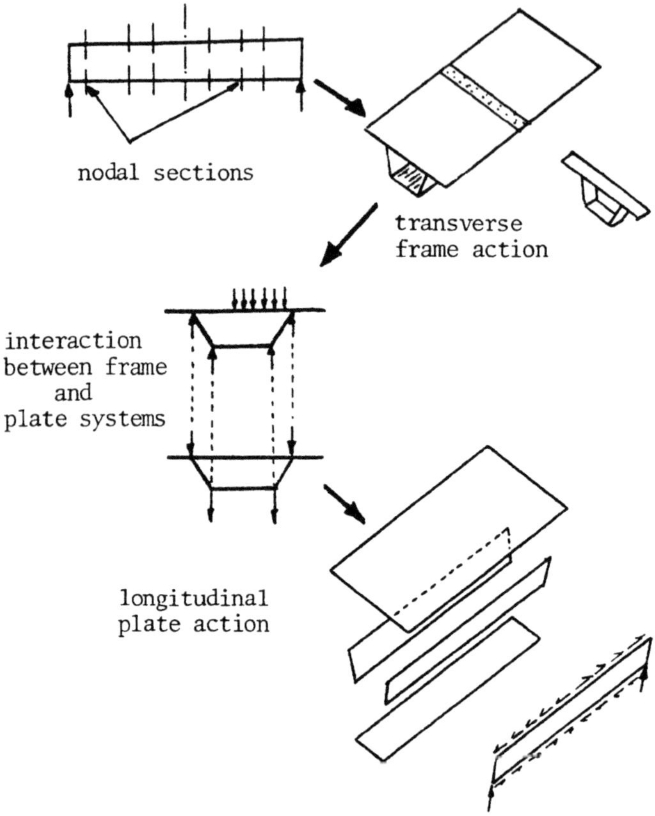

FIG. 1 IDEALIZATION OF STRUCTURAL BEHAVIOUR IN NODAL SECTION METHOD

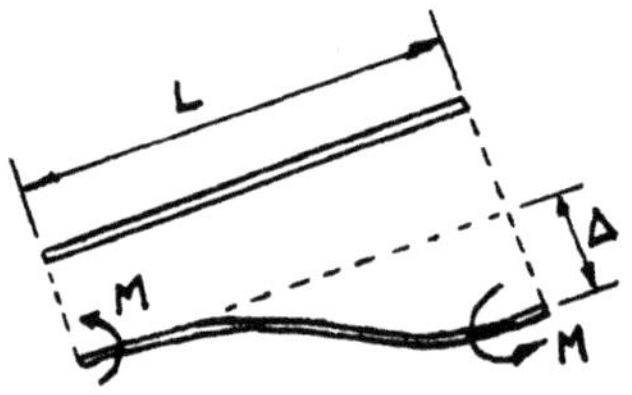

$$M = \frac{6\ EI\Delta}{L^2}$$

FIG. 2 SWAY DISPLACEMENT OF TYPICAL PLATE

deflections are calculated; a new set of incompatibilities will arise from these deflections. The whole process is then repeated and, by using a suitable accelerator, rapid convergence of the iterative solution is guaranteed for virtually all types of cellular cross-sections.

The accuracy of the nodal section approach has been established by several comparisons (2) with other methods. An interesting study was reported by Anderson and Waddell (4) in 1975, wherein the application of the method to the analysis of an actual bridge structure was described. The bridge was divided into 3 parts, each of different cross-sectional form, as shown in Fig. 3. The parts comprised of :

(a) a girder of varying web depth, cantilevering from the supports; this girder was of single-cell construction with edge cantilevers as shown in Fig. 3a.
(b) a twin-cell section with edge cantilevers, as in Fig. 3b, forming the suspended centre span.
(c) a twin-cell section, as in Fig. 3c, forming the 3-span continuous approach viaduct.

Figure 3 also shows the various loading conditions that were considered. Some results, typical of those obtained by Anderson and Waddell, are shown in Fig. 4 for each of the three parts of the bridge. Those quantities that are of most importance in bridge design, viz. transverse bending moments, longitudinal direct stresses and web shears are plotted. The nodal section values obtained in each case are compared to those predicted by other methods (5,6,7) and very good agreement is observed.

The application of the nodal section method to box girders that are curved in plan has been discussed in detail by Al Rifaie and Evans (3) and these authors reported several comparisons of nodal section values with experimental and finite element results for curved girders (8). Figure 5 shows typical results obtained by these authors illustrating the effects of curvature on a single-cell concrete girder. This girder was subjected to a uniformly distributed loading over the complete area of the top flange and the curves show the variation of the transverse moments, longitudinal stresses and vertical deflections at mid-span, with the degree of curvature. Obviously, as the radius of curvature becomes large, the solutions tend towards the symmetrical, straight box condition. The finite element and nodal section curves illustrate the same trends and show a satisfactory degree of agreement.

On the basis of the many comparisons that have been carried out with the results obtained from other methods and from experiments, it may be concluded that the nodal section method yields results of good accuracy. It provides a convenient and economic design tool for analysing the linear elastic behaviour of box girder bridges.

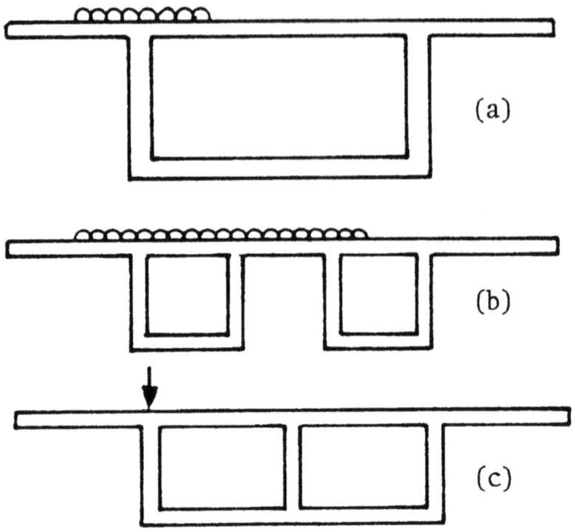

FIG. 3 CROSS-SECTIONS ANALYSED BY ANDERSON AND WADDELL (4)

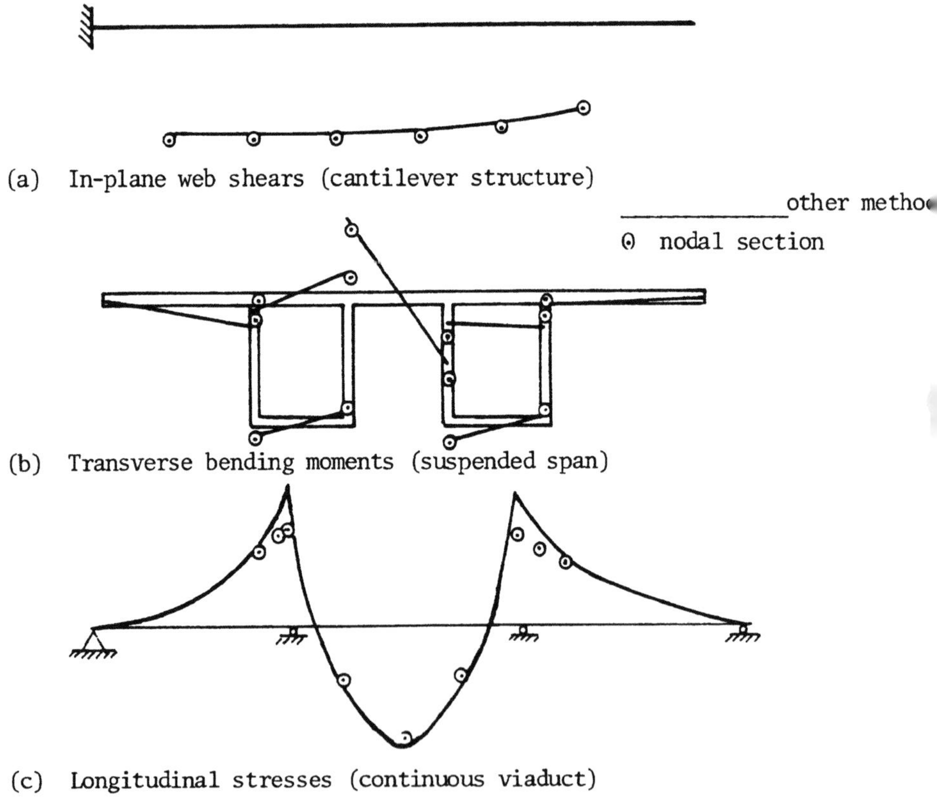

FIG. 4 TYPICAL RESULTS OBTAINED BY ANDERSON AND WADDELL (4)

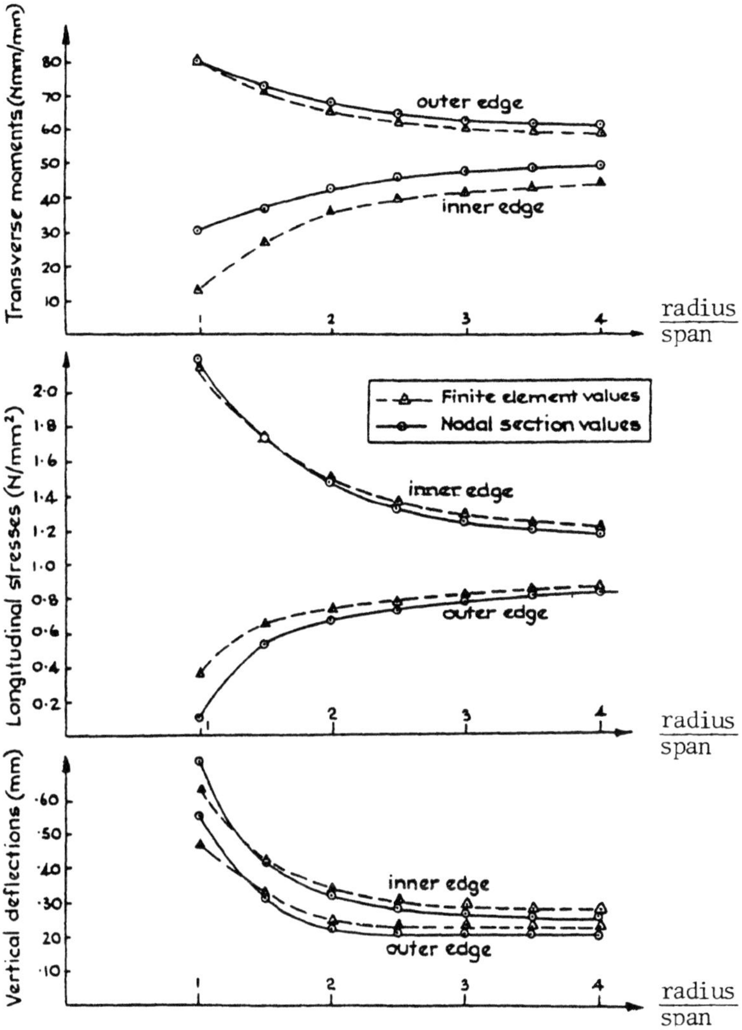

FIG. 5 RESULTS FOR A CURVED BOX GIRDER OBTAINED BY EVANS AND AL-RIFAIE (8)

2.2 The Bar Simulation Method (9,10,11)

When a box girder bends under an applied vertical loading, the longitudinal direct stresses set up within the flanges are assumed to be uniformly distributed across the flange width in normal engineering beam theory. However, in most cases, particularly in a wide flange, the shear deformations developed within the plane of the flange plate lead to a non-uniform distribution of the stresses, as shown in Fig. 6. This phenomenon is known as the "shear lag" effect and its neglect can lead to an underestimation of the stresses in the flange at positions close to the webs and can result in an unsafe design.

The bar simulation method has been developed (9,10,11) to predict this effect in a relatively simple way. The method has been adapted from a technique originally used in the aircraft engineering field (12) and it assumes that a flange plate may be idealized as an assemblage of axial load-carrying "bar" members and shear-carrying "sheet" members.

Figure 7 shows the idealization of the flanges of three typical box girders where the original, unstiffened flange is artificially divided into axial load-carrying and shear-carrying components. The axial load-carrying capacity of the flange is assumed to be concentrated at a number of discrete bar members which are interconnected by a sheet that is assumed to be capable of carrying shears only. The accuracy of the solution will increase with the number of bars taken.

The idealized flange is then analysed by considering the equilibrium of the forces developed in the bars and sheets and the compatibility of bar and sheet displacements. A number of simultaneous differential equations are thus established, the total number of equations being equal to the number of bars assumed. These equations can be solved for any loading and support conditions and explicit solutions have been presented by Evans and Taherian (11) for the conditions of most relevance to bridge analysis.

Typical results obtained by the bar simulation method are compared to finite element values in Figs. 8,9 and 10. Figure 8 shows the influence of the span/flange width ratio of the girder upon the shear lag effect. The curves plotted in Fig. 8 refer to the mid-span stresses in a simply supported girder under a uniformly distributed load and show that the shear lag effect is very dependent upon the span/width ratio. The ratio is plotted on the horizontal axis in the diagram and it is seen that for short, wide girders, the effects of shear lag are very large, whereas they almost disappear for long, narrow girders. The curves plotted in Fig. 8 show that for the single-span girders considered, the results predicted by the bar simulation method lie very close to the finite element values.

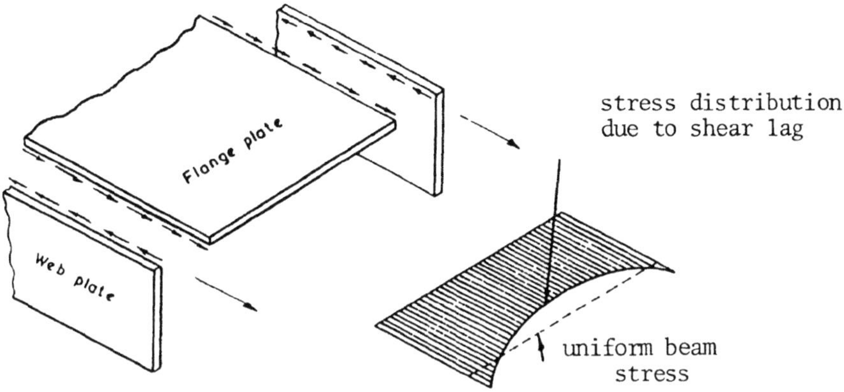

FIG. 6 EFFECT OF SHEAR LAG ON FLANGE STRESSES

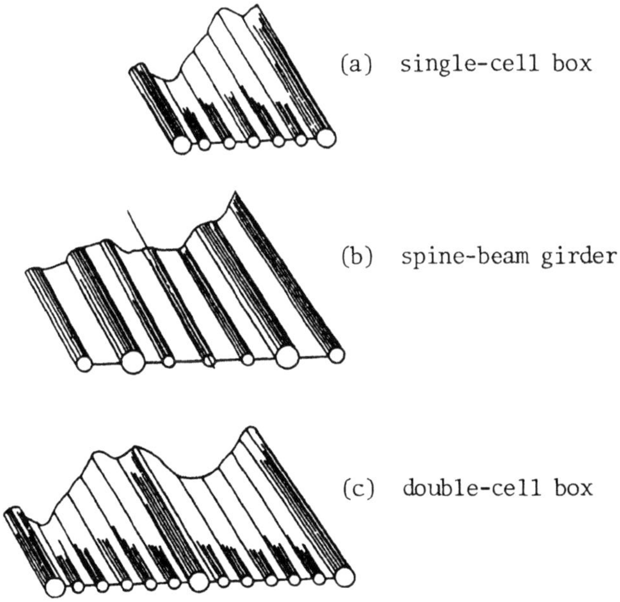

FIG. 7 TYPICAL IDEALIZATIONS OF FLANGE PLATES

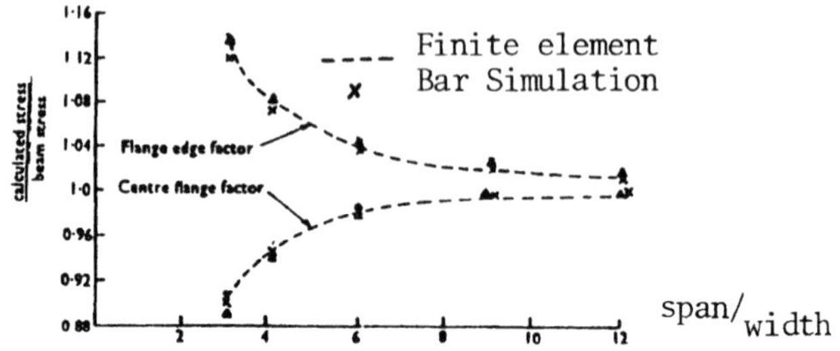

FIG. 8 VARIATION OF SHEAR LAG EFFECT WITH SPAN/WIDTH RATIO

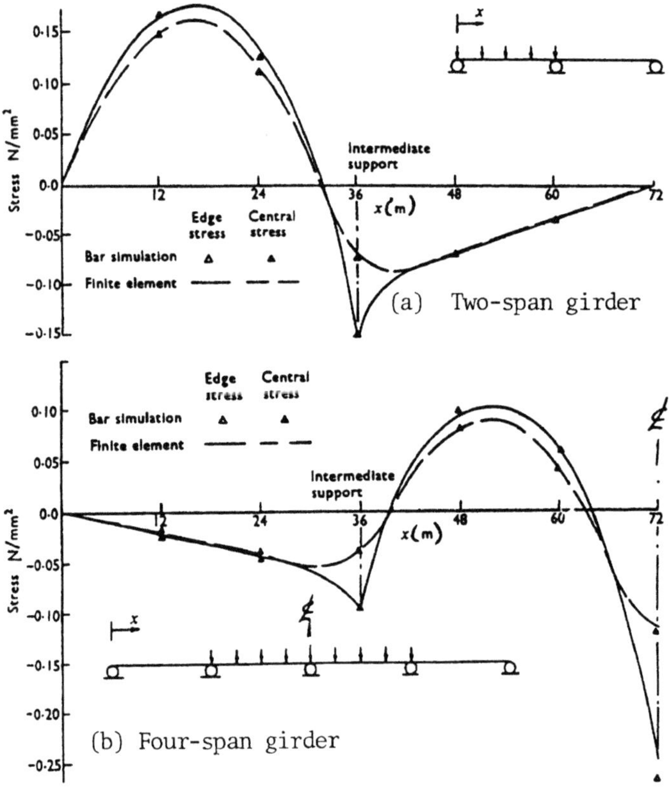

FIG. 9 SHEAR LAG EFFECT IN CONTINUOUS GIRDER

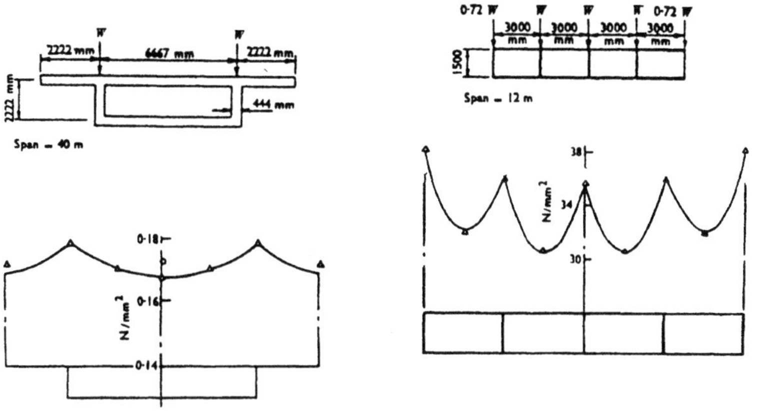

(a) spine-beam concrete box (b) four-cell steel box

FIG. 10 SHEAR LAG EFFECT ON STRESSES IN MULTI-CELL GIRDERS

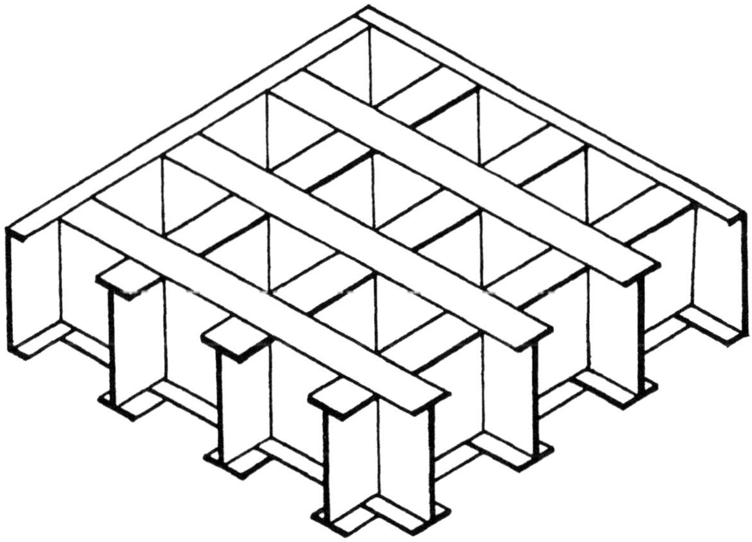

FIG. 11 GRILLAGE IDEALIZATION

Results for continuous single-cell box girders are plotted in Fig. 9. The edge and mid-flange stress variations for a two-span girder, symmetrically supported, but with only one span loaded, are plotted in Fig. 9a; corresponding values for a four-span girder are plotted in Fig. 9b. In both cases, the bar simulation and finite element stress curves show very good agreement at virtually all positions on the span.

Stress distributions across the flange width of two girders of very different cross-sections are plotted in Fig. 10. Stresses for a typical spine beam concrete box are shown in Fig. 10a and corresponding values for a four-cell steel box are given in Fig. 10b. In each case the stress distributions at the mid- and quarter-span cross-sections are plotted and the bar simulation and finite element predictions are again observed to be in good agreement.

These results show that the bar simulation method can be used to determine the shear lag effect in single- and multi-span girders of various cross-sectional types. By adopting the explicit solutions of the differential equations that have been presented (11), the shear lag effect can be calculated very easily at the design stage.

2.3 The Grillage Method (13,14)

A multi-cellular structure may be idealized as a grillage, wherein the real continuous plated structure is represented by a series of discrete, orthogonally intersecting I-beams, as shown in Fig. 11. Once such an idealization of the structure has been established the analysis of the resulting grillage becomes very easy and may be carried out conveniently on a micro-computer.

There are, however, two difficulties associated with the representation of a continuous plated structure by a series of discrete beam members. The first of these is how to deal with the effects of shear lag, see earlier in Fig. 6, in the flanges. It has been shown (13) that shear lag can be taken into account conveniently by adopting an effective width concept where the actual flange width is replaced by a reduced "effective" width, over which the stress is considered to be uniformly distributed. The extent to which the width must be reduced can be determined from the bar simulation method described above or from empirical tables (15). Once the effective width of flange acting with each web is known, the flexural properties of the idealized grillage I-beams can be determined.

The second difficulty that arises from the idealization is to ensure that the torsional stiffness of the closed cells is adequately represented. It has been established (13) that a satisfactory representation can be achieved by assuming that each I-beam possesses a torsional stiffness equal to that of a single closed cell.

The accuracy of such a grillage idealization can be illustrated by considering the typical 16-cell structure shown in Fig. 12a. The structure is square in plan and spans between simple end supports. Figure 12b shows the longitudinal stresses developed at the mid-span cross-section for three very different loading conditions and Fig. 12c shows the corresponding mid-span deflections. Grillage points are compared to finite element curves, showing good agreement in each case.

Cellular structures frequently contain openings in the web and these openings can cause a significant reduction in the shear stiffness of the web plates and in the torsional stiffness of the closed cells. Detailed experimental and finite element studies of these effects have been carried out by Evans and Shanmugam (14,16) and these studies have enabled the effects of openings to be represented by the introduction of two empirical coefficients :

(a) an effective shear area coefficient (K_S) by which the full web area should be multiplied to give an effective shear area.

(b) an effective torsional coefficient (K_T) by which the full torsional stiffness of a cell should be multiplied to give an effective stiffness.

The results of the parametric studies carried out by Evans and Shanmugam (16) were finally reduced to the form of the three curves shown in Fig. 13. The shear area coefficient K_S was found to be largely independent of both the girder width/depth ratio (b/d) and the width/thickness ratio (b/t). Its variation with the diaphragm spacing/depth ratio (a/d) and the web opening ratio is shown in Fig. 13a. The effective torsion coefficient K_T was also found to be independent of the width/thickness ratio (b/t). Its variation with the other two parameters is shown in Fig. 13b.

Once the appropriate values of the reduction coefficients have been determined from these curves, the grillage analysis of a girder with openings may be carried out in the usual way. Results for a typical 16-cell structure, simply supported on all edges are shown in Fig. 14. The girder was subjected to equal loads at all internal nodes, as indicated, and three different extents of web openings were considered. In the first case the webs were solid, then openings representing 12.5% of the original intact web area were introduced and, finally, very large openings of 31.25% were considered.

The grillage values (program GRILEJ) are shown as points for comparison with the finite element curves and very good agreement is observed in Fig. 14 for both deflection and stress values. The plotted deflection values show the very significant loss in stiffness occasioned by the introduction of openings.

Figure 15a shows the variation of the central deflection of the 16-cell girder with web opening ratio for two different end support conditions. For this diagram, because of the computing

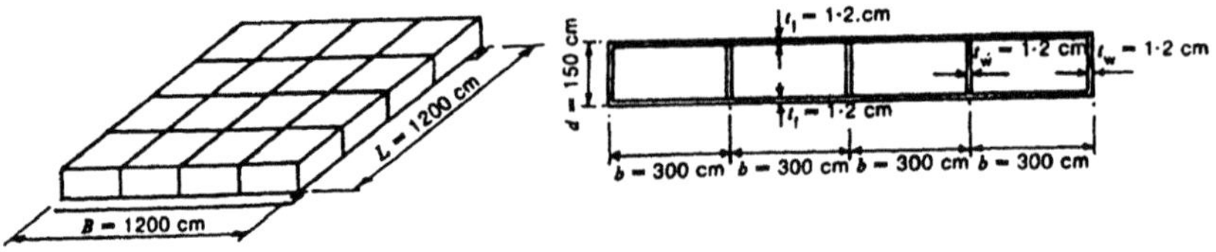

FIG. 12(a) TYPICAL 16-CELL STRUCTURE

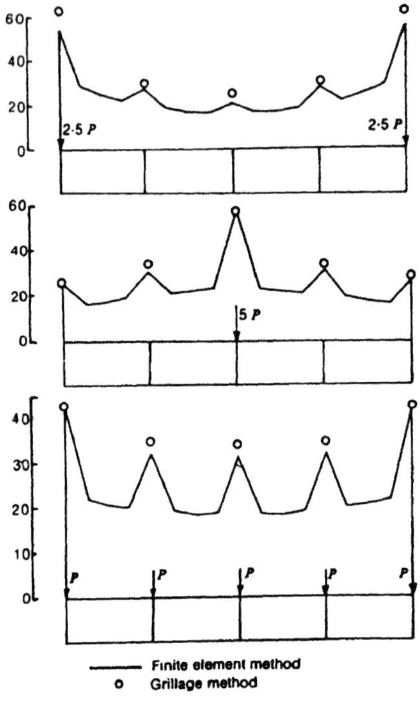

FIG. 12(b) LONGITUDINAL STRESSES (N/mm^2)

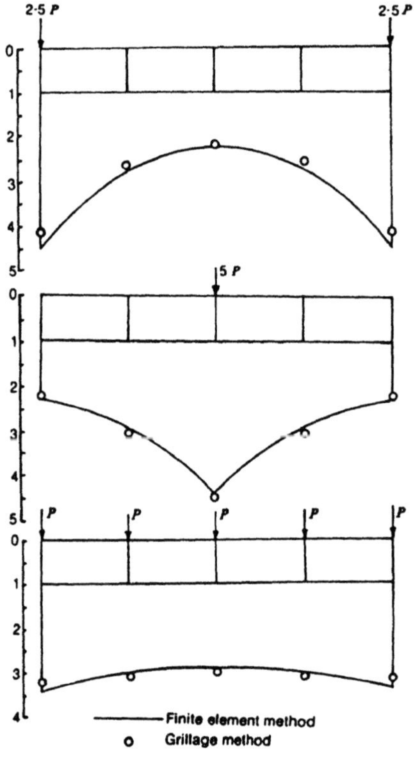

FIG. 12(c) VERTICAL DEFLECTIONS (mm)

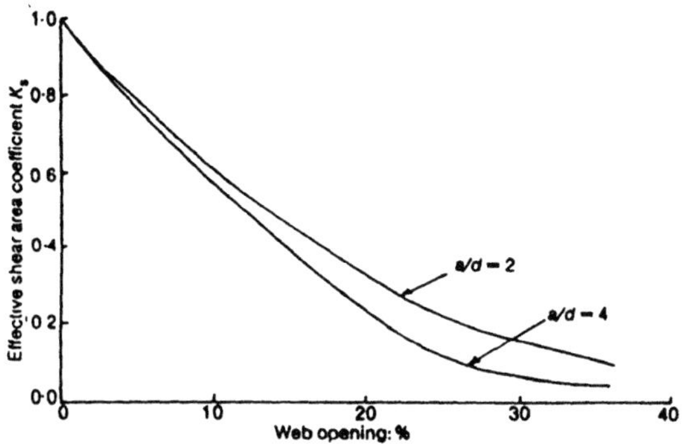

(a) Effective shear area coefficient (K_S)

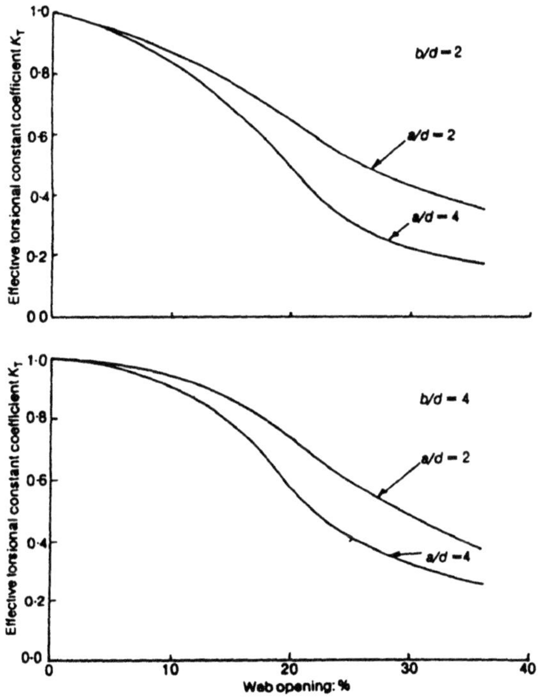

(b) Effective torsional coefficient (K_T)

FIG. 13 DESIGN CURVES TO ALLOW FOR EFFECTS OF OPENINGS

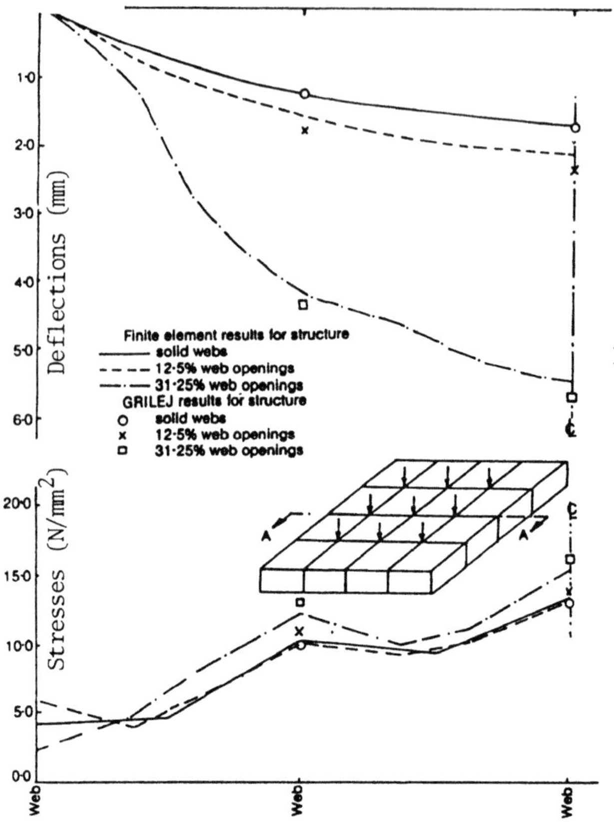

FIG. 14 RESULTS FOR 16-CE
GIRDER

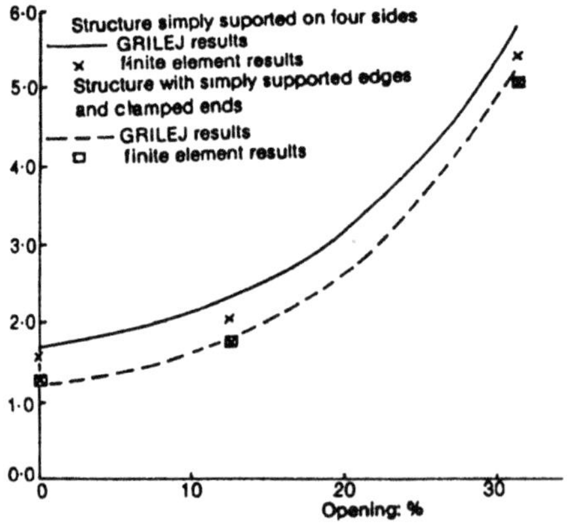

(a) 16-cell girder

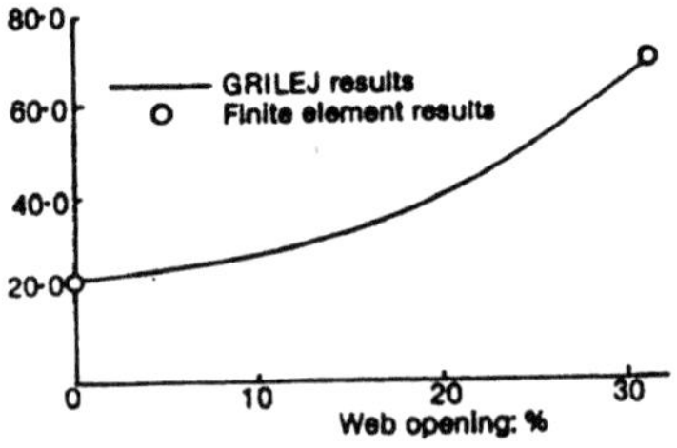

(b) 24-cell girder

FIG. 15 VARIATION OF DEFLECTIONS (mm) WITH OPENING SIZE

time involved, only three finite element solutions were obtained for each case so that these values are shown as points. Several grillage solutions were carried out to enable the continuous curves to be drawn. A similar comparison is shown in Fig. 15b for a 24-cell structure and, in both cases, the finite element points lie close to the grillage curves, indicating good agreement.

These comparisons, and many others, have confirmed the accuracy of the grillage approach for the analysis of multi-cellular girders. Moreover, the grillage solution time of 18 seconds for the 24-cell girder (see Fig. 15b), compared to the corresponding equivalent time of 3 hours for a finite element solution, indicates the usefulness of the grillage approach as a design tool.

3 ANALYSIS OF NON-LINEAR BEHAVIOUR AND COLLAPSE

With the current changes in many design codes from the allowable stress approach to limit state concepts, it is important to be able to calculate the collapse loads of structures. The availability of simplified methods is of even greater importance in this context, since the incremental procedures that must be adopted to allow for the effects of material and geometrical non-linearity make the computing time required for a finite element solution unacceptable for the analysis of anything other than an isolated element. When an individual panel is analysed in this way, it is difficult to prescribe the correct boundary conditions to represent the restraints imposed by adjacent panels.

The extension of the grillage approach to the analysis of the non-linear and collapse behaviour of a multi-cellular structure will now be considered.

3.1 The Prediction of the Non-linear and Collapse Behaviour of a Multi-Cellular Girder (17)

The grillage method discussed in section 2.3 can be extended to analyse the non-linear and collapse behaviour of cellular structures. The high bending moments arising from lateral loading usually cause buckling in the slender compression flanges of these girders. In the post-buckling range, the non-uniform distribution of flange stress, caused by the shedding of load away from the buckled central region of the flange to the edge regions, may be taken into account by adopting an effective width approach.

The reduced effective width in the post-buckling range can be determined from well-established empirical formulae. It will be dependent upon the magnitude of the stress developed in the compression flange and upon the flange buckling stress. Since the effective flange width to be assumed depends upon the magnitude of the flange stresses and since these stresses, in turn, depend upon the effective width assumed, a closed form solution is not possible.

Thus, in the non-linear, post-buckling range the grillage solution has to be carried out incrementally. Starting from the linear elastic range before buckling, the applied loading is increased in steps. For each load, section properties corresponding to the flange stresses calculated for the previous load level are assumed for the members and an elastic grillage analysis is carried out. The procedure is approximate since the assumed member properties are always one step behind the applied loading, so that the magnitude of the chosen loading increment affects the accuracy of the solution. However, it has been found that the solution converges rapidly with the increment size.

At each loading level, every grillage member is considered individually; the average shear stress in the web is combined with the flange stress and the Von Mises yield criterion is applied. If yield is observed to have occurred in any of the members, they are not allowed to carry any further loading and are effectively removed from the girder. Complete failure of the structure occurs when a sufficient number of individual members have failed so that a collapse mechanism can develop.

Typical results are shown in Fig. 16 where the central deflection of a 24-cell girder is plotted agains the applied load; mid-span stress values are also plotted. The non-linear relationship between load and deflection is apparent and the girder is observed to fail at a load of 4900 kN. This value predicted by the grillage method lies within 7% of the load of 5260 kN obtained from a yield line analysis, thus showing satisfactory agreement.

Further results for a typical 28-cell structure are presented in Fig. 17, where the mode of failure of the girder is shown on the schematic plan views on the left, and the table on the right of the diagram indicates the magnitudes of the bending and shearing stresses in the various members as they reach yield.

First yield is observed to have occurred in the transverse members at the centre of the span at a load of 75 kN. These members are close to the longitudinal edges so that shear is the predominant action, as shown by the tabulated values. Before increasing the load to 80 kN, the four yielded members were removed and during this increment, yield is observed to have occurred in four further transverse members, with shear again predominating. Upon removing these four members and applying a further increment of load to 85 kN, widespread yield is observed allowing the formation of a collapse mechanism and consequent failure of the girder.

Unfortunately, little theoretical or experimental evidence exists at present to enable the accuracy of the proposed grillage approach in the non-linear range to be proved conclusively. The complexity of a multi-cellular structure also makes it impossible to obtain finite element results for comparison. The limited

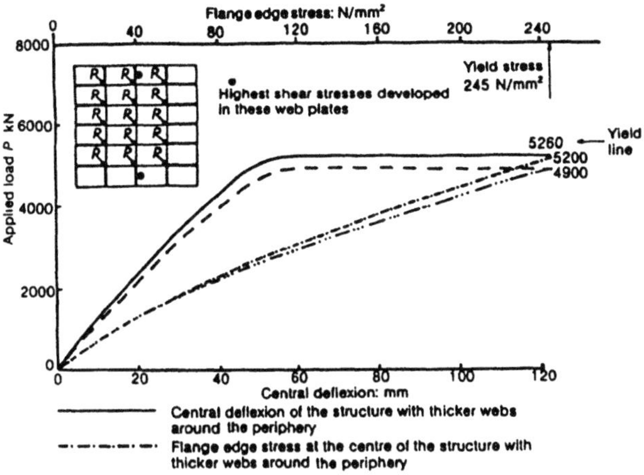

FIG. 16 RESULTS FOR 24-CELL GIRDER

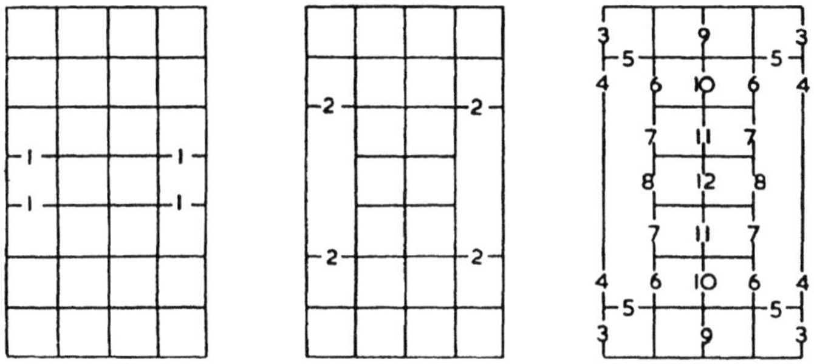

stage 1 load per node - 75 kN

stage 2 load per node = 80 kN

stage 3 load per node = 85 kN

Stage	Element	bending stress ratio σ/σ_Y	shear stress ratio τ/τ_Y
1	1	0.52	0.86
2	2	0.84	1.00
3	3	0.57	1.00
	4	0.78	0.67
	5	1.00	1.00
	6	0.58	1.00
	7	1.00	0.78
	8	1.00	0.00
	9	0.34	1.00
	10	0.66	0.99
	11	1.00	0.57
	12	1.00	0.00

FIG. 17 TYPICAL RESULTS INDICATING COLLAPSE OF 28-CELL GIRDER

comparisons that have been carried out to date have shown encouraging agreement but the author is currently conducting a series of tests on large-scale models to provide further data.

The extension of the grillage approach has been carried out without adding greatly to the complexity of the solution process. Consequently, even in its non-linear form, it may still be regarded as a convenient design aid.

4 CONCLUSION

A number of different methods of analysis for cellular bridge structures have been described in this paper. In each method, an idealized structural behaviour is assumed so as to provide a simplified solution procedure, making each method suitable for use during the preliminary stages of design when repeated analyses may be required to establish optimum proportions. Some typical comparisons with results obtained from other methods have been presented to illustrate the accuracy that can be achieved by the simplified approaches; such comparisons are presented in more detail in the listed references. All the methods presented are suitable for solution on the types of microcomputers that are becoming increasingly available to the design engineer.

REFERENCES

1. Zienkiewicz, O.C. The Finite Element Method, McGraw Hill Ltd, London, 1977.
2. Rockey, K.C. and Evans, H.R. The Nodal Section Method for the Analysis of Box Girders. Proc.Int.Assoc.Bridge and Struct.Eng. Vol. 35, 1975, pp. 185-216.
3. Al Rifaie, W.N. and Evans, H.R. An Approximate Method for the Analysis of Box Girder Bridges that are Curved in Plan. Proc. Int.Assoc.Bridge and Struct.Eng., Feb. 1979, pp. 1-15.
4. Anderson, R.G. and Waddell, J.W. Analysis of Concrete Box Girder Structures by the Nodal Section Method, PTRC Conference, London, July 1975.
5. Highway Engineering Computer Branch of the Department of the Environment. Program for the Analysis of Box Girder Bridges using Quadrilateral Finite Elements, Report HECB/B/14 (QUEST), May 1975.
6. Cheung, Y.K. STRIP Programs for the Analysis of Straight and Curved Box Girder Bridges. Report and User Manual, R. Travers Morgan, Consulting Engineers, London, March 1972.
7. Scordelis, A.C. Box Girder Bridges - A Folded Plate Analysis. Department of the Environment Bridge Design Computer Application Programs (MUPDI) Report HECB/B/11, Nov. 1970.
8. Evans, H.R. and Al Rifaie, W.N. An Experimental and Theoretical Investigation of the Behaviour of Box Girders Curved in Plan. Proc.Instn.Civ.Engrs., Part 2, 1975, 59, June, pp. 323-352.

9. Evans, H.R. and Taherian, A.R. The Prediction of the Shear Lag Effect in Box Girders. Proc.Instn.Civ.Engrs., Part 2, 1977, 63, March, pp. 69-92.
10. Taherian, A.R. and Evans, H.R. The Bar Simulation Method for the Calculation of Shear Lag in Multi-cell and Continuous Box Girders. Proc.Instn.Civ.Engrs., Part 2, 1977, 63, Dec., pp. 881-897.
11. Evans, H.R. and Taherian, A.R. A Design Aid for Shear Lag Calculations. Proc.Instn.Civ.Engrs., Part 2, 1980, 69, June, pp. 403-424.
12. Kuhn, P. Stresses in Aircraft and Shell Structures. McGraw Hill, New York, 1956.
13. Evans, H.R. and Shanmugam, N.E. An Approximate Grillage Approach to the Analysis of Cellular Structures. Proc.Instn. Civ.Engrs., Part 2, 1979, 67, March, pp. 133-154.
14. Evans, H.R. and Shanmugam, N.E. The Elastic Analysis of Cellular Structures Containing Web Openings. Proc.Instn.Civ. Engrs., Part 2, 1979, 67, Dec, pp. 1035-1063.
15. British Standards Institution. Steel Concrete and Composite Bridges; Part 3 - Code of Practice for Design of Steel Bridges. BS5400, 1982.
16. Shanmugam, N.E. and Evans, H.R. An Experimental and Theoretical Study of the Effects of Web Openings on the Elastic Behaviour of Cellular Structures. Proc.Instn.Civ.Engrs., Part 2, 1979, 67, Sept., pp. 653-676.
17. Shanmugam, N.E. and Evans, H.R. A Grillage Analysis of the Non-linear and Ultimate Load Behaviour of Cellular Structures under Bending Loads. Proc.Instn.Civ.Engrs., Part 2, 1981, 71, Sept., pp. 705-719.

III. CONCRETE SLAB AND BOX GIRDER BRIDGES

BERKELEY COMPUTER PROGRAMS FOR THE ANALYSIS OF CONCRETE BOX GIRDER BRIDGES

Alex C. Scordelis

Professor of Civil Engineering
University of California
Berkeley, California USA

SUMMARY

A review of the analytical methods and computer programs, which have been developed at the University of California at Berkeley for the linear and nonlinear analysis of reinforced and prestressed concrete box girder bridges, is presented. Short descriptions of 15 different computer programs based on one of the following methods of analysis are given: (1) folded plate elasticity method; (2) finite strip method; (3) finite element method; (4) finite segment method; and (5) nonlinear analysis method. These various programs provide the capability for the detailed analysis of single and multicell box girder bridges of one or more spans, which are straight, curved, skew or of arbitrary geometry in plan. General loadings and arbitrary boundary conditions may be treated.

The linear analysis programs assume the analytical model to be a linear uncracked homogeneous concrete system. The nonlinear analysis programs utilize analytical models which include the concrete, reinforcing steel and prestressing steel and can predict the structural response of bridges, under initial and time dependent load histories, through their elastic, cracking, inelastic and ultimate ranges.

Five representative numerical examples are presented to illustrate some of the solutions which can be obtained using the Berkeley computer programs. Results obtained are compared with experimental results and those from other analytical solutions.

1. INTRODUCTION

The continuing expansion of the highway network throughout the world is largely the result of the great increase in traffic, population, and the extensive growth of metropolitan urban areas. This expansion has led to many changes in the use and development of various kinds of bridges. In recent years, single or multicell reinforced and prestressed concrete box girder bridges have been proposed and widely used as economic and aesthetic solutions for the overcrossings, undercrossings, separation structures and viaducts found in today's modern highway systems. These structures can be constructed to follow any desired alignment in plan. The very large torsional rigidity of the box girder's closed cellular section provides structural efficiency, while its broad, unbroken soffit, viewed from beneath is more esthetically pleasing than open-web type systems.

In the span ranges of 60 to 100 ft (18 to 31 m), cast-in-place multicell reinforced concrete box girder bridges which are straight, skew, curved or of arbitrary geometry in plan are widely used for overcrossings, undercrossings and interchange structures. For longer spans up to about 200 ft (61 m), post-tensioned, prestressed, cast-in-place multicell box girder bridges are often used. For spans between 200 and 800 ft (61 to 244 m), commonly used in viaducts or water crossings, segmentally erected prestressed concrete box girder bridges of one, two or more cells may be used. The typical box girder bridge is a folded plate structure consisting of top and bottom concrete slabs monolithically connected by a series of vertical, sloping or curved webs (Fig. 1).

Present and past designs of box girder bridges usually have been based on linear elastic analyses of simplified models of uncracked homogeneous systems. Empirical formulas are often used to determine the transverse distribution of longitudinal girder moments and vertical shear forces in the webs due to design truck vehicle wheel loads. Empirical formulas are also used to determine local transverse and longitudinal slab design moments. In many recent bridge specifications, an alternate load factor method of design based on ultimate strength calculations is permitted, but little guidance is given on how this can be applied to complex box girder bridge geometries. Basic knowledge on the nonlinear behavior and ultimate strength of concrete box girder bridges is lacking at the present time.

Because of their extensive use in California, a continuing program of research on box girder bridges has been conducted at the University of California at Berkeley since 1965. A systematic plan was developed to study successively straight, simple and continuous bridges, curved bridges, skew bridges and bridges of arbitrary plan geometry. For each of these configurations the

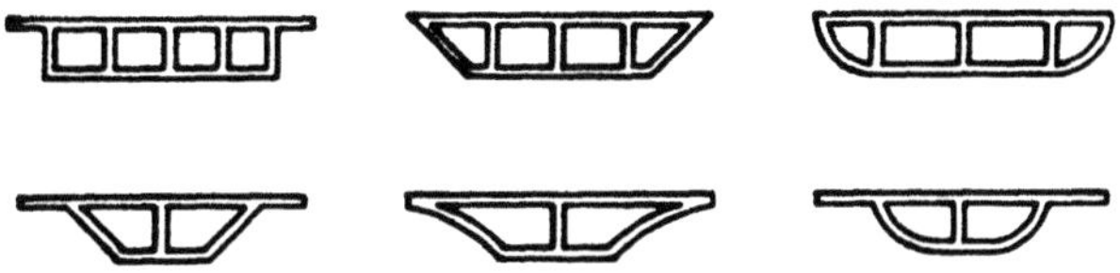

FIG. 1 MULTI-CELL BOX GIRDER BRIDGE CROSS SECTIONS

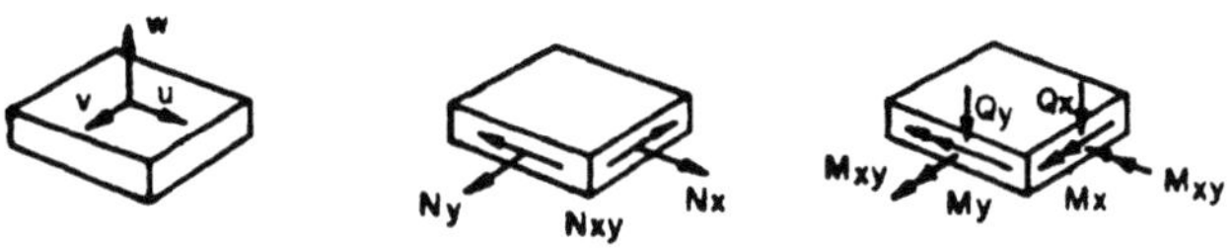

FIG. 2 DISPLACEMENTS AND INTERNAL FORCES ON A DIFFERENTIAL ELEMENT

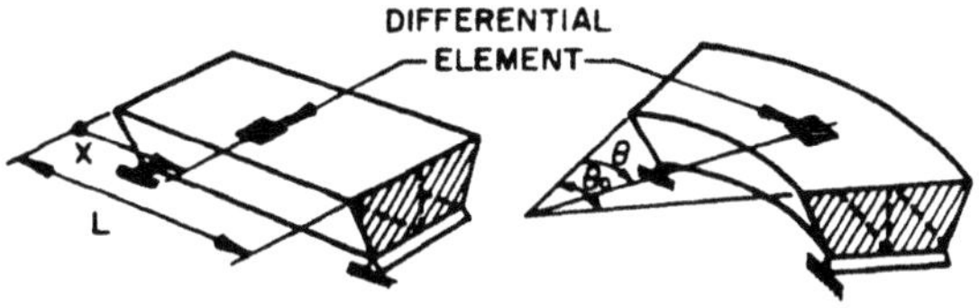

FIG. 3 PRISMATIC BOX GIRDER BRIDGES

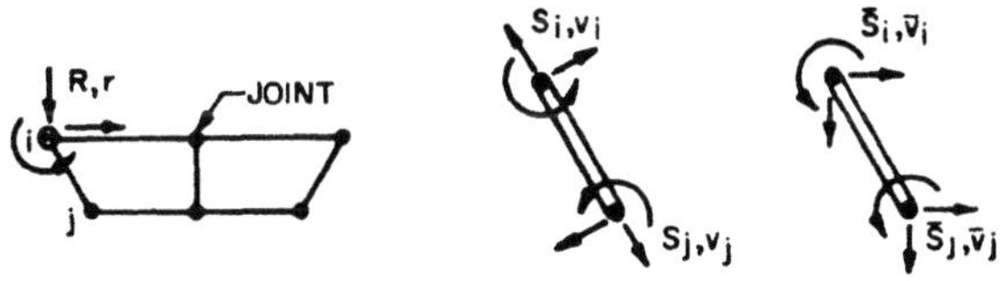

FIG. 4 JOINT AND PLATE EDGE FORCES AND DISPLACEMENTS
IN GLOBAL AND LOCAL COORDINATE SYSTEMS

approach has been to: (1) study the available literature; (2) develop analytical methods and general computer programs; (3) perform experimental studies on elastic models to verify the analytical methods developed if deemed necessary; (4) make analytical parameter studies; (5) test large scale reinforced concrete models or prototypes; and (6) develop recommended design procedures. Initially, reinforced concrete bridges and linear methods of analysis were studied, but in recent years since 1977 the emphasis in this research has shifted to prestressed concrete bridges and to nonlinear methods of analysis. The results of this continuing research program at the University of California have been described in a series of research reports and technical papers which are listed as references in Appendix 2, under A and B respectively.

The purpose of the present paper is to review the analytical methods and computer programs, which have been developed at Berkeley for box girder bridges, and to present numerical results from several examples. Both linear and nonlinear analytical models and methods will be described in the body of the paper, while in Appendix 1 short descriptions of 15 different computer programs for box girder bridges, which are available, will be described. Detailed information on these computer programs and their theoretical basis can be found in the published research reports and technical papers listed as references in Appendix 2.

2. LINEAR ELASTIC ANALYSIS

An analytical solution of the true response of a concrete box girder bridge under load is complicated by the usual factors common to other reinforced or prestressed concrete structural systems. It is a highly indeterminate structure; it is made of two materials, concrete and steel; under increasing load it experiences cracking and material nonlinearities and thus some redistribution of internal forces; and also the internal forces are time dependent because of creep and shrinkage in the concrete. Nevertheless, as for other reinforced or prestressed concrete systems, such as frames, slabs, and shells, it has generally been accepted that for design purposes, the distribution of internal forces, moments and displacements in a box girder bridge due to applied loads can be based on linear elastic analyses of an uncracked homogeneous concrete system.

In a complete analysis of a multi-cell box girder bridge, all of the internal forces and displacements shown on a typical element in Fig. 2 taken from a deck or web plate of the bridge should be determined. The internal forces N_x, N_y and N_{xy} are termed membrane forces while M_x, M_y, M_{xy}, Q_x and Q_y are internal forces due to plate bending. In many approximate analyses certain internal forces are assumed to be negligible and are thus taken as zero.

Of prime interest from a design standpoint are those internal forces which determine the reinforcing or prestressing steel requirements for the bridge. These are N_x for the main longitudinal tension steel; N_{xy} for the diagonal tension steel; and M_y and N_y for the transverse steel.

Many analytical models and methods have been developed for the linear elastic analysis of box girder bridges. Among these are approximate methods based on simplified structural behavior such as the use of an elementary or extended beam method; an equivalent beam grillage or anisotropic slab to represent the system; exact and approximate methods based on folded plate theory; and numerical solutions based on finite strip, finite element, finite segment or finite difference methods. No attempt will be made to review all of these methods here. Instead, only solutions and associated computer programs which have been developed at the University of California at Berkeley will be described. These linear elastic analyses of an uncracked homogeneous structure give complete solutions for all of the internal forces, moments and displacements shown in Fig. 2 for any type or position of loading on the bridge. The linear elastic solutions and computer programs to be described can be classified under the following methods of analysis; (1) folded plate elasticity method; (2) finite strip method; (3) finite element method; and (4) finite segment method. Each of these methods has its advantages and disadvantages and the appropriate method should be carefully selected based on its advantages and applicability for the particular problem to be solved.

3. FOLDED PLATE ELASTICITY METHOD

A multi-cell prismatic box girder bridge (Fig. 3) may be modelled analytically as a folded plate system consisting of an assembly of longitudinal plate elements interconnected at joints along their longitudinal edges and simply supported at the two ends. The known quantities input into the problem include the geometry, dimensions and material properties of the plate elements, the surface and joint loadings and the boundary conditions along the longitudinal joints. Each plate element selected is assumed to extend longitudinally over the entire span and transversely between designated joints on the cross-section (Fig. 4).

An analysis for applied loads with any arbitrary longitudinal distribution for straight bridges or any arbitrary circumferential distribution for curved circular bridges may be performed using a direct stiffness harmonic analysis. The applied loads are first resolved into Fourier series components. An analysis is made for the components of each particular harmonic and then the final results are obtained by summing the results for all harmonics used to represent the load. Once the solution technique, which involves extensive computations, has been developed for a single harmonic, it can be reused for any harmonic, and thus the method is ideally suited to the application of a digital computer.

The analysis for each harmonic load has the advantage that for straight or curved circular bridges such loads will produce displacements of the same variation and vice versa and thus a single characteristic value may be used to describe any force or displacement pattern. For example, the displacement pattern for the *n*th harmonic:

$$r(x) = r_0 \sin \frac{n\pi x}{L} \text{ for straight bridges} \tag{1}$$

$$r(\theta) = r_0 \sin \frac{n\pi\theta}{\theta_0} \text{ for curved circular bridges} \tag{2}$$

may be described by a single value r_0. L and θ_0 define the span for the straight and circular bridges respectively (Fig. 3). This makes it possible to treat an entire joint as a single nodal point and to operate with single forces and displacements instead of functions. If the condition of static equilibrium and geometric compatibility are maintained at a nodal point they will automatically be satisfied along the entire longitudinal joint. Thus the three-dimensional prismatic bridge problem may be treated as a two-dimensional problem in the transverse direction. A direct stiffness method applied to such a system results in a structure stiffness matrix which is extremely well conditioned for solution since the non-zero coefficients are all grouped in a narrow band about the main diagonal.

Each longitudinal joint has four characteristic degrees of freedom; it can displace vertically and horizontally in the plane of the cross-section; it can move longitudinally tangent to the joint; and it can rotate about an axis tangent to the joint. These directions define a global coordinate system for displacements or forces at the joint (Fig. 4a).

The direct stiffness method is described in detail in many textbooks and other publications. Thus it need only be briefly outlined here by the following steps for loads applied at the joints.

1. Determine the 8 x 8 element stiffness matrix k in $S = kv$ for each plate element, relating the edge forces S to the edge displacements v in the local coordinate system (Fig. 4b).

2. Transform the element stiffnesses from a local k (Fig. 4b) to a global $\bar{k}$ (Fig. 4c) coordinate system. Assemble these 8 x 8 matrices into the structure stiffness matrix K, which for the structure in Fig. 4a would be a 24 x 24 matrix.

3. Solve the equilibrium equations, $R = Kr$, where R represents the applied loads, for the unknown joint displacements r.

4. Determine the plate element internal forces and displacements (Fig. 2) by expressions relating these quantities to the joint displacements.

The basic logic of a general computer program for the direct stiffness harmonic analysis is independent of the method chosen for determining the element stiffness matrices.

3.1 Straight Single Span Bridges

For straight bridges, formulas based on elasticity theory for the stiffness matrix coefficients defining membrane and plate bending action for isotropic, linearly elastic plates were first utilized to develop the computer program MULTPL (1965) [A1, B1, B2, B6] for straight single span bridges, simply supported at the two ends. The elasticity equations defining the plane stress problem for membrane loads in the plane of the plate and classical thin plate theory for loads normal to the plate are used to determine the necessary expressions for the fixed edge-forces and moments due to applied loads between the joints and also for the element stiffness matrix k in the local coordinate system needed in the direct stiffness harmonic solution. Results obtained by this elasticity method represent the most exact available solution for a linear elastic analysis of box girder bridges and they are often used to measure the accuracy of results by other methods.

3.2 Straight Continuous Span Bridges

The above solution for a single span bridge without interior supports was extended to continuous span bridges with interior rigid diaphragms in the computer program MUPDI (1966) [A1, B3, B6] and to bridges with interior flexible diaphragms and supporting planar frame bents in the computer program MUPDI3 (1971) [A13, B6, B16]. A force method of analysis is used in combination with the direct stiffness harmonic analysis described above.

For structures supported by flexible planar frame bents the redundants are taken as the interaction forces between the folded plate system and the bents (Fig. 5). The interaction forces are represented by a set of three joint forces at each longitudinal joint (Fig. 5d) consisting of vertical, horizontal and rotational components in the plane of the transverse cross-section. The analysis is carried out in the following sequence of steps:

1. With the redundants X set equal to zero (Fig. 5b),

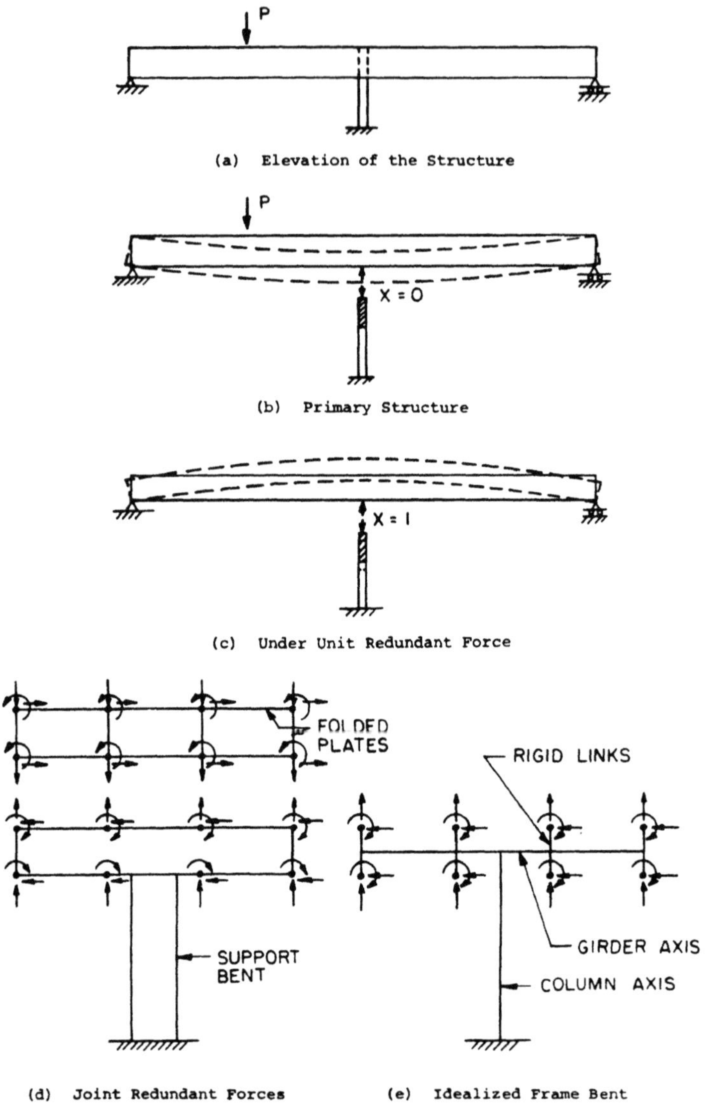

FIG. 5 ANALYSIS OF A BOX GIRDER BRIDGE ON A FLEXIBLE BENT

the folded plate structure is analyzed for the given external applied load using a direct stiffness harmonic analysis as described above. A displacement vector δ_0 (24 x 1 in Fig. 5) is found for this case, which defines the displacements at the points where the redundants are to act.

2. The folded plate structure is next analyzed for unit values of each of the redundant forces X (Fig. 5c) and the corresponding flexibility matrix F_1 (24 x 24 in Fig. 5) for the folded plate system is formed.

3. The idealized planar frame bent (Fig. 5e), composed of one dimensional elements, is then analyzed by the direct stiffness method. The total structure stiffness matrix for the frame bent is formed and a static condensation is carried out to eliminate the degrees of freedom which do not correspond to the redundant forces. Finally, the flexibility matrix F_2 (24 x 24 in Fig. 5) corresponding to the unit redundant forces is found by inverting the frame stiffness matrix.

4. Geometric compatibility requires that

$$\delta_0 + \delta_1 + \delta_2 = 0 \qquad (3a)$$

$$\delta_0 + F_1X + F_2X = 0 \qquad (3b)$$

or

$$\delta_0 + FX = 0 \qquad (3c)$$

where

$$F = F_1 + F_2 \qquad (4)$$

and thus the redundants X may be found from Eq. (3c)

$$X = F^{-1}\delta_0 \qquad (5)$$

5. The simply supported folded plate structure and the planar frame bent can now be analyzed, subjected to the known applied external loading and the known redundant forces, to determine the final stresses and displacements in the structure.

For interior diaphragms which are not externally supported, a similar force method is used to determine the redundants. However, in this case, because the diaphragms can undergo three degrees of rigid body motion in their own plane in addition to the deformation of the diaphragms themselves, additional transformation and condensation of the flexibility matrices F_1 and F_2 are required. These are done automatically in the computer programs and are described in detail in Ref. A13.

4. FINITE STRIP METHOD

When using the direct stiffness harmonic analysis, the direct application of the elasticity theory to determine the stiffness matrix of plate elements which are not isotropic becomes exceedingly complex and resort must be made to simpler approaches. A theory known as the finite strip method is used in these cases. The finite strip method may be thought of as a special form of the finite element method. It approximates the behavior of each plate by an assemblage of narrower longitudinal finite strips for which selected displacement patterns varying as harmonics longitudinally and as polynomials in the transverse direction are assumed to represent the behavior of the strip in the total structure. Local coordinate systems and displacement interpolation functions for a typical finite strip having a longitudinal span $L = 2a$ and a transverse width $B = 2b$ are shown in Figs. 6a and 6b. With these assumptions the displacement at any point in the strip can be expressed in terms of the eight nodal point displacements shown in Fig. 4b. Using successively strain-displacement relationships, a stress-strain law and thence either the principle of virtual displacements or the principle of minimum potential energy, the element stiffness matrix and also the generalized or consistent nodal forces for surface loads can be derived for the finite strip.

4.1 Straight Single Span Bridges

The above method has been used to develop a computer program MULSTR (1970) [A4, B5, B6, B16] for straight single span box girder bridges made up of orthotropic plates with closely spaced eccentric ribs or stiffeners in the longitudinal and transverse directions. The properties of such an orthotropic plate are assumed to be uniform over each longitudinal finite strip, but can vary from strip to strip if desired. The two basic types of eccentric ribs used are designated as torsionally soft ribs and torsionally stiff ribs (Figs. 7a and 7b). The former consists of open slender sections that have little torsional resistance, whereas the latter includes open or closed box sections with considerable torsional resistance. Detailed derivations for the element stiffness matrix and the consistent nodal forces for such finite strips using the basic theory for orthotropic plates with either torsionally soft or stiff eccentric ribs are presented in Ref. A4.

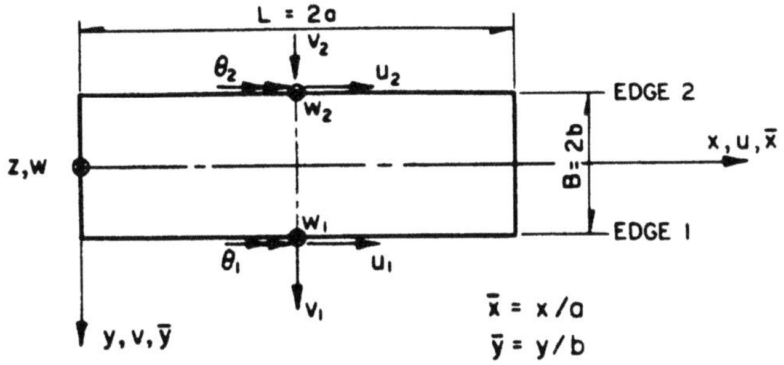

FIG. 6a POSITIVE NODAL POINT DISPLACEMENT AND LOCAL COORDINATE SYSTEM

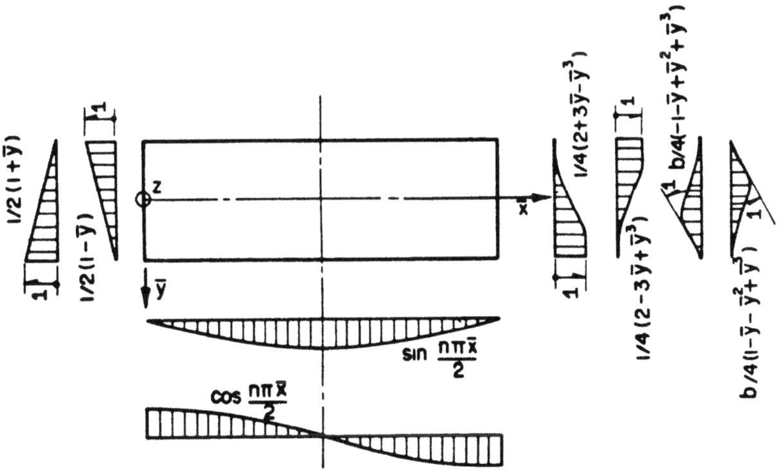

FIG. 6b DISPLACEMENT INTERPOLATION FUNCTIONS FOR A FINITE STRIP

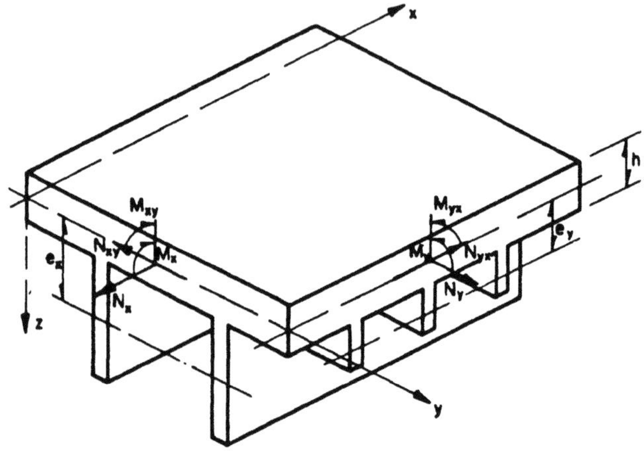

FIG. 7a TYPICAL ELEMENT OF TORSIONALLY SOFT ORTHOTROPIC PLATE

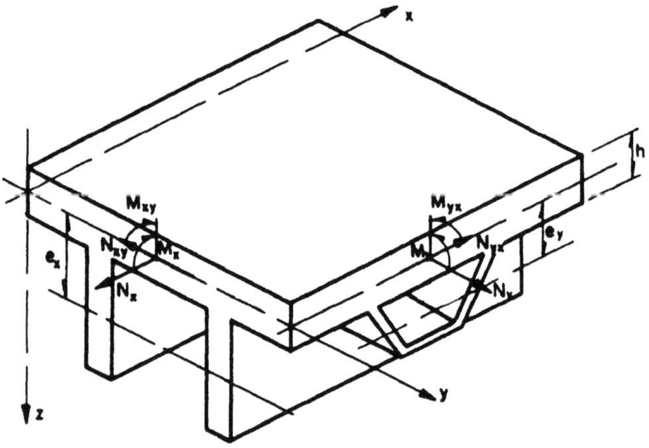

FIG. 7b TYPICAL ELEMENT OF TORSIONALLY STIFF ORTHOTROPIC PLATE

4.2 Circularly Curved Bridges

The finite strip method and direct stiffness harmonic analysis have also been used to develop the computer programs CURSTR (1970) [A6, B6, B8, B14, B16], CURDI (1974) [A20], and CURDIP (1980) [A26] for box girder bridges which are circularly curved in plan (Fig. 8a) and made up of orthotropic plate elements without eccentric stiffeners. Each plate element, which may be a segment from a circular ring plate, a conical shell or a cylindrical shell (Figs. 8b, c, d), is idealized by a number of circumferential finite strips. Displacement patterns in each strip are assumed to vary as harmonics in the circumferential direction and as shown in Fig. 6b in the transverse direction.

CURSTR can be used to analyze curved single span bridges and is similar in concept to MULTPL for straight bridges. CURDI can be used to analyze curved continuous span bridges with interior radial diaphragms or supporting planar frame bents and is similar in concept to MUPDI3 for straight bridges. By setting the radius to a very large value in CURSTR or CURDI a close approximation to a straight bridge is obtained and thus, if desired, results from a finite strip method can be compared with those from the more exact folded plate elasticity method used in MULTPL and MUPDI3.

CURDIP extends CURDI to take into account automatically the effect of post-tensioned prestressing tendons. By defining as input the tendon profiles, jack forces, jacking procedure and friction properties, the corresponding equivalent loads are computed automatically by the program and applied to the structure. This important addition simplifies considerably the analysis of prestressed concrete box girder bridges subjected to general dead and live loadings.

In order to describe a general prestressing tendon profile in CURDIP the real tendon is divided into piecewise linear segments (Fig. 9). To minimize the necessary amount of input data each tendon is first divided into portions which are thence subdivided into various numbers of segments of equal length in each portion. The tendons are assumed to be in the mid-thickness plane of the finite strips. Cylindrical coordinates defining the tendon points at each end of a segment are input.

These coordinates are transformed to a global Cartesian coordinate system so that vector algebra may be used for the necessary calculations to determine the equivalent loads due to prestressing.

This process involves the following steps (Fig. 10a): (1) the length and direction of each tendon segment is determined; (2) the angle θ_i between segments at each tendon point is calculated;

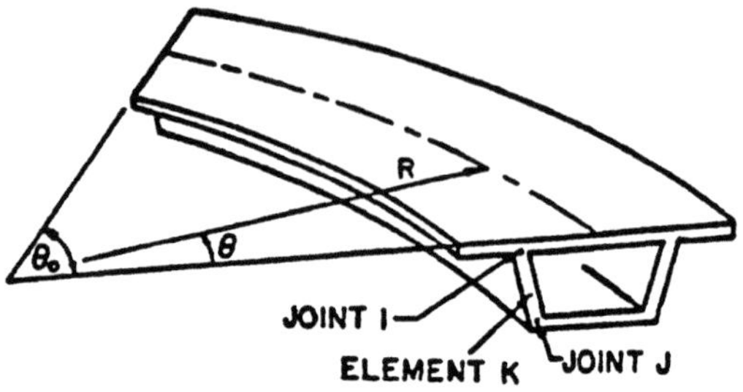

(a) Circularly Curved Bridge

(b) Ring Plate Element

(c) Conical Element

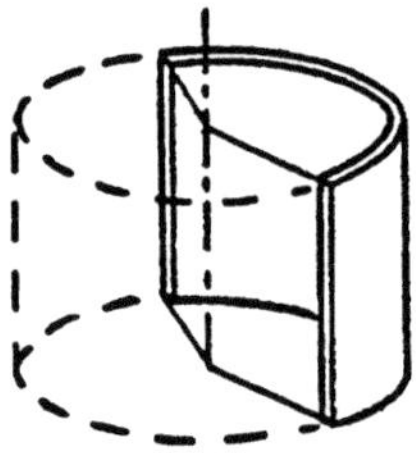

(d) Cylindrical Element

FIG. 8 CURVED BOX GIRDER BRIDGE ELEMENTS

(3) the standard friction formula is used to calculate the prestressing tendon force F_i existing in each tendon segment.

$$F_i = F_o \; e^{-(\mu\alpha + kS)} \tag{6}$$

where F_o = tendon force at jacking end; μ = coefficient of friction; α = total cummulative angle change from jacking end; k = wobble friction coefficient; and S = cummulative distance along tendon from jacking end; (4) with the force in each tendon segment known, the concentrated equivalent load vector at each tendon segment point can be found (Fig. 10b) in a global coordinate system by vector algebra; (5) this load vector R is transformed back to a cylindrical coordinate system and applied as statically equivalent nodal joint loads R_i and R_j to the finite strip in which the tendon point occurs (Fig. 11).

5. ADVANTAGES AND DISADVANTAGES OF DIRECT STIFFNESS HARMONIC ANALYSIS

Both the folded plate elasticity method and the finite strip method make use of the direct stiffness harmonic analysis procedure described earlier. Advantages of this procedure are:

1. It is well suited for computer programming and can yield a complete and accurate solution in a reasonable amount of computer time.

2. Any desired theory can be used to determine the response of the individual plate or strip elements.

3. Both surface and joint loadings of arbitrary longitudinal variation as well as prestressing can be treated.

4. Any combination of displacement and force boundary conditions along the longitudinal joints can be used.

Disadvantages of this procedure are:

1. It is restricted to prismatic structures which may have interior supports; but must be simply supported at the extreme ends.

2. The material and geometric properties of each plate or strip element making up the cross-section must be constant in the longitudinal direction.

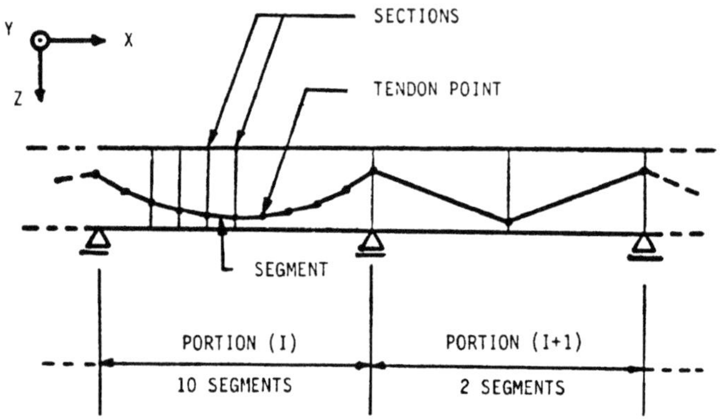

FIG. 9 PRESTRESSING TENDON SUBDIVISION INTO PORTIONS AND SEGMENTS

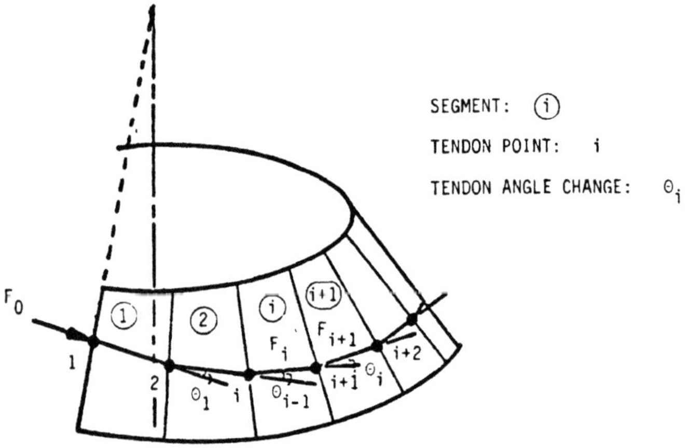

FIG. 10a LENGTHS, ANGLE CHANGES, AND TENDON SEGMENT FORCES

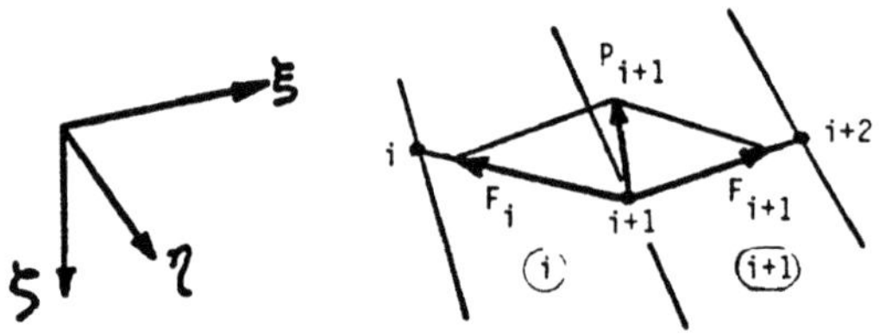

FIG. 10b CONCENTRATED EQUIVALENT LOAD AT TENDON POINT

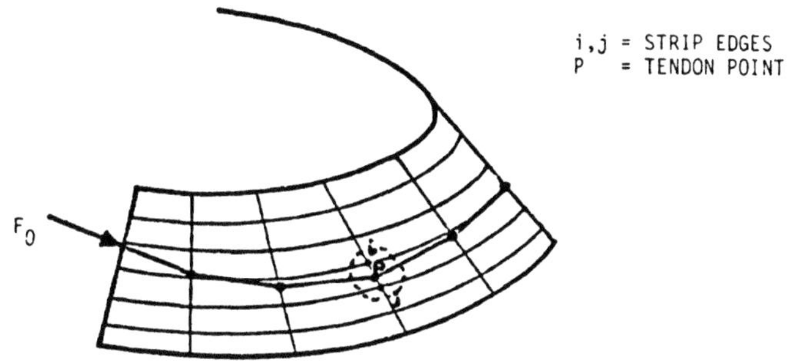

FIG. 11a TENDON PROFILE ON SIX FINITE STRIPS

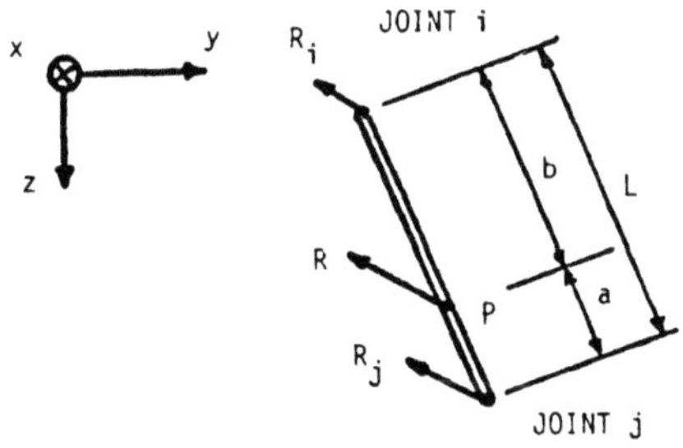

FIG. 11b STATICALLY EQUIVALENT JOINT LOADS

6. FINITE ELEMENT METHOD

The finite element method is described extensively in textbooks and in the technical literature. In the finite element method the actual continuum is replaced by an assembly of finite elements interconnected at nodal points (Fig. 12). For a general box girder bridge system, the finite elements may consist of two-dimensional shell or plate elements and transverse or longitudinal one dimensional frame type elements. Stiffness matrices which approximate the behavior in the continuum, are developed for the finite elements based on assumed displacement or stress patterns, after which an analysis based on the direct stiffness method may be performed to determine nodal point displacements and thence the internal stresses in the finite elements. The accuracy obtained is dependent on the assumptions used in deriving the stiffness matrices and on the fineness of mesh used in subdividing the structure. The results often satisfy compatibility, but not necessarily equilibrium in the continuum until a sufficiently fine mesh is used.

A number of investigators have developed general shell programs which could be used for analyzing box girder bridges. However, if available, it is better to use special purpose programs which take advantage of the repetitive and special nature of these structures. These should provide the required accuracy in the results with a minimum of the following: (1) required amount of input preparation; (2) execution time and core storage in the computer; and (3) amount of output data reduction necessary for meaningful interpretation. Three such special purpose programs developed at the University of California for the linear elastic analysis of box girder bridges are CELL (1970) [A7, B6, B7, B10, B16), FINPLA (1967) [A2, A5, A6] and FINPLA2 (1971) [A12, B6, B16].

6.1 Bridges with Arbitrary Plan Geometry and Constant Depth

Many multicell box girder bridges used in separation and interchange structures fall into this category. The computer program CELL was written especially for this type of bridge in which the bridge is made up of top and bottom horizontal deck slabs and vertical longitudinal webs and transverse diaphragms (Fig. 14).

The deck slabs are idealized by quadrilateral elements having a total of 5 degrees of freedom (DOF) per node, 3 translations and 2 rotations (Fig. 13a, b). The in-plane action of the quadrilateral elements is represented by the plane stress mixed model Q8D11 having 2 translational DOF at each external corner node and 3 internal DOF (Fig. 13a). The mixed model is constructed using separate expansions for the displacement and strain fields. The variations of the u and v components of the displacement field are approximated by the standard bi-linear expansion for the 8 corner node DOF and by bi-quadratic expansions for 2 of the internal DOF.

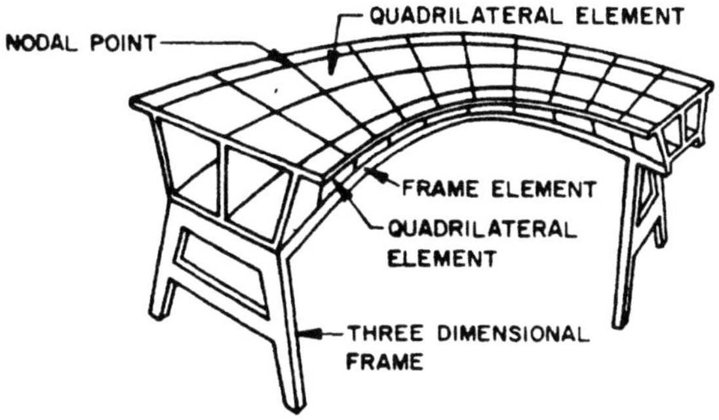

FIG. 12 BOX GIRDER BRIDGE OF GENERAL GEOMETRY

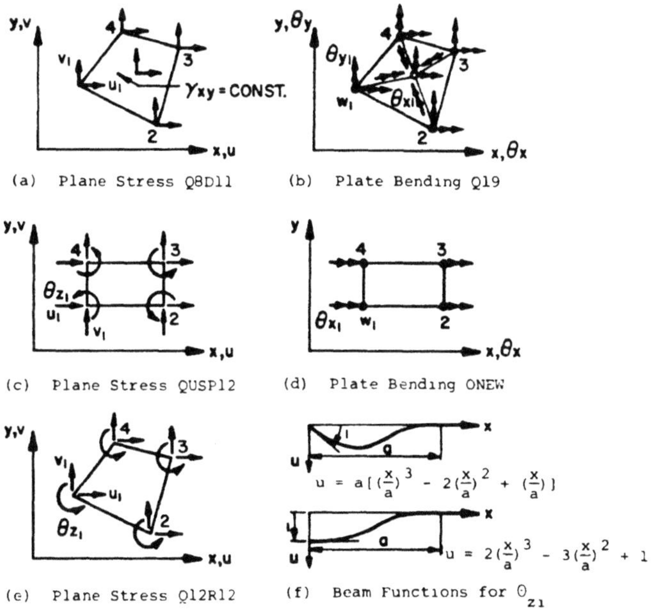

FIG. 13 FINITE ELEMENTS USED IN CELL AND FINPLA2 PROGRAMS

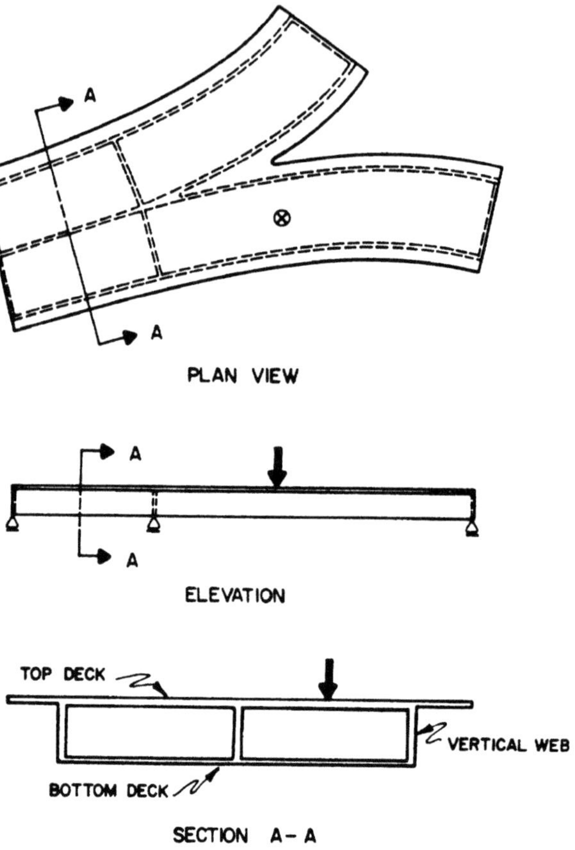

FIG. 14 BOX GIRDER BRIDGE OF CONSTANT DEPTH WITH VERTICAL WEBS

The third internal DOF is used to enforce a constant shear-strain variation over the entire element, which produces a more flexible and better element. After the element stiffness is formed the 3 internal DOF are eliminated by an internal static condensation process. The quadrilateral plate bending element Q19 (Fig. 13b) used for the deck slabs is a compatible element made up of four subtriangles, each of which has 11 DOF associated with full cubic expansions of the w-displacement field and an enforced linear variation of the normal slope along one edge. In combining the four sub-elements, a quadrilateral with 19 DOF is obtained. However, the 7 internal DOF are eliminated by static condensation leaving the essential 3 DOF at each corner node, 2 rotations and a translation (Fig. 13b).

The veritcal webs of the bridge are idealized by special rectangular spar elements having a total of 5 DOF at each corner node, 3 translations and 2 rotations. A single element over the entire depth of the bridge can be used to capture the essential behavior of the web. The in-plane action of these elements is represented by the model QUSP12 (Fig. 13c). A bilinear expansion for u and v is associated with the 2 translational DOF at each node and a cubic variation in the x-direction of v is defined by the rotation $\theta_{zi} = \partial v/\partial x$ at each node. The plate bending is represented by a simple one-way bending element ONEW having 2 DOF at each node, a rotation and a translation (Fig. 13d). Orthotropic plate properties and arbitrary loadings and boundary conditions can be treated in CELL. Automatic element and coordinate generation options which minimize the required input data, have made CELL a widely used program.

6.2 Bridges of General Geometry with Arbitrary Integrated Three-Dimensional Frame

A structure of this type (Fig. 12) is made up of quadrilateral two dimensional elements and one dimensional frame elements, each of which is assumed to have 6 DOF at each node, 3 translations and 3 rotations. Although 5 DOF have often been used for general shell programs, special techniques are required to account for the missing DOF, the rotation normal to the shell or plate surface. The use of 6 DOF for the two dimensional elements eliminates this complication and also makes the integration of the standard 6 DOF per node frame elements into the system a simple matter.

The quadrilateral elements for both the decks and the web may have an arbitrary orientation in space. They are taken as flat plate elements which give the best least squares fit through the actual location of the corner nodes. For box girder bridges of general plan and elevation geometry this assumption should be quite good. The special feature of the plane stress element, Q12R12, is that in addition to the commonly used 2 translational DOF per node, each node is also assigned a rotational DOF (Fig. 13e). This is defined as the average rotation about the element z-axis. The internal element displacements u and v are assumed to vary linearly with the 2 translational DOF and as beam functions (Fig. 13f) with the rotational DOF. The assumed nodal rotations introduce small angular discontinuities so that the element is not fully compatible.

The plate bending model used for the quadrilaterals is identical to that described in the preceding section (Fig. 13b) for the CELL program. The stiffness matrix for the standard one dimensional frame elements can be found in any textbook.

The above elements were first incorporated in 1967 into the program FINPLA to analyze straight box girder bridges made up of rectangular plate finite elements, reinforced by one dimensional transverse or longitudinal beam elements, and supported by single or multiple column rigid frame bent supports. The transverse center-line dimensions of the box girder system must be constant along the span in FINPLA, however, the thickness and orthotropic material properties of the rectangular plate elements may vary along the width and length of the bridge. The program has been used extensively for box girder bridges and requires a minimum of input data to define the structure and the loading.

In 1970 FINPLA was extended to analyze box girder bridges of general geometry (Fig. 12) in a computer program of much greater complexity called FINPLA 2. FINPLA 2 can analyze bridges of varying width and depth made up of quadrilateral planar finite elements (Fig. 13b, f) and one dimensional frame elements, so that almost any three dimensional bridge system can be studied. The form of the input data was developed primarily to simplify its use for box girder bridges. The structure is discretized by dividing it longitudinally into a number of segments by vertical sections and by subdividing each segment into finite elements. The structure alignment is described by a longitudinal reference line, which may be a straight line, a circular curve or an arbitrary planar string polygon and cross-sections are defined with respect to this line. Because of its complexity FINPLA2 has not been used extensively and thus complete checking, documentation and elimination of errors in all options and applications of the program cannot be assured.

7. ADVANTAGES AND DISADVANTAGES OF FINITE ELEMENT METHOD

Advantages of this method are:

1. It is the most general method available and can treat arbitrary loadings, boundary conditions, varying material and dimensional properties and cutouts.

3. One dimensional frame type elements can be readily incorporated as an integral part of the structural system.

Disadvantages of this method are:

1. It requires a greater amount of computer time than a direct stiffness harmonic analysis to obtain a solution of comparable accuracy.

2. A refined mesh size must be used to achieve accurate results in the vicinity of steep stress gradients.

3. Static equilibrium is not automatically satisfied for the displacement models normally used, but is approached as the mesh size is refined. Judgement must be used in selecting an appropriate mesh layout and in interpreting the results.

4. Automatic mesh and load generation schemes need to be incorporated into the computer program used to avoid the large manual input of data otherwise required.

8. FINITE SEGMENT METHOD

The name finite segment method is often used to designate solutions in which larger elements than those used in the classical finite element method are employed. Typically, these larger elements capture the dominant structural behavior and satisfy equilibrium. A computer program SIMPLA (1967) [A2, B4] based on this method has given excellent solutions for straight box girder bridges with constant transverse centerline dimensions along the span and arbitrary boundary and support conditions. The basic structural element used in this method is formed by dividing each longitudinal plate into a number of finite segments along the span. These finite segments each have a width equal to the transverse distance between the longitudinal joints of the plate. Thickness and material properties of each finite segment may be different. In the analysis, the finite segments are first interconnected transversely at one end of the bridge to form a full transverse segment of the entire bridge cross-section. The solution proceeds by a segment progression method along the span to connect one transverse segment to the next until the far end of the bridge is reached. The boundary conditions at the two ends of the bridge provide sufficient equations to determine all the unknowns needed in the solution of the problem.

Each plate segment is assumed to obey the ordinary theory of folded plates. Thus slab action is determined by the behavior of one way slab strips spanning between longitudinal joints, so that torsional and longitudinal slab moments are neglected. In-plane membrane action is defined by elementary beam theory applied to individual finite segments which results in a linear valuation of longitudinal stresses over the width of each segment between longitudinal joints.

Advantages of the finite segment method are:

1. A reduced number of elements is needed to define the structure and thus the storage and computer time required for solution is reduced.

2. The internal forces found satisfy equilibrium.

Disadvantages of this method are:

1. The segment progression solution used may be numerically sensitive requiring so-called "stop-overs" along the span in the solution.

2. Because of the larger elements used, highly localized stresses or steep stress gradients cannot be captured.

9. NONLINEAR ANALYSIS

Present design procedures for reinforced and prestressed concrete box girder bridges by either allowable stress or load factor design methods are generally based on internal forces determined from linear elastic analyses of simplified models of uncracked homogeneous concrete systems. There is a need to develop rational methods of nonlinear analysis from a fundamental viewpoint which can accurately predict the response of these complex bridge systems under initial and time-dependent load histories, expecially through their cracking, inelastic and ultimate ranges, so as to determine their nonlinear behavior, ultimate strength and mechanisms of failure.

An important problem has arisen with respect to the many hundreds of existing box girder bridges. Many bridge agencies are attempting to determine the overload capacities of these bridges because of the demand for permission to use heavier and heavier loads on the highway system. These bridges were originally designed using linear elastic analyses of simplified analytical models. Their true ability to carry the heavier loads being proposed can only accurately be determined by means of analyses which take into account the dominant nonlinearities under overloads up to failure.

In this presentation, the most recent research at the University of California in the nonlinear analysis of reinforced and prestressed concrete box girder bridges, which was completed in three separate Ph.D. dissertations under the supervision of the writer, will be reviewed. Each dealt with seperate bridge types as follows:

(1) Straight, Cast-In-Place, Reinforced or Prestressed Concrete Box Girder Bridges.

(2) Curved, Segmentally Erected, Single Cell, Prestressed Concrete Box Girder Bridges.

(3) Cast-In-Place, Multicell, Reinforced Concrete Box Girder Bridges of Arbitrary Plan Geometry.

The analytical models, solution techniques and computer programs developed as part of each of these extensive research studies will be described briefly. Detailed information on these studies may be found in the research reports and technical papers given as references in Appendix 2.

9.1 Straight Cast-In-Place, Reinforced or Prestressed Concrete Box Girder Bridges

The nonlinear analysis of bridges, which can be modeled as planar frames composed of one dimensional elements, can be performed using the computer program PCFRAME (1977) [A29, B22, B23, B26]. This program was developed as part of the Ph.D. dissertation completed by Y.J. Kang [A29] in 1977.

The analytical model consists of a series of interconnected one dimensional elements between joints lying in a plane (Fig. 15). Each joint has three degrees of displacement freedom, two translational and one rotational. Each element is assumed to have a prismatic cross section, which has an axis of symmetry, but the shape of the cross section may differ from element to element. The element cross section is divided into a descrete number of concrete and steel layers, (Fig. 15). These layers are defined by their cross-sectional area and distance from the element reference plane, which contains the longitudinal axis of the member. Different material properties in the layers due to cracking of the concrete and nonlinear material stress-strain relationships are monitored during the analysis. Element stiffnesses and properties are evaluated by a layer integration procedure.

Pre-tensioned, post tensioned bonded and unbonded prestressing tendons each having a given profile, initial tensioning force and

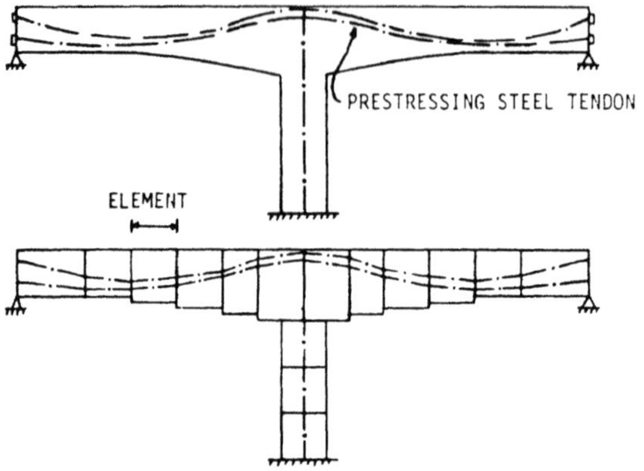

(a) Actual and Idealized Prestressed Concrete Frame

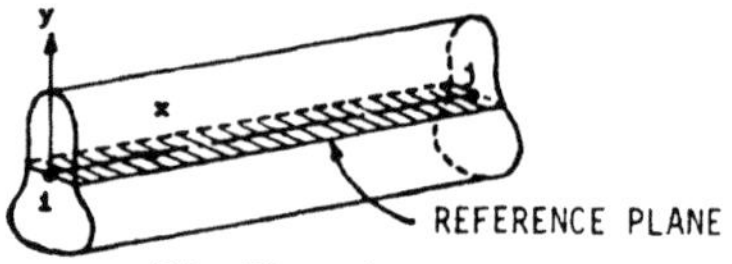

(b) Element

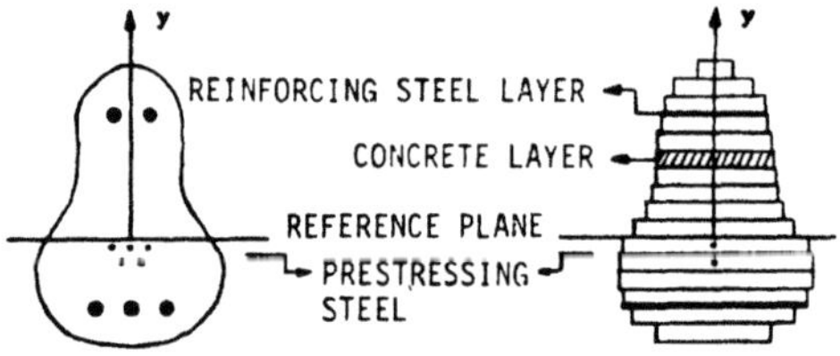

(c) Actual and Idealized Cross Section

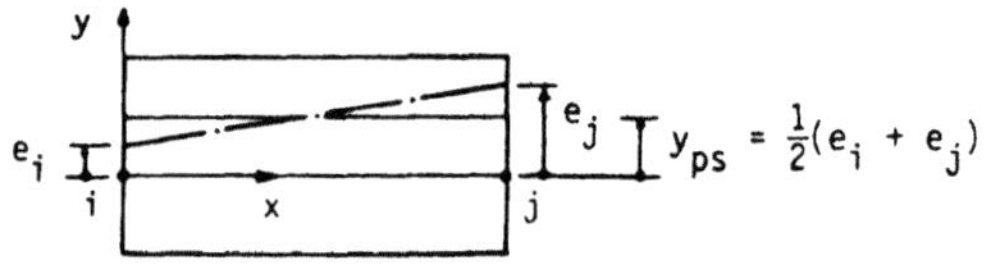

(d) A Prestressing Steel Segment in an Element

(e) Elevation of an Element and Displacement Componenents

FIG. 15 ANALYTICAL MODEL FOR COMPUTER PROGRAM PCFRAME

and a constant cross-sectional area along its length can be included in the analytical model. The actual curved prestressing tendons is idealized (Fig. 15) by a series of segments each of which is straight, spans an element and is assumed to have a constant prestress force.

An efficient numerical procedure is used for the material and geometric nonlinear analysis which includes the time-dependent effects due to load history, temperature history, creep, shrinkage and aging of the concrete and friction and relaxation of the prestress. The procedure is capable of predicting the response of planar structures throughout their service load history as well as throughout elastic, cracking, inelastic and ultimate load ranges.

A step forward integration is performed by dividing the time domain into a discrete number of intervals for the quasi-static time dependent analysis. For each time interval, nonlinear equilibrium equations which are valid for the current geometry and material properties are set up and solved by the finite element method based on the displacement formulation. An incremental load method combined with unbalanced load iterations for each load increment is used for the solution of the nonlinear equilibrium equations.

Time-dependent variation of concrete properties is recognized. Concrete strain is assumed to consist of the mechanical component and the non-mechanical component due to creep, shrinkage, aging and temperature variations. Parabolic-linear, bilinear and multi-linear approximations of the stress-strain curves are utilized for concrete, reinforcing steel and prestressing steel, respectively. A simple model for any inelastic load reversal is incorporated.

Creep strain is evaluated by an efficient numerical procedure based on the age and temperature dependent integral formulation. In the procedure, the history effect is incorporated by updating the values of only two variables instead of the storage of all the previous stress or strain histories. Nonlinear creep effect at high stress levels is also considered.

Output from the computer program PCFRAME for each load or time step includes the following: (1) joint displacements and rotations; (2) support reactions; (3) element axial forces and moments; (4) strains, stresses and forces for each prestressing steel segment; and (5) strains and stresses for each concrete and reinforcing steel layer.

9.2 Curved, Segmentally Erected, Single Cell, Prestressed Concrete Box Girder Bridges

Deflection and stress prediction are important in the design of segmentally erected bridges. A method is presented for the analysis of curved segmentally erected prestressed concrete box girder bridges including time-dependent effects which can be performed using the computer program SEGAN (1978) [A21, B19]. This program was developed as part of a Ph.D. dissertation completed by S. F. Van Zyl in 1978.

The bridge cross section must consist of a single box section with vertical or inclined webs and can have cantilever flanges. Width and depth dimensions can vary along the bridge span (Fig. 16a). The segments can be either precast or cast-in-place and can be erected at various specified times.

The analytical model consists of a series of joints interconnected by a series of special straight one dimensional elements. These are skew-ended finite elements (Fig. 16b) with eight degrees of freedom at each of the two end nodes. In addition to the usual three translational and three rotational degrees of freedom, a transverse distortional and a longitudinal warping degree of freedom are used at each end. Because the element end cross section need not be normal to the longitudinal axis, it can be made to coincide with the end of the adjacent element meeting it on angle. Diaphragm and support spring stiffnesses are added to the structure stiffness matrix before solving for the displacements by the direct stiffness method.

Time-dependent material changes such as creep and shrinkage of concrete as well as environmental effects such as humidity and temperature changes are accounted for. However, other concrete material nonlinearities such as cracking or inelastic behavior are not included in the analysis. A super-position method where the previous strain history is contained in a limited number of variables is used in the creep analysis which is similar to that used in PCFRAME.

Prestressing in the longitudinal direction can be included in the analysis. Tendons can be located in either the webs or flanges and can have a linear or parabolic profile. Stressing operations encountered in segmental construction such as stress and release operations, slip-in of tendons, prestressing and removal of tendons are included. Prestress losses due to creep, shrinkage and elastic deformations of the concrete as well as friction and relaxation of prestressing steel are computed and the effect included in the analysis.

A procedure is presented for specifying the structural configuration and the erection sequence. Operations used in segmental construction such as addition of segments, prestressing, changing of support boundary conditions, can be analyzed. At each construction stage the current structure is analyzed by the direct

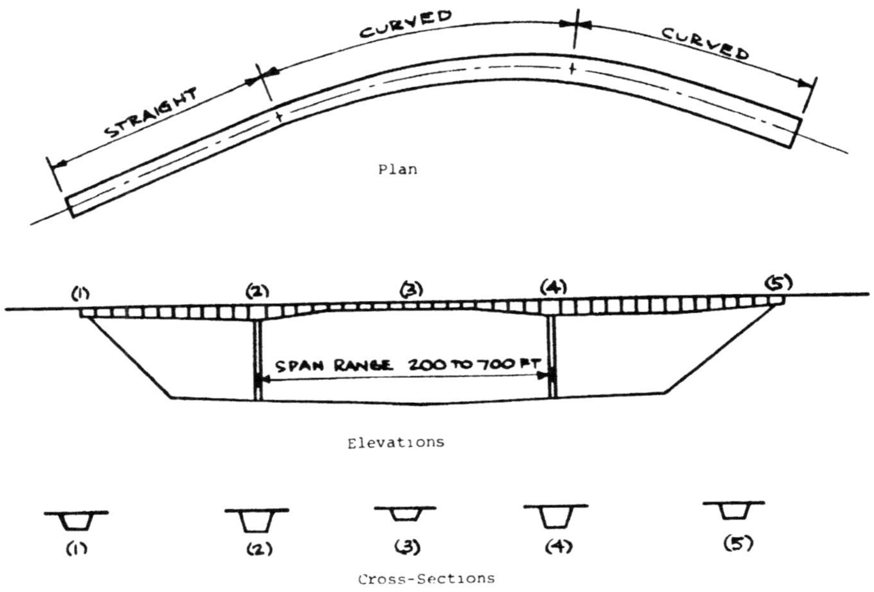

FIG. 16a CURVED SEGMENTAL BRIDGE

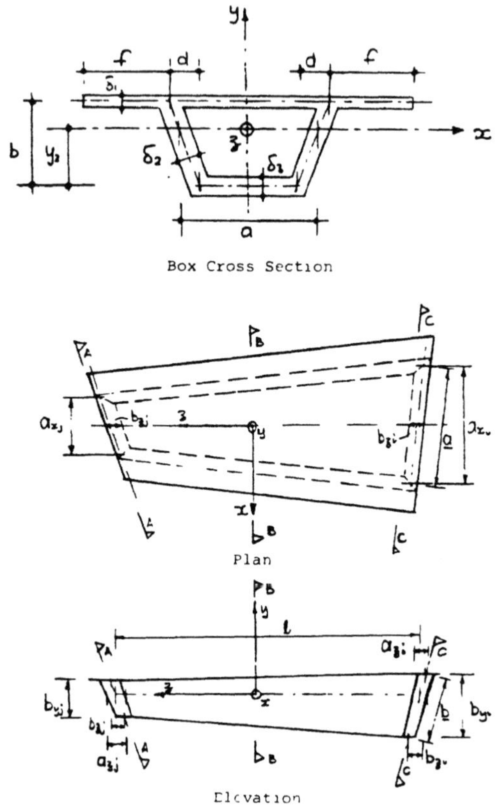

FIG. 16b SKEW ENDED FINITE ELEMENT FOR COMPUTER PROGRAM SEGAN

stiffness method and the incremental displacements, stresses and strains are obtained. By superposition the total displacements, stresses and strains are obtained. The complete structure also can be analyzed for a variety of live loads and time steps.

9.3 Cast-in-place, Multicell, Reinforced Concrete Box Girder Bridges of Arbitrary Plan Geometry

A numerical method of analysis has been developed by F. Seible, in a Ph.D. dissertation [A28] completed in 1982, to trace the nonlinear response of multicell reinforced concrete box girder bridges under stepwise increasing static loads. This analysis forms the basis for the computer program NOBOX (1982) [A28, B27, B28]. Nonlinearities considered are material nonlinearities inherent in reinforced concrete structural members under short term loading, such as cracking of the concrete, yielding of the reinforcement and formation of plastic hinges due to shear and flexure.

The analytical model is developed for multi-cellular structures of arbitrary plan geometry and constant height. The analytical method can trace the complete nonlinear behavior of the bridge beyond the working stress range all the way up to the ultimate failure and collapse of the structure. Both collapse mechanisms and failure loads can be determined.

A three-dimensional grillage model (Fig. 17a) comprised of special elements, has been chosen in order to minimize the computational effort. All the elements used in this displacement model satisfy external and internal equilibrium at all times. Individual primary actions in a multi-cellular structure such as longitudinal bending and shear, transverse bending and torsion are, with some engineering judgment, directly assigned to the individual elements comprising the model which minimizes the number of DOF's necessary to capture the complex cellular behavior. The structure is modeled by longitudinal beam elements, transverse bending frames and special torsional shear panels. The DOF's necessary for this three-dimensional grillage model are shown in Fig. 17a. No axial deformations are considered, i.e., all translational DOF's (Fig. 17a) are assumed to act uniformly over the height of the webs or the width of the top and bottom slabs of the structure.

The longitudinal beam elements along the girder lines (Fig. 17b) represent the longitudinal bending and shear resistance in the bridge structure. Transverse bending frames (Fig. 17c) comprised of individual beam elements are arranged transversely at discrete intervals (Fig. 17a) in order to model the transverse bending in top and bottom slabs, and in the vertical webs, as well as the transverse distortion in the box section. Geometric

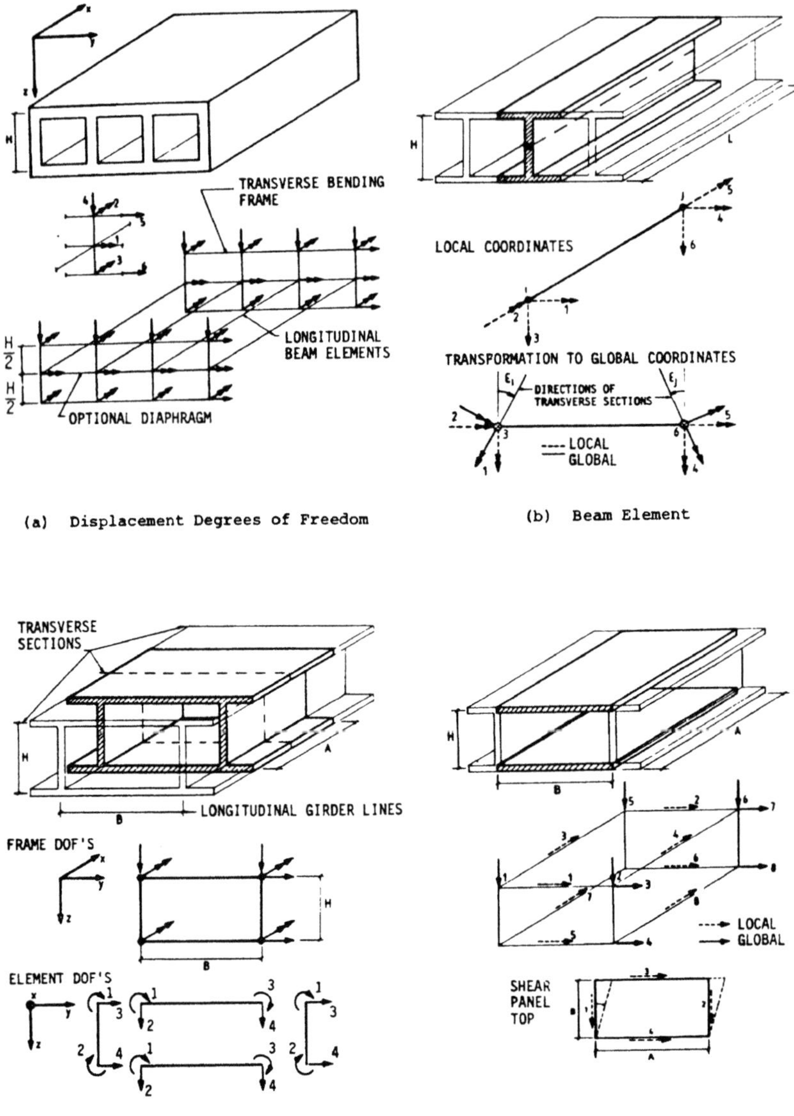

(a) Displacement Degrees of Freedom

(b) Beam Element

(c) Transverse Bending Frame

(d) Torsional Shear Panels

FIG. 17 3-D GRILLAGE MODEL AND STRUCTURAL ELEMENTS FOR COMPUTER PROGRAM NUBOX compatibility between adjacent longitudinal beam elements or adjacent transverse bending frames is established through shear panels, which can only represent a uniform state of shear. These are introduced between girder lines and bending frames at the top and bottom slab levels (Fig. 17d). These shear panels are also used to model the torsional properties of the box section by simply requiring an equal and opposite shear flow in top and bottom slabs at a certain section. Optional transverse diaphragms in the form of transverse beam elements (Fig. 17a) can also be introduced in the mid-plane of the structural model.

Material nonlinearities in the form of member force-deformation relationships for reinforced concrete members can be characterized from numerous experimental investigations by three major events: (1) cracking of the concrete; (2) yielding of the steel reinforcement; and (3) crushing of the concrete. These are approximated in each of the three types of elements used by simple trilinear force-deformation relationships.

Based on a mixed model formulation, an analytical scheme is developed which can treat the perfectly plastic branch of the force deformation characteristics by physically introducing mechanisms into the analytical model. A simple separation of the force and displacement unknowns in the mixed model leads to an analytical scheme in which the nonlinear element states are considered by a modification of the structural system, while the nonlinear element history is accounted for in a corrected load vector. Plastic deformations in the introduced mechanisms are readily available at the element level from the mixed model formulation.

The discretization of the analytical model as a grillage, consisting of structural elements which satisfy internal equilibrium and capture the dominant structural behavior, allows an economical solution due to the minimization of the number of degrees of freedom.

The computer program NOBOX can be used to give directly as output the displacements, the longitudinal girder moments and shears, transverse slab moments and torsional shear forces in the top and bottom slabs at any stage of loading.

10.0 NUMERICAL EXAMPLES

In order to demonstrate, test and verify the analytical solutions and computer programs developed for the linear and nonlinear analysis of concrete box girder bridges, a large number of numerical studies have been performed. Complete details on these studies for each of the computer programs developed may be found in the pertinent research reports given as references in Appendix

A.2. In general, the numerical results from the computer programs have been verified by one or more of the following ways: (1) comparison with experimental results from physical model studies; (2) comparison with numerical results from other analytical models; and (3) performing statical checks.

Five representative numerical examples will be presented here to illustrate some of the solutions which can be obtained using the Berkeley computer programs. Additional numerical examples may be found in the references given in Appendix A2.

10.1 Example 1 -- Small Scale Aluminum Model Studies

Extensive experimental studies have been conducted on a series of small scale elastic aluminum models (Fig. 18) [A14, A15, A16, B10, B14]. All models had the same cross section (Fig. 18e) which represented a 1:29 scale model or a typical two lane bridge in California.

Instrumentation of the models made it possible to determine experimental values of reactions, deflections, longitudinal membrane forces and transverse plate bending moments (Fig. 2), which could then be compared to theoretical values found using the computer programs CELL for the straight and skew bridges and CURSTR for the curved bridges. Each model was subjected to a series of vertical point loads at various locations.

For the straight and skew bridges, ten models were tested. The model reference numbers and plan geometries including spans and angle of skew are shown in Fig. 18a. Starting with the largest span rectangular bridge (Model OB), the model was cut back in five stages to study two spans at 30° skew (Models 2 and 4), two spans at 45° skew (Models 1 and 3), and a final short span rectangular model (Model 5). Model references A and B were, respectively, models tested without and with a midspan right transverse diaphragm. All models had end diaphragms and had only four reaction supports R1 to R4 at the corners (Fig. 18b).

For the curved bridges, six models were tested (Figs. 18c, d) using a similar cutting back procedure. The radius of curvature selected represented the sharpest curvature normally used in California bridges.

Comparisons of theoretical and experimental results for all bridges tested under a variety of load positions permit the following conclusions: (1) the computer programs CELL and CURSTR accurately predict the magnitude and distribution of longitudinal membrane forces N_x, transverse plate bending moments M_y, and reactions R1 to R4; (2) the transverse distribution of N_x at midspan, for a given load position, has essentially a constant form of

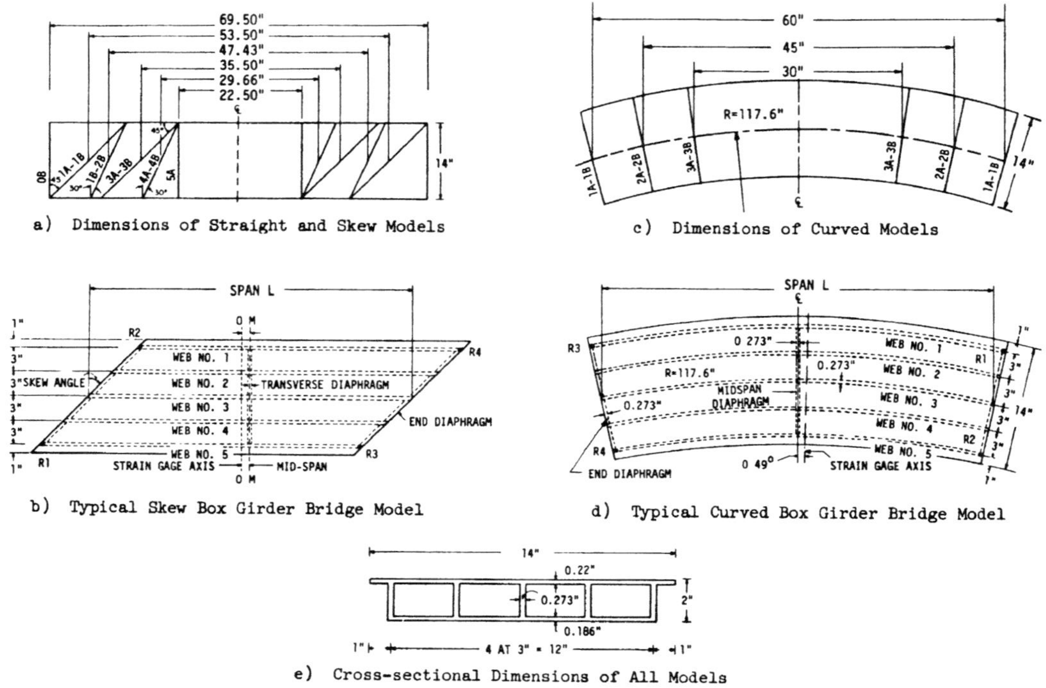

a) Dimensions of Straight and Skew Models

c) Dimensions of Curved Models

b) Typical Skew Box Girder Bridge Model

d) Typical Curved Box Girder Bridge Model

e) Cross-sectional Dimensions of All Models

FIG. 18 EXAMPLE 1 - DIMENSIONS OF SMALL SCALE ALUMINUM BOX GIRDER BRIDGE MODELS

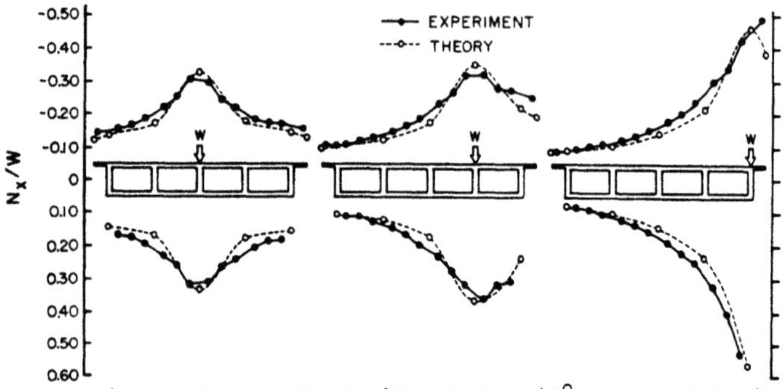

a) Skew Bridge Model 3A (Skew Angle = 45°, Span = 35.5 in)

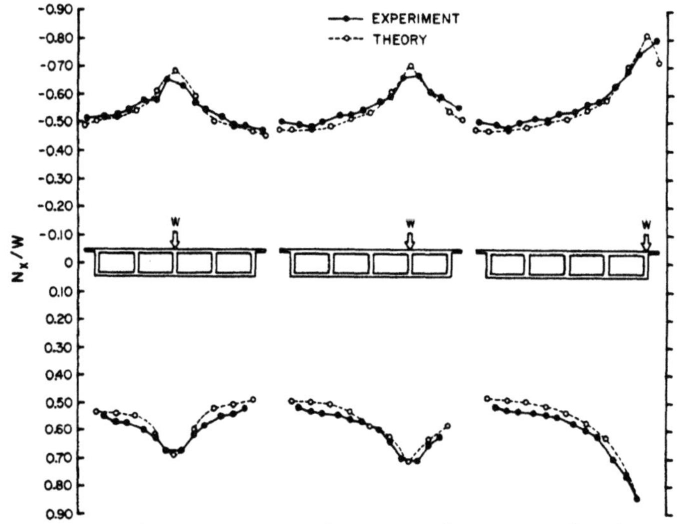

b) Curved Model 1A (Radius = 116 in, Span = 60 in)

FIG. 19 EXAMPLE 1 - TRANSVERSE DISTRIBUTION OF LONGITUDINAL MEMBRANE FORCES N_x/W AT MIDSPAN distribution that is a function of span, skew angle or curvature; and (3) the transverse distribution of M_y at midspan, for a given load position, is essentially independent of span, skew angle or curvature.

Typical theoretical and experimental results for the transverse distribution of N_x at midspan for three midspan load positions are presented in Fig. 19 for skew Model 3A having a skew angle = 45° and a span = 35.5 in. (902 mm) Fig. 18a) and for curved Model 1A having a radius of 118 in. and a span = 60 in. (1524 mm) (Fig. 18c). Note the good agreement between theoretical and experimental values. It can also be seen that the distribution curves for the two bridges are eseentially parallel, as indicated in conclusion (2) above, indicating that the departure from a uniform transverse distribution is primarily a function of the local plate bending moments in the bridge.

10.2 Example 2 -- Large Scale Reinforced Concrete Model Studies

Extensive studies of the structural behavior of three reinforced concrete 1:2.82 scale models of a typical two-lane bridge in California having two 101.5 ft (31 m) spans with a single column bent at the center have been performed at the University of California. The elevations, plan and section dimensions of the three models are shown in Fig. 20. All three had the same cross-sectional and longitudinal span dimensions, but in plan one was straight, one was curved and one was skew. Amounts of steel reinforcement used in the straight and curved models were somewhat similar, but in the skew model the amount was considerably less.

The 72 ft. (21.1 m) long, 12 ft. (3.7 m) wide, and 1 ft., 8-9/16 in. (0.52 m) deep models shown in Fig. 20 were built and tested in the Structural Engineering Laboratory of the University of California, Berkeley, using standard high-strength [60 ksi (410 MN/m^2) yield] deformed bars as reinforcement and concrete with a 3/8 in. (10 mm) aggregate instead of a mortar mix as the model material.

A detailed presentation of the investigation on each of these models including their description, instrumentations, construction, automatic data recording and reduction, test programs and comparisons of theoretical and experimental results may be found in a series of publications listed in Appendix A2. The experimental program was divided into two parts to study the response of the bridges under: (1) dead load and working loads; and (2) overloads including loading to failure.

The straight bridge was loaded to failure in 1970 [A9, A10, A11, B7, B11, B12, B13], the curved bridge in 1973 [A17, A18, A19, B17, B18, B20] and the skew bridge in 1979 [A22, A23, A24, A25,

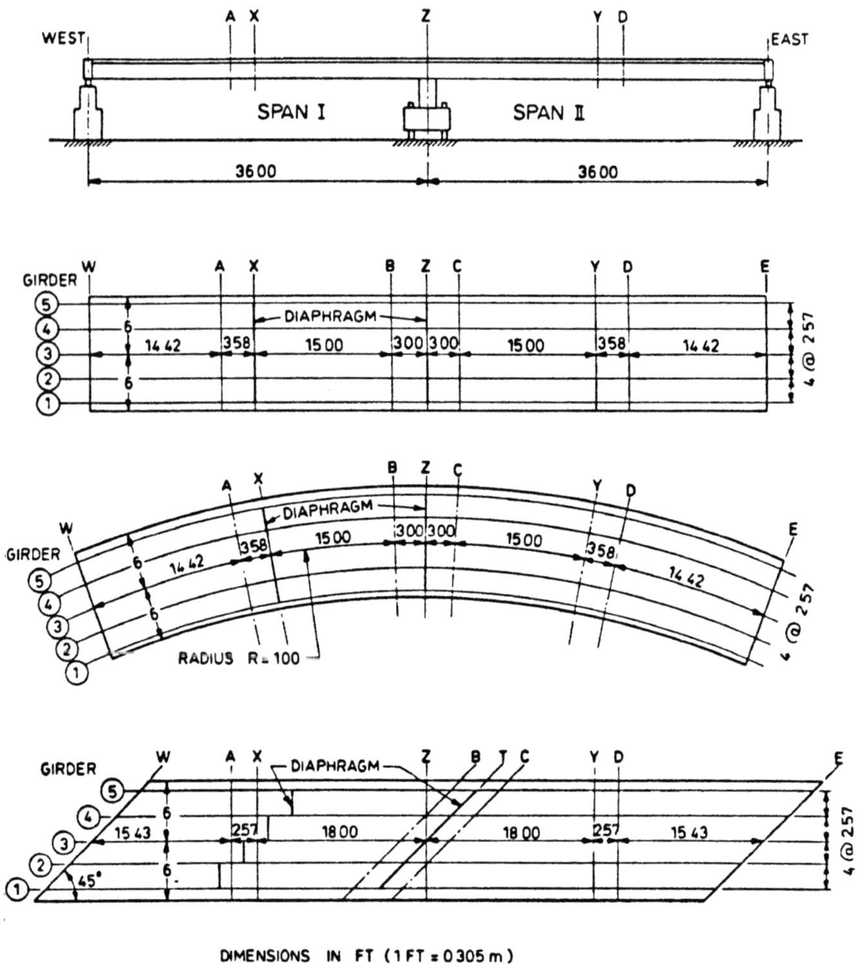

FIG. 20a EXAMPLE 2 - ELEVATION AND PLAN DIMENSIONS OF STRAIGHT, CURVED AND SKEW BRIDGE MODELS

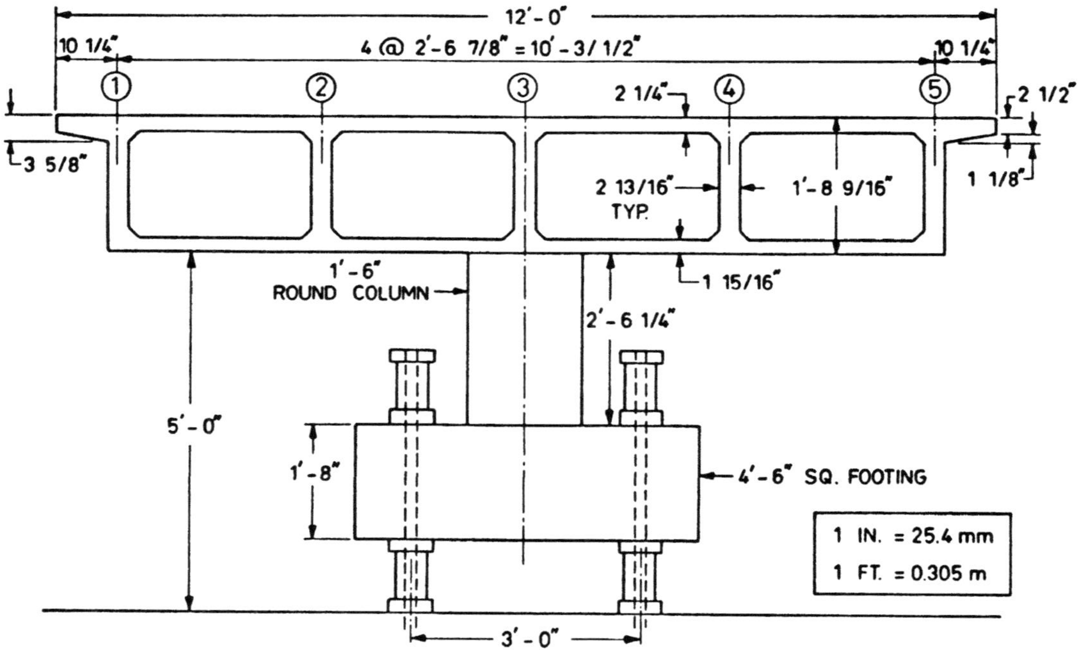

FIG. 20b EXAMPLE 2 - TYPICAL SECTION OF STRAIGHT, CURVED AND SKEW BRIDGE MODELS

B21, B24, B25]. Extensive correlative studies of experimental values for reactions, deflections, internal forces and moments with theoretical values predicted by various computer programs are presented in the above references.

Linear elastic analyses were carried out for the straight bridge using FINPLA2 and for the curved and skew bridges using CELL. Only a typical comparison of theoretical results for the three bridge models [A23] is given in Table 1 for midspan deflections and in Table 2 for longitudinal membrane forces N_x at Section A near midspan, for midspan point loads of 100 kips (445 kN).

The transverse distribution of deflections at midspan sections X and Y given in Table 1 indicates that for point loads on the longitudinal centerline at 3X + 3Y the skew bridge deflections are generally smaller than the corresponding deflections for the straight and the curved bridge. For a point load on the acute side of the span the skew bridge deflections at the loaded point are higher than for the straight or curved bridge, while for loads on the obtuse side of the bridge the opposite is the case.

From the transverse distribution of longitudinal membrane forces N_x at Section A given in Table 2, it can be observed that for loads along center girder 3, the membrane forces in the longitudinal direction in the skew bridge are of the same order of magnitude as for the straight and the curved bridges. It is interesting to note that higher membrane forces exist at Section A on the acute side of the span than on the obtuse side for the skew bridge. For loads along girder 1 the membrane forces N_x at Section A are considerably higher for the skew bridge than the straight or the curved bridge models, while for loads along girder 5 the membrane forces at Section A of the skew bridge are much smaller than in the corresponding straight and curved bridge cases. This emphasizes the sensitivity of the skew bridge to transverse load position as compared to the straight or curved bridge.

Nonlinear analyses were also carried out for the straight and curved bridge models using NOBOX [A28, B27, B28]. For both box girder bridge models a longitudinal collapse load analysis was performed first in order to obtain an upper bound on the ultimate collapse load and an indication of the ultimate collapse mechanism. A complete nonlinear analysis was then done for both examples and results were compared with the experimental findings. Finally eccentric midspan point loads were applied in both cases as opposed to the more uniform midspan loadings used for the experimental ultimate load test, and again longitudinal collapse load analyses and complete nonlinear analyses were performed.

Only typical results for load vs. midspan deflections are presented in Fig. 21 and 22 for the curved bridge model. Fig. 21

TABLE 1 COMPARISON OF TRANSVERSE DISTRIBUTION OF DEFLECTIONS ($FT \times 10^{-2}$) AT MIDSPAN SECTIONS X AND Y FOR STRAIGHT, CURVED AND SKEW BRIDGE MODELS

LOAD CASE	BRIDGE MODEL	SECTION X					SECTION Y				
		GIRDERS					GIRDERS				
		1	2	3	4	5	1	2	3	4	5
1X + 1Y	STRAIGHT	6.86	5.27	3.88	3.59	1.61	8.00	5.30	3.50	2.25	1.36
	CURVED	5.79	4.61	3.41	2.40	1.50	4.77	4.77	3.18	2.01	1.33
	SKEW	11.69	8.30	5.81	3.99	1.32	5.75	2.63	0.33	-1.48	-3.04
3X + 3Y	STRAIGHT	3.86	3.98	4.13	3.98	3.86	3.50	3.89	4.90	3.89	3.50
	CURVED	3.44	3.72	4.04	4.07	4.12	3.15	3.67	4.81	3.95	3.71
	SKEW	3.78	3.77	3.78	3.16	2.56	2.36	3.06	4.36	3.65	3.55
5X + 5Y	STRAIGHT	1.61	2.68	3.88	5.28	6.86	1.36	2.25	3.50	5.30	8.00
	CURVED	1.56	2.78	4.15	5.72	7.49	1.27	2.30	3.69	5.63	8.46
	SKEW	-2.82	-1.18	.55	2.42	5.12	1.10	3.16	5.55	8.50	12.33
1X	STRAIGHT	8.29	6.97	5.85	4.93	4.17	-1.43	-1.71	-1.99	-2.28	-2.55
	CURVED	7.14	6.11	5.25	4.59	4.06	-1.16	-1.49	-1.81	-2.15	-2.50
	SKEW	12.46	9.18	6.80	4.60	2.56	-.76	-1.88	-3.02	-4.18	-5.37
3X	STRAIGHT	5.86	5.96	6.10	5.96	5.86	-1.98	-1.97	-1.97	-1.97	-1.98
	CURVED	5.25	5.61	6.01	6.13	6.31	-1.84	-1.91	-1.97	-2.06	-2.16
	SKEW	6.81	6.27	5.77	4.66	3.58	-.99	-1.48	-1.99	-2.51	-3.03
5X	STRAIGHT	4.17	4.94	5.86	6.97	8.30	-2.55	-2.28	-1.99	-1.71	-1.44
	CURVED	4.06	5.11	6.31	7.73	9.36	-2.55	-2.37	-2.18	-2.02	-1.87
	SKEW	2.59	3.01	3.58	4.30	5.90	-1.24	-1.12	-1.02	-.91	-.78

1 FT = 0.305 m

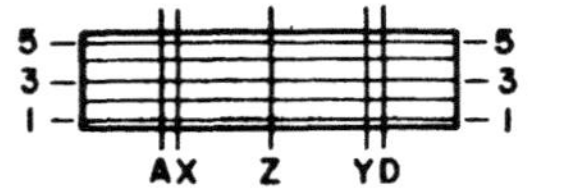

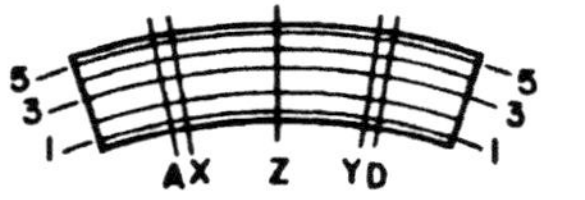

TABLE 2 COMPARISON OF TRANSVERSE DISTRIBUTION OF LONGITUDINAL MEMBRANE FORCES N_x (KIPS/FT) AT SECTION A FOR STRAIGHT CURVED AND SKEW BRIDGE MODELS

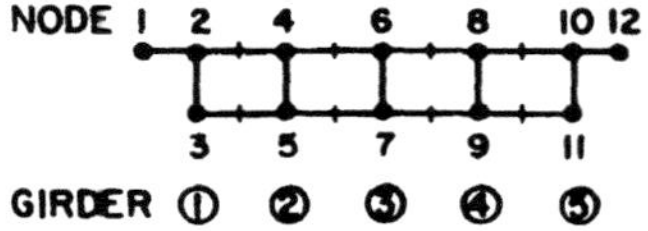

1 KIP = 4.448 kN
1 FT = 0.305 m

LOAD CASE	LOCATION	BRIDGE MODEL	TOP NODES							BOTTOM NODES				
			1	2	4	6	8	10	12	3	5	7	9	11
1X + 1Y	PLATE	STRAIGHT	-28.1	-26.2	-24.1	-22.1	-20.7	-21.6	-23.2	32.3	27.0	21.6	17.6	15.6
		CURVED	-41.6	-32.3	-24.7	-20.8	-18.0	-19.4	-22.5	32.5	26.7	21.8	18.6	17.2
		SKEW	-42.2	-37.2	-34.2	-32.5	-25.7	-28.0	-33.4	34.8	38.1	34.6	27.6	26.1
	WEB	STRAIGHT		-25.6	-24.5	-22.5	-20.9	-20.9		47.5	39.9	32.1	22.8	21.7
		CURVED		-32.0	-29.1	-24.6	-21.2	-19.3		48.0	39.5	32.2	27.0	25.4
		SKEW		-41.7	-46.2	-44.1	-34.7	-31.5		52.3	57.8	52.9	41.8	40.8
3X + 3Y	PLATE	STRAIGHT	-21.6	-22.1	-23.0	-23.8	-23.0	-22.1	-21.6	21.7	22.8	23.1	22.8	21.7
		CURVED	-27.2	-24.2	-20.1	-23.0	-21.8	-23.1	-24.7	22.8	23.3	23.5	23.0	21.2
		SKEW	-30.1	-25.4	-24.0	-20.4	-18.7	-18.6	-19.2	26.4	25.7	21.2	20.1	17.4
	WEB	STRAIGHT		-21.4	-22.9	-23.6	-22.9	-21.4		31.8	32.2	33.5	33.2	31.8
		CURVED		-23.9	-25.6	-26.4	-25.3	-22.9		33.4	34.0	34.2	33.6	31.1
		SKEW		-28.5	-31.4	-26.1	-24.1	-20.7		39.4	37.9	30.7	29.2	25.7
5X + 5Y	PLATE	STRAIGHT	-23.2	-21.5	-20.7	-22.1	-24.1	-26.2	-28.1	15.6	17.6	21.6	27.0	32.3
		CURVED	-24.3	-19.7	-18.0	-20.7	-23.8	-25.9	-37.3	18.1	18.6	21.5	25.6	29.3
		SKEW	-6.5	-5.2	-5.6	-5.3	-12.7	-26.0	-34.2	4.0	5.5	5.9	15.1	28.6
	WEB	STRAIGHT		-20.9	-20.9	-22.5	-24.5	-25.6		23.2	26.1	32.1	39.9	47.5
		CURVED		-19.6	-21.2	-24.6	-28.2	-29.3		26.8	27.6	32.0	38.0	43.3
		SKEW		-5.8	-7.4	-6.7	-16.6	-29.0		-6.0	7.9	8.1	21.9	42.9

compares theoretical values from NOBOX with experimental values obtained in the test of the bridge to failure during which three increasingly equal point loads on girders 2, 3 and 4 were applied at both midspans X and Y in addition to the dead load. Very close agreement can be observed between theoretical and experimental results. Also note the smaller deflections at the inner girder 1X as compared to the outer girder 5X, under this loading.

Fig. 22 compares only the theoretical values from NOBOX of midspan deflections for three different point loadings, symmetrically placed longitudinally at the two midspan sections X and Y: (1) the actual failure loading used in the test program of pointloads on girders 2, 3 and 4 (note that a different horizontal scale has been used for the deflection in Fig. 22 compared to that in Fig. 21); (2) an eccentric point load on inner girder 1; and (3) an eccentric point load on outer girder 5. A comparison of the midspan deflection under the exterior girders for all three load cases for which a complete nonlinear analysis was performed (Fig. 22) shows the enormous influence of the transverse stiffness deterioration in the bridge model. Not only is the ultimate load level and the ultimate deflection for the eccentric loadings drastically different from the more uniform experimental failure loading in the curved bridge model, but the eccentric loading cases feature also a torsional rotation of the bridge model at midspan, i.e., a negative deflection or uplift of the unloaded side of the span. This uplift would not occur without the transverse stiffness deterioration, since the downward loading would be distributed transversely all the way to the other side of the span. This can be seen for the lower load levels in Fig. 22 and also from the results of the linear analyses given in Tables 1 and 2.

10.3 Example 3 -- Two Cell Aribtrary Quadrilateral Box Girder

A linear analysis of a single span, two cell box girder of arbitrary plan geometry with diaphragms at the ends only (Fig. 23a) was performed using the grillage model of NOBOX (termed BOX for a linear analysis) and the finite element model of CELL. The same discretization (Fig. 23a) was used in both cases. The analysis was performed for an eccentric point load of 1 kip (4.448 kN) on girder 3 at Section 6 rather than at midspan Section 7 in order to avoid possible disturbances at the point of load application in the finite element model.

The vertical deflections of the three girder lines obtained by NOBOX and CELL (Fig. 23b) are in close agreement. A similar conclusion can be reached for the transverse bending moments in the top slab at midspan Section 7 (Fig. 24a) and the percentage of the total midspan longitudinal moment distributed to each of the three girders (Fig. 24b).

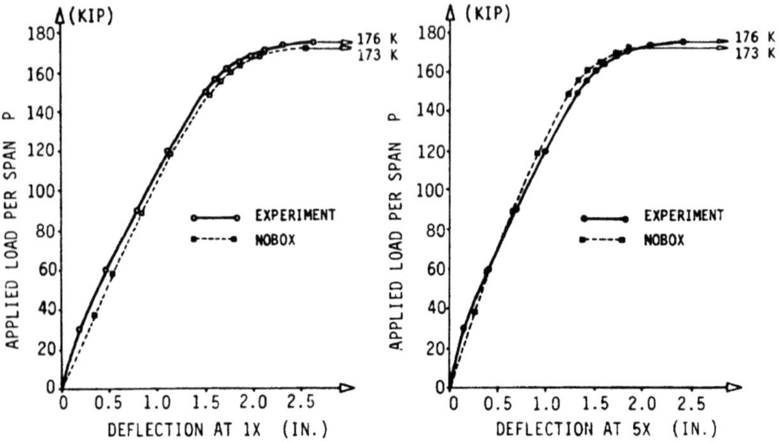

FIG. 21 EXAMPLE 2 - MIDSPAN DEFLECTIONS AT EXTERIOR GIRDERS OF CURVED BRIDGE DURING FAILURE LOADING

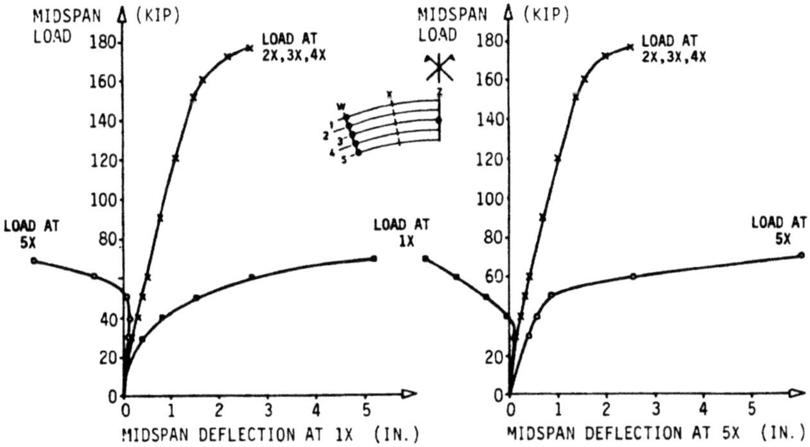

FIG. 22 EXAMPLE 2 - THEORETICAL MIDSPAN DEFLECTIONS AT EXTERIOR GIRDERS OF CURVED BRIDGE UNDER DIFFERENT LOADINGS

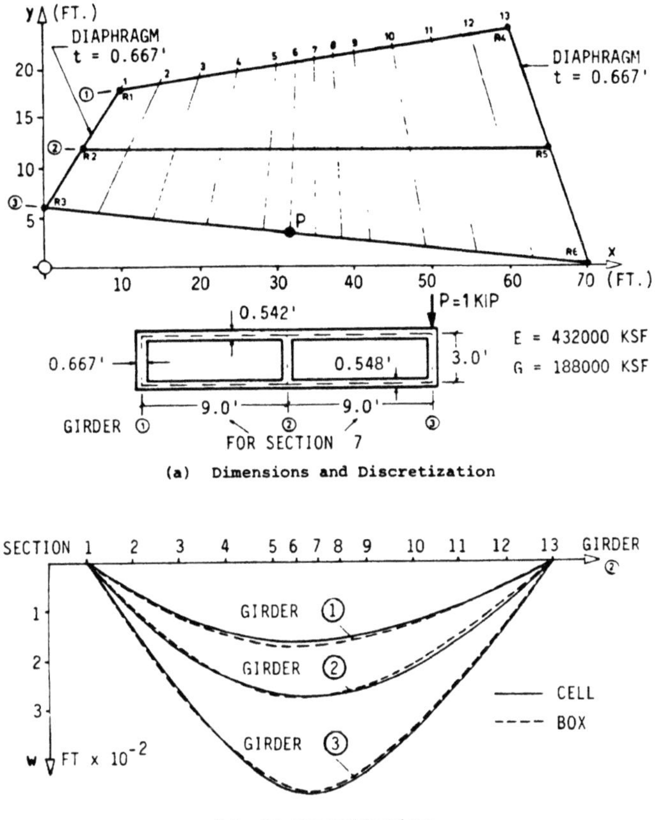

(a) Dimensions and Discretization

(b) Girder Deflections

FIG. 23 EXAMPLE 3 - DIMENSIONS, DISCRETIZATION AND DEFLECTIONS OF ARBITRARY QUADRILATERAL TWO-CELL BOX GIRDER BRIDGE

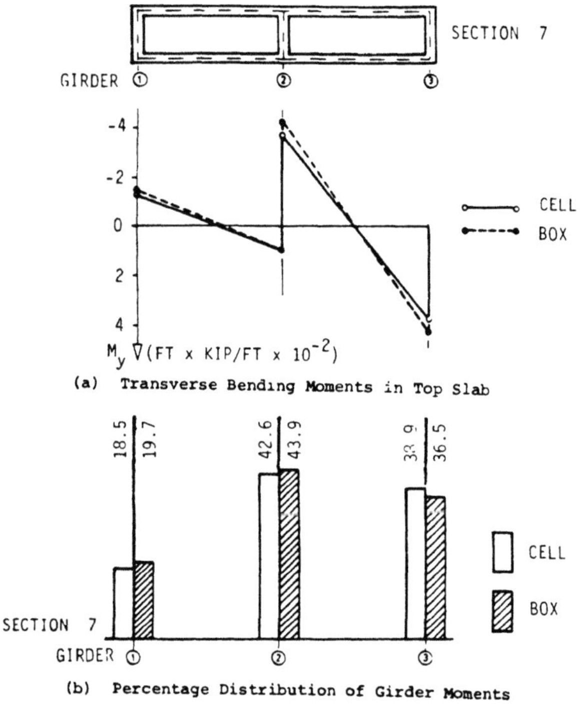

(a) Transverse Bending Moments in Top Slab

(b) Percentage Distribution of Girder Moments

FIG. 24 EXAMPLE 3 - TRANSVERSE SLAB MOMENTS AND LONGITUDINAL GIRDER MOMENTS AT MIDSPAN SECTION 7

A large difference exists, however, in the computer time necessary for the solution by the grillage model of NOBOX as compared to the finite element model of CELL. CELL requires about six times the central processor time required by NOBOX. In addition, it should be noted that NOBOX yields directly design quantities such as girder moments and shears, which have to be calculated in an additional step from the internal nodal forces when CELL is used.

10.4 Example 4 -- Single Cell, Curved, Two Span, Prestressed Box Girder Model

Linear analyses due to external loading; and prestress, including friction, of a curved bridge model (Fig. 25, 26), have been performed using the skewed finite segment model of SEGAN [A21, B19] and the finite strip model of CURDIP [A6]. The model was constructed by Hood of New Zealand [A21], who also conducted an extensive experimental and analytical investigation of its behavior. It was made of an epoxy resin. The cross section of the actual experimental model as well as those assumed in the analyses by CURDIP and SEGAN are shown in Fig. 26. The two span bridge had vertical supports at the two ends and at its center (Fig. 25a). It had single prestressing tendons in each web with the profile shown in Fig. 25b. The tendon geometry formed part of the input into the computer analyses.

Only typical analytical and experimental results for the case of prestress alone are presented in Fig. 27. Friction along the tendon was accounted for in the analyses. Fig. 27a depicts the longitudinal (tangential) stress distribution at a midspan section between the end and center support. The stresses shown are membrane stresses, therefore the average through the thickness of each plate. The agreement of the analytical results by CURDIP and SEGAN with those found experimentally are quite good. A check of the statical moment at the cross-section reveals that the internal moment is within 2 to 3 percent of the external moment, which is quite satisfactory.

In Fig. 27b, the transverse (radial) stresses at the outer and inner faces of the top slab at the midspan section are plotted. Note that because of the analytical models used (Fig. 26) only two plotted points are available from SEGAN, while five are available from CURDIP. The agreement between analytical and experimental results is satisfactory.

10.5 Example 5 -- Single Cell, Straight, Three Span, Prestressed Box Girder Bridge

A nonlinear material and time dependent analysis has been applied to a three span, continuous, post-tensioned, single cell,

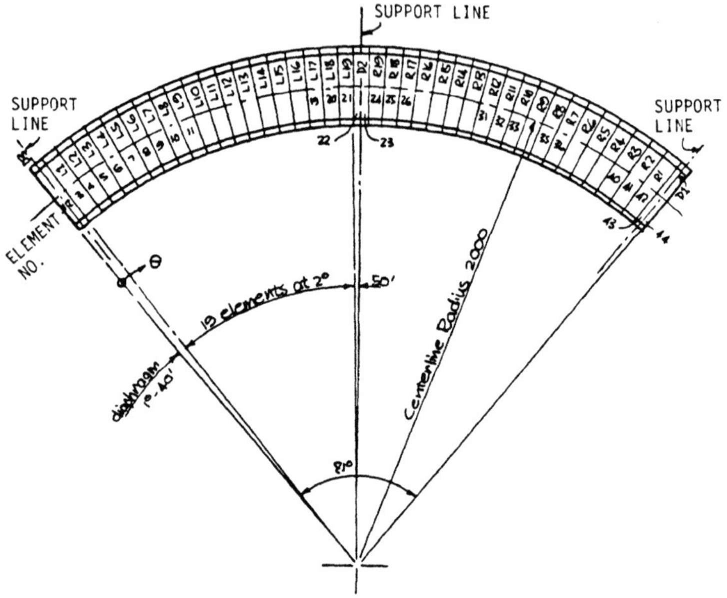

FIG. 25a EXAMPLE 4 - PLAN OF CURVED, TWO SPAN, PRESTRESSED SINGLE CELL BOX GIRDER MODEL TESTED BY HOOD

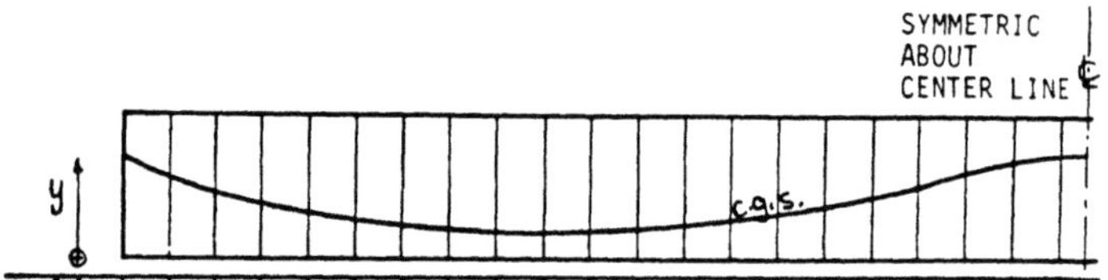

y soffit to cgs	50.00	44.04	37.69	32.12	27.37	23.43	20.29	17.95	16.44	15.73	15.82	16.72	18.43	20.96	24.28	28.44	33.38	39.14	45.70	51.81	54.72
θ	0°	1°-40'	3°-40'	5°-40'	7°-40'	9°-40'	11°-40'	13°-40'	15°-40'	17°-40'	19°-40'	21°-40'	23°-40'	25°-40'	27°-40'	29°-40'	31°-40'	33°-40'	35°-40'	37°-40'	39°-40'

FIG. 25b EXAMPLE 4 - PRESTRESSING LAYOUT FOR SINGLE TENDON IN EACH WEB

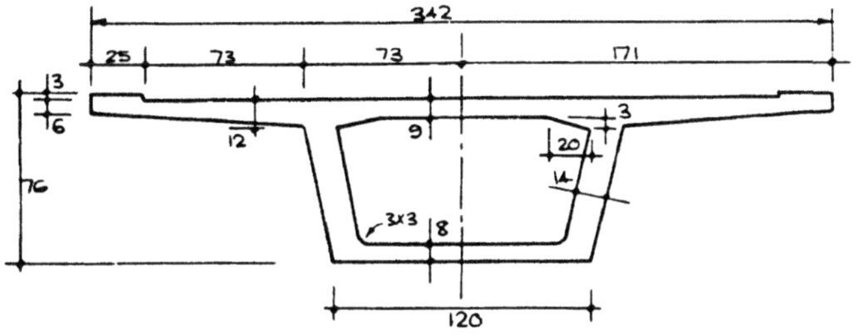

(a) Actual Experimental Model

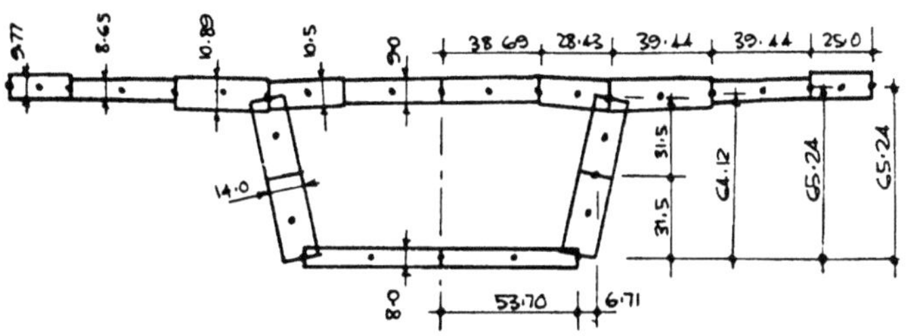

(b) Finite Strip Analysis by CURDIP

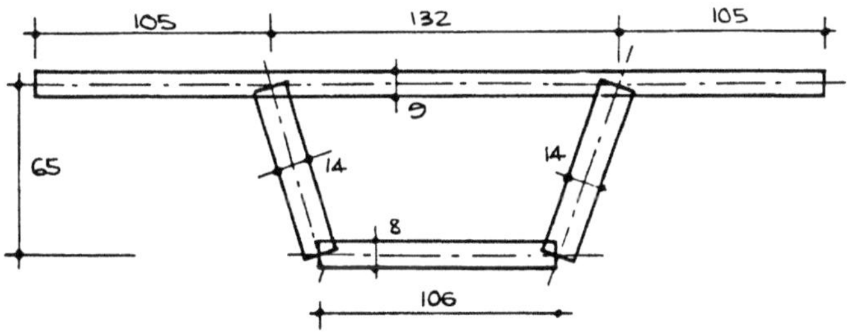

(c) Finite Element Analysis by SEGAN

FIG. 26 EXAMPLE 4 - CROSS-SECTIONAL DIMENSIONS OF ACTUAL MODEL AND THOSE USED IN ANALYSES (mm)

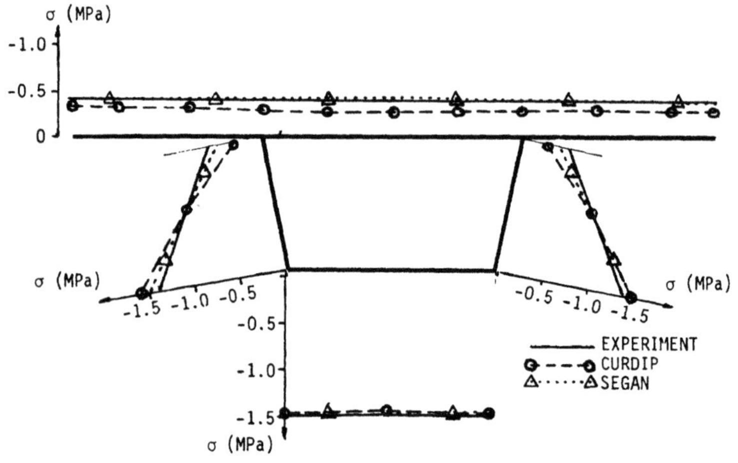

(a) Longitudinal (Tangential) Membrane Stresses (MPa) at Middle Plane of Plates

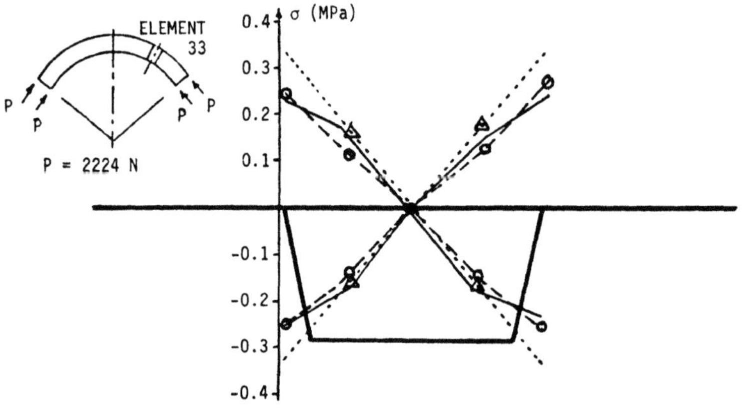

(b) Transverse (Radial) Stresses (MPa) at Outer and Inner Surface of Top Slab

FIG. 27 EXAMPLE 4 - COMPARISON OF ANALYTICAL (CURDIP AND SEGAN) AND EXPERIMENTAL RESULTS AT CENTER OF ELEMENT 33 DUE TO PRESTRESSING ALONE (P = 2224N IN EACH WEB)

concrete box girder bridge (Fig. 28) using the computer program PCFRAME [A29, B23] in order to determine its response to time dependent effects and to increasing truck overloads up to failure. The cross section (Fig. 28), with a two lane roadway width, is assumed constant over the entire length of the bridge. It is assumed that simple support conditions exist at the ends of the structure and vertical bearing supports exist at the two interior supports.

The cast-in-place, post-tensioned bridge is designed using California highway design criteria. One prestressing (PS) post-tensioned tendon in each vertical web is assumed to have the profile shown in Fig. 28. Two bridges are analyzed for comparison, one without and one with additional longitudinal mild reinforcing steel (RS), uniformly distributed over the cross-section. This steel, not required for strength, is provided for construction purposes.

For PCFRAME analysis the cross section is discretized into ten concrete layers, two for each of the top and bottom slabs and six for the webs. Where included the mild RS is modeled by three steel layers, one in each of the top and bottom slabs and one at the midheight of the web.

The time dependent analysis traces the response of the bridge under dead load only from its initial prestressing at 28 days through a period of 10,000 days (27 years). Time dependent variations of deflections, curvatures, moments, and stresses as well as prestressing forces are traced. Only typical results for the vertical deflection (Fig. 29a) and the prestressing tendon stress (Fig. 29b) versus time are presented here.

Notably evident are the smaller deflections for the bridge with, as compared to without, mild RS (Fig. 29a). The ratios of the dead load deflection after 10,000 days to the initial deflection at 28 days are about 3.0 and 4.2 for the two cases. As expected the prestressing tendon stress decreases with time (Fig. 29b) due to creep and shrinkage of the concrete and relaxation of the stress in the prestressing steel. Most of the prestress loss occurs within the first 100 days. The analysis by PCFRAME predicts considerably smaller losses than code values, probably because the latter are a conservative estimate for design.

In order to study the response of the bridge to increasing load levels up to ultimate failure, a nonlinear material and time dependent analysis was performed with PCFRAME in which the bridge was loaded to failure using increasing levels of truck live loads. The truck live loads were applied at two different ages after the initial application of dead load plus prestress, first at an early age of 30 days and second at a much later age of 10,000 days (27

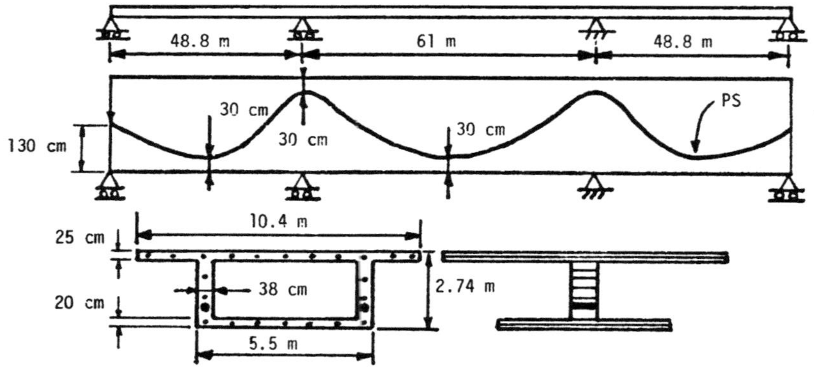

FIG. 28 EXAMPLE 5 - DIMENSIONS OF THREE SPAN PRESTRESSED CONCRETE BOX GIRDER BRIDGE

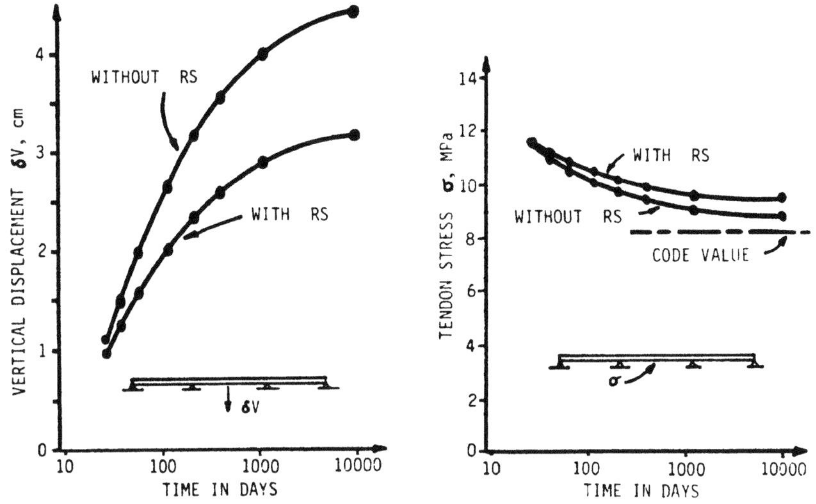

(a) Vertical Displacement δv vs. Time (b) Tendon Stress σ vs. Time

FIG. 29 EXAMPLE 5 - PCFRAME RESULTS FOR TIME DEPENDENT ANALYSIS

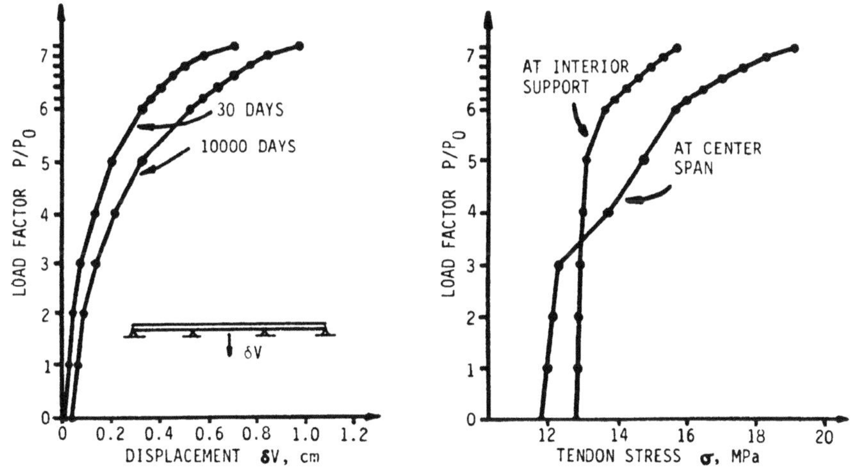

(a) Vertical Displacement δv vs. Overload P/P_0 (b) Tendon Stress σ vs. Overload P/P_0

FIG. 30 EXAMPLE 5 - PCFRAME RESULTS FOR OVERLOAD ANALYSIS TO FAILURE years).

The truck live load consisted of an overload vehicle requiring a special permit to be used on California highways. Total length of the vehicle is 108 ft (33m), total weight P_o= 107 tons (952 kN), which is carried on 13 axles. In the overload analysis one truck only is positioned in the middle of the center span of the bridge, and the structural load vector representing P_o is incremented until ultimate failure occurs. Only the bridge without mild RS was analyzed since its contribution to the ultimate capacity is small.

In general, overload response of the bridge at the two loading ages proved to be quite similar. Typical results are presented for the vertical deflection (Fig. 30a) for both loading ages and for the prestressing tendon stress (Fig. 30b) loaded only at an age of 30 days, since loading at 10,000 days results in similar curves, but shifted slightly.

The load-deflection curves (Fig. 30a) show that the structure is slightly stiffer when loaded at 30 days than when loaded at 10,000 days. This is a result of the reduction in prestressing force, due to time dependent effects, and the decrease in the cracking load of the bridge. The large overload capacity of the bridge is evident with ultimate failure occuring under dead load plus 7.2 times the live load due to one vehicle. It can also be seen in Fig. 30a that the ultimate load does not vary significantly with age at loading.

Fig. 30b plots tendon stresses at two critical locations along the girder under increasing vehicle overload. Response is shown only for loading at the age of 30 days. The initiation of cracking at the critical locations is readily identified in this figure. After dead load plus three times the load of one vehicle, cracking is initiated at the center of the main span, resulting in a large increase in tendon stress at that location. After dead load plus five times the load of one vehicle, cracking is initiated over the interior support, resulting in a similarly large increase in tendon stress at that location, and a further increase in tendon stress at center span due to a redistribution of moment. Tendon stress continues to increase under increasing overload until ultimate failure of the bridge occurs due to rupture of the tendons at the center of the main span.

11.0 SUMMARY AND CONCLUSIONS

A review of the analytical methods and computer programs which have been developed at the University of California at Berkeley, for the linear and nonlinear analysis of concrete box girder bridges has been presented. Short descriptions of 15 different computer programs and pertinent references, where details on their theoretical basis, input-output capabilities and numerous numerical examples can be found, are given in Appendix 1 and 2.

In the present paper, five numerical examples were presented to illustrate the wide range of problems, which can be solved and the type of output results obtained.

It can be concluded that the available computer programs described are valuable computational aids for the analysis of single or multi-cell, reinforced or prestressed, concrete box girder bridges which are straight, curved, skew or of arbitrary geometry in plan.

12.0 ACKNOWLEDGEMENTS

Many graduate students, listed as authors or co-authors in the publications referenced in Appendix 2, participated extensively in the development of the computer programs listed in Appendix 1. Their excellent contributions to this research effort are gratefully acknowledged.

Most of the computer programs for the linear analysis of box girder bridges were developed as part of a continuing research program sponsored by the Division of Highways, Department of Transportation, State of California and the Federal Highway Administration, United States Department of Transportation.

Much of the work on the computer programs for the nonlinear analysis of box girder bridges was done as part of a continuing research program sponsored by the National Science Foundation.

The support of these sponsoring agencies is acknowledged with many thanks. However, the content of this paper reflects the views of the author and not necessarily the sponsors.

APPENDIX 1. COMPUTER PROGRAMS FOR BOX GIRDER BRIDGES

This Appendix 1 contains a list of the computer programs developed at the University of California since 1965 for the analysis of single or multi-cell box girder bridges. All of the programs, except the last three (SEGAN, PCFRAME, NOBOX), are based on linear elastic analyses of uncracked homogeneous structures. The last three are nonlinear analysis programs, which trace the response of bridge systems under various nonlinear effects.

The programs are grouped in the list under the following methods of analysis:

A. Folded Plate Elasticity Method
B. Finite Strip Method
C. Finite Element Method
D. Finite Segment Method
E. Nonlinear Analysis Method

For each computer program a short computer program name of four to seven letters is given followed by a brief description of the program. Figures in parentheses indicate the year the program was first used and figures in brackets indicate references in Appendix 2 where detailed descriptions of the programs and their theoretical development may be found.

Anyone interested in any of the programs listed should first read the detailed references indicated and then if they wish to obtain a copy of the program itself, they should write for further information to:

Professor A. C. Scordelis
Department of Civil Engineering
729 Davis Hall
UNIVERSITY OF CALIFORNIA
Berkeley, California 94720

COMPUTER PROGRAMS FOR THE ANALYSIS OF BOX GIRDER BRIDGES DEVELOPED AT THE UNIVERSITY OF CALIFORNIA, BERKELEY, UNDER THE SUPERVISION OF PROFESSOR A. C. SCORDELIS

Note: All of the computer programs listed below are written in FORTRAN IV language.

A. FOLDED PLATE ELASTICITY METHOD

MULTPL Program (1965) [A1, B1, B2, B6]

This program provides a rapid solution for open or cellular folded plate structures having single simple spans. A direct stiffness harmonic analysis is performed in the solution. The elasticity theory is used for the isotropic plate elements. Uniform or partial surface loads as well as line loads and concentrated loads may be applied anywhere on the structure. Up to 100 non-zero terms of the appropriate Fourier series may be used to approximate the loading. Internal forces, moments and displacements in each plate element at any desired point may be found.

MUPDI Program (1966) [A1, B3, B6]

This program extends the MULTPL program to continuous span bridges and can be utilized to analyze open or cellular folded plate structures simply supported at the two ends and having up to four interior rigid diaphragms or supports between the two ends. A direct stiffness harmonic analysis based on the elasticity theory is used for the folded plate system. Compatibility at the interior rigid diaphragms or supports is accomplished by a force method of analysis. Loads and redundant forces may be approximated by up to 100 non-zero terms of the appropriate Fourier series.

MUPDI3 Program (1971) [A13, B6, B16]

This program extends the original MUPDI program such that up to twelve interior diaphragms or supports may be used and they need no longer be rigid. Diaphragms may be defined by flexible beams and supports may be defined by two dimensional planar rigid frame bents. Options permit evaluation of internal forces and moments in the diaphragms and bents as well as in the plate elements of the bridge.

B. FINITE STRIP METHOD

MULSTR Program (1970) [A4, B5, B6, B16]

This program is capable of analyzing straight prismatic folded plates made up of orthotropic plate elements with eccentric stiffeners. The structures must be simply supported at its two ends. Each plate element is idealized by a number of longitudinal finite strips in which the properties of the longitudinal and transverse stiffeners are distributed uniformly over the area of each strip and are accounted for in the analysis. The finite strip method is used to determine the strip stiffness. The displacement patterns are assumed to vary as harmonics longitudinally. In the transverse direction, a linear variation of the in-plane displacements and a cubic variation of the normal displacements are chosen. A direct stiffness harmonic analysis is used to analyze the assembled structure.

CURSTR Program (1970) [A6, B6, B8, B14, B16]

This program is capable of analyzing prismatic folded plate structures which are circular in plan and made up of orthotropic plate elements. The structure must be simply supported by radial diaphragms at its two ends. Each plate element, which may be a segment from a general cone, a cylindrical shell, or a circular ring plate, is idealized by a number of circumferential finite strips. The finite strip method is used to determine the strip stiffness. The displacement patterns are assumed to vary as harmonics in a circumferential direction. In the transverse direction, a linear variation of the in-plane displacements and a cubic variation of the normal displacements are chosen. A direct stiffness harmonic analysis is used to analyze the assembled structure.

CURDI Program (1974) [A20]

This program extends the CURSTR program to continuous span bridges and can be utilized to analyze circularly curved structures with up to twelve interior radial diaphragms or supports. Diaphragms may be defined by flexible beams and supports may be defined by two dimensional planar rigid frame bents. A direct stiffness harmonic analysis is used for the folded plate systems. Compatibility at the interior diaphragms or supporting bents is accomplished by a force method of analysis.

Loads and redundant forces may be approximated by up to 100 non-zero circular harmonics of the appropriate Fourier series. Internal forces, moments and displacements in the folded plate elements as well as the one dimensional rigid frame elements may be found at any desired point.

CURDIP Program (1980) [A26]

This program extends the CURDI program to automatically include prestressing as well as external loads. Up to ten post-tensioned longitudinal prestressing tendons of general three dimensional profiles may be specified with a minimum of input data. Longitudinal tendons may exist in the top or bottom slabs and the webs of the bridge. Friction between the prestressing tendons and the ducts is accounted for and jacking may be specified from one or both ends of the tendon. Equivalent loads due to prestressing are calculated automatically in the program using vector algebra.

COSLAB Program (1980) [A27]

This program provides a rapid solution for circularly curved orthotropic slabs simply supported at the two ends and having up to twelve interior column supports. Uniform or partial surface loads, line loads and concentrated loads may be applied anywhere on the structure. Up to ten longitudinally post-tensioned prestressing tendons with profiles over the depth of the slab made up of parabolic and straight portions may be defined with a minimum of input data. Resulting joint displacements and the internal forces, moments and displacements in the finite strip elements are calculated.

A direct stiffness harmonic analysis is used for the slab system. Compatibility at the column supports is accomplished by a force method of analysis. The finite strip method is used to evaluate plate edge forces, stiffnesses and final internal forces, moments and displacements. Fourier series are used for the loads. Geometry and tendon forces are given as input for the prestressing tendons, and equivalent loads are computed. Friction between tendon and duct is neglected. Column supports are approximated as vertical one-dimensional springs, thus neglecting the influence of the bending properties of the columns.

C. FINITE ELEMENT METHOD

CELL Program (1970) [A7, B6, B9, B10, B16]

This program analyses cellular structures of constant depth with arbitrary plan geometry. The structure must be made up of top and bottom decks and vertical webs. Two different finite element types are used to capture the main behavior of the deck and web components. The top and bottom deck slabs are idealized by quadrilateral elements having a total of five degrees of freedom (DOF) at each corner node. These elements possess both membrane and plate bending stiffnesses. The vertical webs are idealized by special rectangular spar elements having a total of five DOF at each corner mode. A single element over the entire depth of the bridge can be used to capture the essential in-plane behavior and transverse bending of the web.

Orthotropic plate properties and arbitrary loadings and boundary conditions can be treated. Automatic element and coordinate generation options minimize the required input data.

FINPLA Program (1967) [A2, A5, B6]

This program utilizes the finite element method to analyze straight prismatic folded plate structures made up of rectangular plate finite elements reinforced by one dimensional transverse or longitudinal beam elements and supported by single or multiple column rigid frame bent supports. Internal diaphragms can also be included. Arbitrary loadings and boundary conditions can be treated.

The rectangular plate finite elements possess both membrane and plate bending stiffnesses. They have six degrees of freedom (DOF) at each corner mode. The beam and frame one dimensional elements possess the standard six DOF at each end and may be eccentric with respect to the plate system. The program has been written primarily for box girder bridge structures and requires a minimum of input data to define the structure and the loading. Displacements and internal forces and moments are output for all elements specified.

FINPLA 2 Program (1971) [A12, B6, B16]

This program utilizes the finite element mothod to analyze general nonprismatic box girder bridge structures of varying width and depth and may have an integrated three-dimensional frame. The structure is discretized by dividing it longitudinally into a certain number of structure segments by vertical sections, and by subdividing each such segment into finite elements. The structure alignment is described by a longitudinal reference line which may be a straight line, a circular curve or an arbitrary planar string polygon and cross sections are defined with respect to this line.

A structure of this type is made up of quadrilateral two dimensional elements and one dimensional frame elements, each of which is assumed to have six DOF at each node, three translations and three rotations.

The quadrilateral elements for both the decks and the webs possess both membrane and plate bending stiffnesses and they may have an arbitrary orientation in space. They are taken as flat plate elements which give the best least squares fit through the actual location of the corner nodes. For box girder bridges of general plan and elevation geometry this assumption should be quite good.

Orthogropic plate properties and arbitrary loadings and boundary conditions can be treated. Automatic element and coordinate generations options minimize the required input data.

D. FINITE SEGMENT METHOD

SIMPLA Program (1967) [A2, B4]

The program provides a complete analysis of straight prismatic cellular or open folded plate structures. The structure may be composed of one or more (up to 15) types of plates. The folded plate structure is defined transversely by the cross section in terms of the dimensions of its plate elements and their joint interconnections, and longitudinally by the number of segments. Arbitrary support conditions can be accounted for. The input data is so arranged that only the properties of a typical cross section need be specified, and any repeating segment can be indicated by a simple input format.

The solution is based on the finite segment concept in which each plate element is divided into a finite number of segments longitudinally. Compatibility and equilibrium conditions are then satisfied along the four edges of each segment. Each segment of the plate is assumed to obey the ordinary theory of folded plate analysis. This means, slab action is defined by the behavior of a transverse one-way slab spanning between longitudinal joints, and membrane stresses produced in each plate by longitudinal plate action are calculated by the elementary beam theory. A transfer matrix method is used to solve the resulting set of linear equations for the unknown displacements and then the internal plate forces and moments can be found.

E. NONLINEAR ANALYSIS METHOD

PCFRAME Program (1977) [A29, B23, B26]

This program provides an efficient numerical procedure for the material and geometric nonlinear analysis of planar reinforced and prestressed concrete frames including the time dependent effects due to load history, temperature history, creep, shrinkage and aging of concrete and relaxation of prestressing tendons. The analysis is capable of predicting the response of these structures throughout their service load history as well as throughout the elastic cracking, inelastic and ultimate load ranges.

The analytical model consists of a series of interconnected beam elements with three degrees of freedom at each end. Varied material properties within a beam element are accounted for by a composite concrete and reinforcing steel layer system. Pre-tensioned or post-tensioned bonded or unbonded tendons of general profile can be specified and results for stresses in the concrete, reinforcing and prestressing steel and displacements can be obtained before, at and anytime after the transfer of prestress.

<u>SEGAN Program</u> (1978) [A21, B19]

This program permits the analysis of curved segmentally erected prestressed concrete box girder bridges, including time dependent effects, so that deflections and stresses can be computed at any stage during the construction or service life of the bridge.

The segments can be either precast or cast in place. The bridge cross section must consist of a single box section with vertical or inclined webs and have cantilever flanges. Width and depth dimensions can vary along the bridge span. Time-dependent material changes, such as creep and shrinkage of concrete, as well as environmental effects such as humidity and temperature changes, are accounted for. However, other concrete material non-linearities such as cracking or inelastic behavior and geometric nonlinearities are not included in the analysis.

Prestressing in the longitudinal direction can be included in the analysis. Tendons can be located in either the webs or flanges and can have a linear or parabolic profile. Stressing operations encountered in segmental construction, such as stress and release operations, slip in of tendons, and restressing and removal of tendons are included. A procedure is included for specifying the structural configuration and the erection sequence. Operations used in segmental construction, such as addition of segments, prestressing, changing of support boundary conditions, application or removal of construction loads, and prescribed displacements can be analyzed. At each construction stage, the current structure is analyzed and the incremental and total displacements, stresses, and strains are obtained. The complete structure can also be analyzed for a variety of live loads and time steps.

<u>NOBOX Program</u> (1982) [A28, B27, B28]

This program can be used to trace the nonlinear response of multi-cell reinforced concrete box girder bridges under stepwise increasing static loads. Nonlinearities considered are material nonlinearities inherent in reinforced concrete structural members under short term loading such as cracking of the concrete, yielding of the reinforcement and formation of plastic hinges due to shear and flexure.

The analytical model is developed for multi-cellular structures of arbitrary plan geometry and constant height. The analytical method can trace the complete nonlinear behavior of the bridge beyond the working stress range all the way up to the ultimate failure and collapse of the structure. Both collapse mechanisms and failure loads can be determined.

A three-dimensional grillage, consisting of longitudinal beam elements, transverse diaphragm elements, transverse bending frames and specially developed torsional shear panels, forms the basis for the analytical model. Material nonlinearities in the form of member end force-deformation relationships for reinforced concrete members subjected to flexure, shear or torsion, as obtained from numerous experimental investigations, are approximated by simple trilinear elastic-perfectly plastic force-deformation characteristics.

APPENDIX 2. REFERENCES ON BOX GIRDER BRIDGE RESEARCH PERFORMED AT THE UNIVERSITY OF CALIFORNIA, BERKELEY

This Appendix 2 contains a list of research reports (part A) and technical papers (part B) on analytical and experimental research investigations performed in the Division of Structural Engineering and Structural Mechanics of the University of California, since 1965.

The technical papers (part B) are publications in technical society journals and proceedings, which can be found in most engineering libraries.

The research reports (part A) are published as part of a Structural Engineering and Structural Mechanics (UC-SESM) report series of the University of California. Information on their availability from the University of California may be obtained by writing to:

Professor A. C. Scordelis
Department of Civil Engineering
729 Davis Hall
University of California
Berkeley, California 94720

Unfortunately, the original supply of some of the research reports has been exhausted. However, most of them have been placed on file with the U.S. Department of Commerce and may be obtained on request for cost of reproduction by writing to:

National Technical Information Service
Operations Division
Springfield, Virginia 22161

The accession number (shown in parentheses on reference list A) should be specified when ordering a particular report.

REFERENCES ON BOX GIRDER BRIDGE RESEARCH PERFORMED AT THE UNIVERSITY OF CALIFORNIA, BERKELEY

A. Research Reports

1. Scordelis, A.C., "Analysis of Simply Supported Box Girder Bridges," Structural Engineering and Structural Mechanics Report No. SESM 66-17, University of California, Berkeley, October 1966 (PB 175 646).

2. Scordelis, A. C., "Analysis of Continuous Box Girder Bridges," Structural Engineering and Structural Mechanics Report No. SESM 67-25, University of California, Berkeley, November 1967 (PB 178 355).

3. Scordelis, A.C., and Meyer, C., "Wheel Load Distribution in Concrete Box Girder Bridges," Structural Engineering and Structural Mechanics Report No. SESM 69-1, University of California, Berkeley, January 1969 (PB 183 923).

4. Willam, K. J., and Scordelis, A.C., "Analysis of Orthotropic Folded Plates with Eccentric Stiffeners," Structural Engineering and Structural Mechanics Report No. SESM 70-2, University of California, Berkeley, February 1970 (PB 191 051).

5. Meyer, C., and Scordelis, A.C., "Computer Program for Prismatic Folded Plates with Plate and Beam Elements," Structural Engineering and Structural Mechanics Report No. SESM 70-3, University of California, Berkeley, February 1970 (PB 191 050).

6. Meyer, C., and Scordelis, A.C., "Analysis of Curved Folded Plate Structures," Structural Engineering and Structural Mechanics Report No. UC SESM 70-8, University of California, Berkeley, June 1970 (PB 193 535).

7. Willam, K. J., and Scordelis, A.C., "Computer Program for Cellular Structures of Arbitrary Plan Geometry," Structural Engineering and Structural Mechanics Report No. UC SESM 70-10, University of California, Berkeley, September 1970 (PB 196 143).

8. Meyer, C., "Analysis and Design of Curved Box Girder Bridges," Structural Engineering and Structural Mechanics Report No. UC SESM 70-22, University of California, Berkeley, December 1970 (PB 197 289).

9. Bouwkamp, J.G., Scordelis, A.C., and Wasti, S.T., "Structural Behavior of a Two Span Reinforced Concrete Box

Girder Bridge Model, Volumn I, "Structural Engineering and Structural Mechanics Report No. UC SESM 71-5, University of California, Berkeley, April 1971 (PB 199 187).

10. Scordelis, A.C., Bouwkamp, J.G., and Wasti, S.T., "Structural Behavior of a Two Span Reinforced Concrete Box Girder Bridge Model, Volume II," Structural Engineering and Structural Mechanics Report No. UC SESM 71-16, University of California, Berkeley, October 1971 (PB 210 431).

11. Scordelis, A. C., Bouwkamp, J. G., and Wasti, S.T., "Structural Behavior of a Two Span Reinforced Concrete Box Girder Bridge Model, Volume III," Structural Engineering and Structural Mechanics Report, No. UC SESM 71-17, University of California, Berkeley, October 1971.

12. Meyer, C., and Scordelis, A.C., "Computer Program for Non-Prismatic Folded Plates with Plate and Beam Elements," Structural Engineering and Structural Mechanics Report No. UC SESM 71-23, University of California, Berkeley, December 1971 (PB 220 197).

13. Lin, C.S., and Scordelis, A.C., "Computer Program for Bridges on Flexible Bents," Structural Engineering and Structural Mechanics Report No. UC SESM 71-24, University of California, Berkeley, December 1971 (PB 210 171).

14. Godden, W. G., and Aslam, M., "Model Studies of Skew Box Girder Bridges," Structural Engineering and Structural Mechanics Report No. UC SESM 71-26, University of California, Berkeley, December 1971 (PB 223 120).

15. Comartin, C.D., and Scordelis, A.C., "Analysis and Design of Skew Box Girder Bridges," Structural Engineering and Structural Mechanics Report No. UC SESM 72-14, University of California, Berkeley, December 1972. (PB 226 793).

16. Godden, W. G., and Aslam, M., "Model Studies of Curved Box Girder Bridges," Structural Engineering and Structural Mechanics Report No. UC SESM 73-5, University of California, Berkeley, March 1973 (PB 226 842/AS).

17. Scordelis, A.C., Bouwkamp, J.G., and Larsen, P.K., "Structural Behavior of a Curved Two Span Reinforced Concrete Box girder Bridge Model, Volume I," Structural Engineering and Structural Mechanics Report No. UC SESM 74-5, University of California, Berkeley, September 1974 (PB 242 523/AS).

18. Scordelis, A.C., Bouwkamp, J.G., and Larsen, P.K.,

"Structural Behavior of a Curved Two Span Reinforced Concrete Box Girder Bridge Model, Volume II," Structural Engineering and Structural Mechanics Report No. UC SESM 74-6, University of California, Berkeley, September 1974 (PB 242 524/AS).

19. Scordelis, A.C., Bouwkamp, J.G., and Larsen, P. K., "Structural Behavior of a Curved Two Span Reinforced Concrete Box Girder Bridge Model, Volume III," Structural Engineering and Structural Mechanics Report No. UC SESM 74-7, University of California, Berkeley, September 1974 (PB 242 525/AS).

20. Kabir, A.F., and Scordelis, A.C., "Computer Programs for Curved Bridges on Flexible Bents," Structural Engineering and Structural Mechanics Report No. UC SESM 74-10, University of California, Berkeley, September 1974 (PB 242 470/AS).

21. Van Zyl, S., "Analysis of Curved Segmentally Erected Prestressed Concrete Box Girder Bridges," Structural Engineering and Structural Mechanics, Report No. UC/SESM 78-2, University of California, Berkeley, January 1978 (PB 284 939/AS).

22. Scordelis, A.C., Bouwkamp, J.G., Wasti, S.T., and Anicic, D. "Structural Behavior of A Skew Two Span Reinforced Concrete Box Girder Bridge Model," Vol. I, Structural Engineering and Structural Mechanics Report No. UC SESM 80-1, University of California, Berkeley, June 1980.

23. Scordelis, A.C., Bouwkamp, J.G., Wasti, S.T., and Anicic, D., "Structural Behavior of a Skew Two Span Reinforced Concrete Box Girder Bridge Model," Vol. II, Structural Engineering and Structural Mechanics Report No. UC SESM 80-2, University of California, Berkeley, June 1980.

24. Scordelis, A.C., Bouwkamp, J. G. Wasti, S.T., and Anicic, D., "Structural Behavior of a Skew Two Span Reinforced Concrete Box Girder Bridge Model," Vol. III, Structural Engineering and Structural Mechanics Report No. UC SESM 80-3, University of California, Berkeley, June 1980.

25. Scordelis, A.C., Bouwkamp, J.G., Wasti, S.T., and Anicic, D., "Structural Behavior of a Skew Two Span Reinforced Concrete Box Girder Bridge Model," Vol. IV, Structural Engineering and Structural Mechanics Report No. UC SESM 80-4, University of California, Berkeley, June 1980.

26. Sorensen, S.I., and Scordelis, A.C., "Computer Program

for Curved Prestressed Box Girder Bridges," Structures and Materials Research Report No. UC SESM 80-10, Department of Civil Engineering, University of California, Berkeley, December 1980.

27. Sorensen, S.I., and Scordelis, A.C., "Computer Program for Curved Prestressed Slabs," Structures and Materials Research Report No. UC SESM 80-11, Department of Civil Engineering, Universitv of California, Berkeley, December 1980.

28. Seible, F., "Nonlinear Analysis and Ultimate Strength of Multi-Cell Reinforced Concrete Box Girder Bridges," Structural Engineering and Structural Mechanics Report No. UC SESM 82-2, University of California, Berkeley, February 1982.

29. Kang, Y.J., "Nonlinear Geometric, Material and Time Dependent Analysis of Reinforced and Prestressed Concrete Frames," Structural Engineering and Structural Mechanics Report No. UC SESM 77-1, University of California, Berkeley, January 1977 (PB 272 930/AS).

B. Technical Papers

1. DeFries-Skene, A., and Scordelis, A.C., "Direct Stiffness Solution for Folded Plates," Journal of the Structural Division, Proceedings of American Society of Civil Engineers, Volume 90, No. ST-3, June 1964.

2. Scordelis, A.C., Davis, R.E., and Lo, K.S., "Load Distribution in Concrete Box Girder Bridges," ACI Proceedings of First International Symposium on Concrete Bridge Design, Toronto, Canada, April 1967, ACI Publication SP-23, 1969.

3. Scordelis, A.C., and Davis, E.R., "Stresses in Continuous Concrete Box Girder Bridges," ACI Proceedings of Second International Symposium on Concrete Bridge Design, Chicago, April, 1969, ACI Publication SP-26, 1971.

4. Lo, K. S., and Scordelis, A.C., "Finite Segment Analysis of Folded Plates," Journal of the Structural Division, Proceedings of American Society of Civil Engineers, Volume 95, No. ST5, May 1969.

5. Willam, K. J., and Scordelis, A.C., "Analysis of Eccentrically Stiffened Folded Plates," Proceedings of IASS Symposium on Folded Plates and Prismatic Structures,

Vienna, September 1970.

6. Scordelis, A.C., "Analytical Solutions for Box Girder Bridges," Proceedings, Conference on Modern Developments in Bridge Design and Construction, Cardiff, Great Britain, April, 1971.

7. Bouwkamp, J.G., Scordelis, A.C., and Wasti, S.T., "Structural Behavior of a Reinforced Concrete Box Girder Bridge," Proceedings, Conference on Modern Developments in Bridge Design and Construction, Cardiff, Great Britain, April 1971.

8. Meyer, C., and Scordelis, A.C., "Analysis of Curved Folded Plate Structures," Journal of the Structural Division, Proceedings of American Society of Civil Engineers, Volume 98, No. ST1, January 1972.

9. Willam, K. J., and Scordelis, A.C., "Cellular Structures of Arbitrary Plan Geometry," Journal of the Structural Division, Proceedings of American Society of Civil Engineers, Volume 98, No. ST7, July 1972.

10. Godden, W.G., and Aslam, M., "Model Studies of Skew Multicell Girder Bridges," Journal of the Engineering Mechanics Division, Proceedings of the American Society of Civil Engineering, Volume 99, No. EM-1, February 1973.

11. Scordelis, A.C., Bouwkamp, J. G., and Wasti, S.T., "Study of AASHO Loadings on a Concrete Box Girder Bridge," Highway Research Record No. 428, Highway Research Board, Washington, D.C. 1973.

12. Scordelis, A.C., Bouwkamp, J.G., and Wasti, S.T., "Structural Response of a Concrete Box Girder Bridge," Journal of the Structural Division, Proceedings of the American Society of Civil Engineers, Vol. 99, No. ST10, October 1973.

13. Scordelis, A.C., Bouwkamp, J.G., and Wasti, S.T., "Ultimate Strength of a Concrete Box Girder Bridge," Journal of the Structural Division, Proceedings of the American Society of Civil Engineers, Vol. 100, ST1, January 1974.

14. Aslam, M., Godden, W.G., "Model Studies of Multicell Curved Box Girder Bridges," Journal of the Engineering Mechanics Division, Proceedings of the American Society of Civil Engineering, Vol. 101, No. EM3, June 1975.

15. Scordelis, A.C., "Folded Plates for Bridges," Bulletin of the International Association for Shell and Spatial Structures, Madrid, No. 57, April 1975.

16. Scordelis, A.C., "Analytical and Experimental Studies of Multicell Concrete Box Girder Bridges," Bulletin of the International Association for Shell and Spatial Structures, Madrid, No. 58, August 1975.

17. Scordelis, A.C. and Larsen, P.K., "Structural Response of Curved RC Box Girder Bridges," Journal of the Structural Division, Proceedings of the American Society of Civil Engineers, Vol. 103, No. ST8, August 1977.

18. Scordelis, A.C., Larsen, P.K. and Elfgren, L.G., "Ultimate Strength of Curved RC Box Girder Bridge," Journal of the Structural Division, Proceedings of the American Society of Civil Engineers, Vol. 103, No. ST8, August 1977.

19. Van Zyl, S.F., and Scordelis, A.C., "Analysis of Curved Prestressed Segmental Bridges," Journal of the Structural Division, Proceedings of the American Society of Civil Engineers, Vol. 105, No. ST 11, November 1979.

20. Scordelis, A.C., Elfgren, L.G., and Larsen, P.K., "Time Dependent Behavior of Concrete Box Girder Bridges," Journal of the American Concrete Institute, Title No. 76-9, January 1979.

21. Bouwkamp, J.G., Scordelis, A.C., and Wasti, S.T., "Failure Study of a Skew Box Girder Bridge Model," Final Report, 11th Congress, Vienna, International Association for Bridge and Structural Engineering, August 31 - September 5, 1980.

22. Hellesland, J., and Scordelis, A.C., "Analysis of RC Bridge Columns Under Imposed Deformations," Final Report, IABSE Colloquium on Advanced Mechanics of Reinforced Concrete, Delft, June 1981.

23. Ketchum, M.A., and Scordelis, A.C., "Nonlinear Analysis of a Prestressed Concrete Bridge," Final Report, IABSE Colloquium on Advanced Mechanics of Reinforced Concrete, Delft, June 1981.

24. Scordelis, A.C., Wasti, S.T., and Seible, F., "Structural Response of Skew RC Box Girder Bridges," Journal of the Structural Division, Proceedings ASCE, Vol. 108, No. ST1, January 1982.

25. Scordelis, A.C., Bouwkamp, J.G., Wasti, S.T., and Seible, F., "Ultimate Strength of Skew RC Box Girder Bridge," Journal of the Structural Division, Proceedings ASCE, Vol. 108, No. ST 1, January 1982.

26. Kang, Y.J., and Scordelis, A.C., "Nonlinear Analysis of Prestressed Concrete Frames," Journal of the Structural Division, Proceedings of the American Society of Civil Engineers, Vol. 106, No. ST 2, February 1980.

27. Seible, F., and Scordelis, A.C., "Nonlinear Behavior and Failure Analysis of Multicell RC Box Girder Bridges," Proceedings, CSCE International Conference on Short and Medium Span Bridges, Toronto, Canada, August 1982.

28. Seible, F., and Scordelis, A. C., "Nonlinear Analysis of Multicell Reinforced Concrete Box Girder Bridges," Journal of Engineering Structures, IPC Science and Technology Press Limited, January 1983.

COMPARATIVE STRUCTURAL BEHAVIOR OF STRAIGHT, CURVED AND SKEW REINFORCED CONCRETE BOX GIRDER BRIDGE MODELS

S. Tanvir Wasti

Professor of Civil Engineering
Middle East Technical University
Ankara, Turkey

Alex C. Scordelis

Professor of Civil Engineering
University of California
Berkeley, California USA

SUMMARY

Comparisons of theoretical and experimental values of selected structural quantities, such as reactions, deflections and moments are presented for three large scale (1:2.8), two span, four cell, reinforced concrete box girder bridge models tested under dead load, working loads, and overloads, including loading to failure. All three models had the same cross-sectional and longitudinal span dimensions, but in plan one was straight, one was curved and one was skew. The models were 72 ft (21 m) long along the longitudinal centerline, 12 ft (3.7 m) wide and 1.71 ft (0.52 m) deep. Amounts of steel reinforcement used in the straight and curved model were similar, but in the skew model the amount was considerably less.

General agreement was found between theoretical and experimental results. The structural response of the straight and curved models to point loads at midspan sections was similar, however that of the skew bridge was different and highly dependent on the transverse position of the applied point load. All three bridge models exhibited excellent behavior with high overload capacities before ultimate failure.

1. INTRODUCTION

Box girder bridges are extensively used in highway construction because of their structural efficiency and because they offer an economical alternative in many cases. Although plan geometries may be straight, curved or skew because of lack of space or the growing number of complex intersections, design procedures, e.g., the AASHTO and State of California wheel load formulas, do not incorporate any parameters to account for curvature or skew. The purpose of this paper is to present comparisons of theoretical and experimental values for selected structural quantities such as reactions, deflections and moments to assess the effects of curvature or skew on the behavior of otherwise similar large scale reinforced concrete box girder bridge models.

2. DESCRIPTION OF BOX GIRDER BRIDGE MODELS

An in-depth program of analytical and experimental research on box girder bridges has been conducted at the University of California, Berkeley, since 1965. Especially relevant to the present paper are a series of technical reports and papers [1-9] that describe detailed experimental and analytical studies of 1:2.8 scale continuous two span, four cell straight, curved and skew reinforced concrete box girder bridge models. In each case the scale was chosen so as to allow the representation of a typical No. 11 reinforcement bar of area 1.56 sq. in. (1006 mm^2) in the prototype by a No. 4 reinforcement bar of area 0.196 sq. in. (126 mm^2) in the model. Construction of the model was a replica of prototype construction in the field, with the bottom slab and webs being cast first and the top slab being cast subsequently.

In order to obtain proper simulation of prototype behavior and ensure that strains and stresses measured in the model would represent those in the prototype, it was necessary in each case to add extra dead weight which, along with the self-weight of the model, would result in 2.8 times the weight of the model itself. Various schemes for implementing the placing of this extra weight (equivalent in each of the three models to about 95 kips or 423 kN) were examined. For the straight and curved bridge models, it was decided to place steel billets inside the cells and to distribute their weight by means of wooden supports and layers of sand. For the skew bridge model it was decided to use 80 concrete blocks on the top deck of the bridge as the required extra weight.

Main longitudinal reinforcement for the straight and curved models comprised 55 No. 4 rebars of area 10.8 sq. in. (6968 mm^2) in the midspan regions of the bottom slabs and 82 No. 4 rebars of area 16.1 sq. in. (10387 mm^2) in the top slab over the center bent. Comparative figures for the skew model were 50 No. 4 rebars of area

9.8 sq. in. (6323 mm^2) in the bottom slabs and 68 No. 4 rebars of area 13.4 sq. in. (8645 mm^2) in the top slab over the center bent. The reductions represent smaller design moments for skew bridges because of the observed decrease in dead load moments in skew bridges [7]. It should, however, be pointed out that the main longitudinal reinforcement quantities comprise only a part of the total steel in box girder bridges because of the additional requirements for transverse steel and web reinforcement.

The elevation and plan views of the three bridge models are given in Fig. 1 with basic dimensions and designations of significant longitudinal and transverse sections. The common transverse section is shown in Fig. 2. In each model A, B, C and D represent the instrumented sections, W and E the West and East abutment sections and X and Y the midspan sections. Z is the central right section showing the orientation of the center bent diaphragm for the straight and curved models, whereas the skew center bent section along the diaphragm for the skew model is designated as T. Span I for each of the models possesses a midspan diaphragm which, for the skew model, is staggered as shown in Fig. 1. In the straight and curved models it may be observed that the distances between sections A and X and sections D and Y are 3.58 ft (1.09 m) and that sections B and C are in each case 3 ft (0.91 m) on either side of section Z. For the skew model it was decided to divide the sections in multiples of 2.57 ft (0.78 m). In each case it was considered preferable to locate the instrumented sections A and D close to but not at the midspan loaded sections.

For all models, locations were designated by the intersection of the girder and the associated transverse (right or skew) section. Thus 1X refers to the location where girder 1 and the midspan section X cut each other. For the skew bridge model, the acute side and obtuse side refer to the bridge portions on either side of girder 3; these portions are skew-symmetrically placed in the two spans of the bridge.

3. EXPERIMENTAL PROGRAM

The model had a loading frame at each of the midspan sections X and Y, enabling point loads to be applied at each of the girders 1-5 either singly or in combination. Instrumentation and data acquisition were planned to allow the measurement of girder and central column reactions, deflections at the sections of interest and internal tensile and compressive strains.

Each phase of the live load program for each bridge model comprised firstly the application of equal loads called "conditioning loads" on all girders at both midspans to produce the same nominal total steel stresses (24, 30, 40, 50 and 60 ksi = 166, 207,

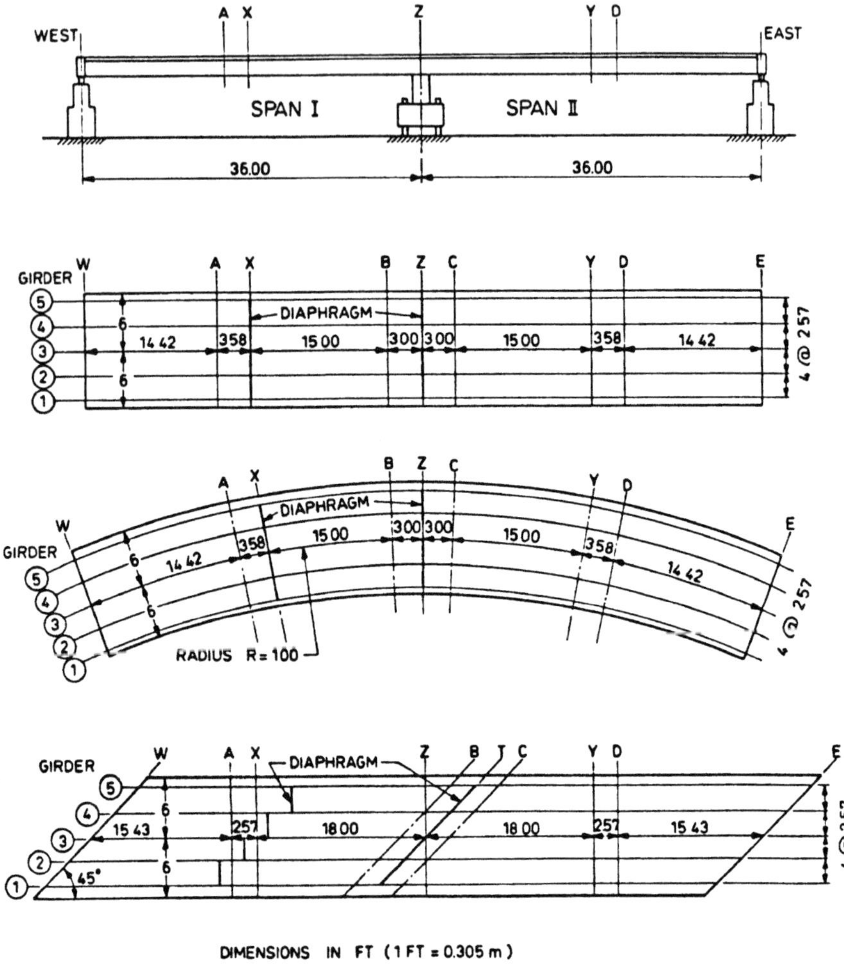

FIG. 1 SCHEMATIC REPRESENTATION OF ELEVATION AND PLANS OF STRAIGHT CURVED AND SKEW BOX GIRDER BRIDGE MODELS

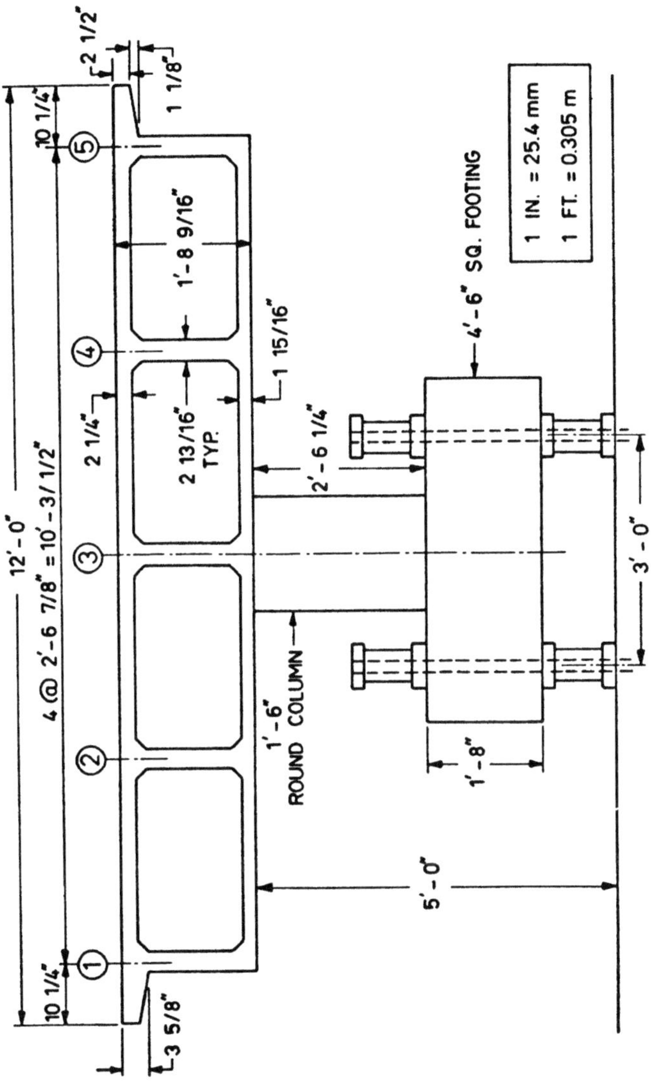

FIG. 2 TYPICAL SECTION OF BOX GIRDER BRIDGE MODELS

276, 345 and 414 MPa) at sections of maximum positive and negative moment. After the removal of the conditioning loads, point loads were applied in each phase in several combinations, to produce total stresses in the tensile reinforcement of each girder (considered separately) of the order of the working stresses, i.e., 24 and 30 ksi (166 and 207 MPa). The conditioning loads represented the successive deterioration of the bridge models under the effects of overloading.

Details of the very extensive loading program for each model, incorporating different support restraints, truck and construction vehicle loads and also loading to failure shall not be dealt with herein, but may be studied in the above mentioned references.

4. THEORETICAL ANALYSES

Theoretical results for the straight bridge model were obtained using a finite element computer program entitled FINPLA2 [10] whereas both the curved and skew bridge models were analyzed using the finite element program CELL [11]. Both FINPLA2 and CELL assume that the plates forming the box girder are elastic, isotropic and homogeneous and that a linear relationship exists between forces and deformations.

For each model the results obtained from the chosen finite element programs were compared in detail with those from several other analytical models. Extensive comparisons indicated the superiority of the specially developed FINPLA2 and CELL programs in predicting reactions, deflections and moments for box girders and also the distribution of the section moments along the different girders.

The FINPLA2 model for the straight bridge consisted of a relatively coarse mesh that resulted in a total of 180 top deck, 120 bottom deck and 150 web elements. For the curved bridge, the analytical model comprised a total of 534 elements. The CELL finite element model for the skew bridge consisted of 576 deck elements each for the top and bottom slabs and 312 vertical web elements.

5. COMPARISONS OF THEORETICAL AND WORKING STRESS EXPERIMENTAL RESULTS

5.1 Preliminary Results

Pertinent comparisons of theoretical and experimental values for end abutment and center footing reactions and for moments (total section moment and distribution to girders) will be presented for the straight, curved and skew bridge models. The load cases treated here will be restricted for lack of space mainly to the following: 1X, 3X, 5X, 1X + 1Y, 3X + 3Y and 5X + 5Y. The experimental values will comprise results recorded for these point loadings after the 30 ksi (207 MPa) conditioning load level. Because the analytical models do not allow for cracking, comparisons of deflections will be limited to experimental results only.

In order to enable an assessment of the response of the three bridge models to overload stresses, some results of reactions, deflections and moments at the 60 ksi (414 MPa) stress level will be included in Section 6.

5.2 Reactions

The theoretical (T) and experimental (E) values of the vertical reactions R_W for the west end, R_F for the center footing and R_E for the east end with the associated bending moments M_W, M_F and M_E and torques T_W, T_F and T_E are listed in Table 1 for the six load cases 1X, 3X, 5X, 1X + 1Y, 3X + 3Y and 5X + 5Y.

For all three models the boundary conditions were in general similar with the exception that flexible vertical end supports were used for the skew bridge instead of the rigid vertical end supports used for the straight and curved bridges and a horizontal transverse restraint at the end supports was used for the straight bridge. These differences, however, are not likely to influence the results considerably.

An observation of the results in Table 1 indicates that good agreement exists between the theoretical and experimental values in virtually all cases.

The total vertical reactions R_W, R_F and R_E change only very insignificantly for point loads moving transversely across the straight or curved bridge. The skew bridge, however, shows significant changes in the vertical reactions depending on the transverse positioning of the load. While for load cases where the acute side of a span is loaded, the adjacent end reaction is smaller than for the corresponding straight and curved bridge cases, this reaction increases to a value much larger than for the straight and curved bridge models as the loads shift from the acute to the obtuse side of the skew span. For the load cases where only the center girder 3 is loaded, the skew bridge features slightly higher end reactions and slightly lower vertical center bent reactions than the curved or straight bridge. In addition to this, however, a significant negative end moment can also be noted for the skew bridge which decreases the total positive midspan moment. This negative end moment increases in magnitude as the loads move towards the obtuse side of the span, but decreases and even changes sign for loads

TABLE 1

COMPARISON OF REACTIONS FOR STRAIGHT, CURVED AND SKEW BRIDGE MODELS

LOAD CASE	MODEL		REACTIONS (KIPS AND FT. KIPS)								
			WEST END			CENTER FOOTING			EAST END		
			R_W	M_W	T_W	R_F	M_F	T_F	R_E	M_E	T_E
1X	ST	T	39.2	0	-355	69.0	96	- 42	-8.2	0	- 95
		E	38.0	0	-300	69.0	135	- 66	-8.3	0	-122
	CU	T	41.2	0	-283	64.5	110	- 83	-5.7	0	- 99
		E	41.2	0	-280	68.1	24	- 78	-7.1	0	- 96
	SK	T	32.5	193	-193	72.9	117	-196	-5.4	125	-125
		E	33.6	181	-181	74.0	133	-216	-6.0	126	-126
3X	ST	T	39.3	0	0	68.8	86	0	-8.1	0	0
		E	38.1	0	5	70.3	135	3	-8.3	0	-2
	CU	T	38.8	0	57	69.1	122	0	-7.9	0	- 17
		E	40.0	0	67	69.7	103	9	-7.2	0	- 20
	SK	T	42.5	-93	93	64.4	91	- 69	-6.9	35	- 35
		E	42.2	-78	78	69.1	107	- 65	-8.1	10	- 10
5X	ST	T	39.2	0	335	69.0	96	42	-8.2	0	95
		E	37.9	0	306	69.5	128	72	-8.3	0	116
	CU	T	36.2	0	395	74.1	133	82	-10.2	0	63
		E	36.4	0	400	76.0	145	74	-9.7	0	45
	SK	T	52.6	-390	390	55.9	65	51	-8.5	-73	73
		E	57.0	-373	373	62.3	91	68	-9.1	-83	83

1 KIP = 4.448 kN	1 FT = 0.305 m	1 FT - KIP = 1.356 kN - m

TABLE 1 (CONT)

COMPARISON OF REACTIONS FOR STRAIGHT, CURVED AND SKEW BRIDGE MODELS

LOAD CASE	MODEL		REACTIONS (KIPS AND FT KIPS)								
			WEST END			CENTER FOOTING			EAST END		
			R_W	M_W	T_W	R_F	M_F	T_F	R_E	M_E	T_E
1X + 1Y	ST	T	31.0	0	-432	138.0	0	- 84	31.0	0	-431
		E	30.1	0	-374	139.2	14	-140	30 4	0	-458
	CU	T	35.3	0	-382	129.3	0	-168	35.3	0	-382
		E	35.5	0	-381	132.5	-54	-149	33.2	0	-346
	SK	T	24.1	264	-264	128.7	54	-247	47.3	517	-517
		E	24.5	244	-244	132.3	58	-277	47 2	505	-505
3X + 3Y	ST	T	31.2	0	0	137.7	0	0	31.2	0	0
		E	29.7	0	2	138.4	9	5	30.5	0	1
	CU	T	30.9	0	39	138.3	-3	1	30.8	0	39
		E	31.3	0	37	141.4	72	7	30.1	0	35
	SK	T	35.6	-118	118	128.9	-2	1	35.5	117	-117
		E	36.0	- 41	99	137.6	3	9	35.9	85	- 85
5X + 5Y	ST	T	31.0	0	432	138.0	0	84	31.0	0	431
		E	30.3	0	389	140.2	-2	144	31.5	0	468
	CU	T	26.0	0	458	147.9	3	166	26.1	0	458
		E	28.7	0	461	151.4	-31	162	27.1	0	417
	SK	T	47.1	-515	515	128.9	-52	249	24.0	-265	-265
		E	47.8	-475	475	140.1	-49	300	24.0	-277	277

1 KIP = 4.448 kN	1 FT = 0.305 m	1 FT - KIP = 1.356 kN - m

moving towards the acute side of the span. Note that there is no end moment for the simply supported straight or curved bridges and that the end moment for the skew bridge is only introduced by the skewness of the simply supported end abutment.

The torque at the end abutments of the skew bridge has the same magnitude but opposite sign as the end moment and is smaller for loads on the acute side but larger for loads on the obtuse side of the skew span than in corresponding straight or curved bridge cases.

5.3 Deflections

In Figs. 3, 4 and 5 the experimental vertical deflections along girders 1, 3 and 5 have been plotted for the straight, curved and skew bridge models for a total of nine different load cases. The load values have been normalized to 100 kips (445 kN). The number of experimental points for the skew bridge model is much greater than for the other two cases because several more linear potentiometers were used for deflection measurements. No modifications have been made in the plotted experimental results to reflect the fact that the modulus of elasticity of the concrete was slightly different for the straight, curved and skew bridge models.

Considering the deflection under the application point of the load, Figs. 3, 4 and 5 show that the skew bridge features the highest deflection at the acute side of a span and the lowest deflection of all three bridge types at the obtuse side of the span.

The influence of the midspan diaphragm in Span I can also be seen by comparing the deflections at section X and section Y. Thus for the load cases 1X + 1Y and 5X + 5Y it is observed that the maximum deflection in the diaphragmed span (between 0 and 36 ft) is less for both the straight and curved bridges than the maximum deflection in the undiaphragmed span.

For the load at 1X it is normal to expect the curved bridge to have the smallest deflection under the load because of the smaller length of girder 1. Conversely, for the load at 5X, the curved bridge should feature the largest deflection. This is borne out by Figs. 3, 4 and 5. Finally, the measured deflections for the load cases 3X, 3Y and 3X + 3Y which comprise loads on the center girder only are observed not to be significantly different for the three types of bridge models.

While no theoretical values for deflections have been plotted in the above figures because the analytical model assumes a homogeneous material with completely linear elastic behavior, the observed deflections lie in all cases between the limits of 1.0 to

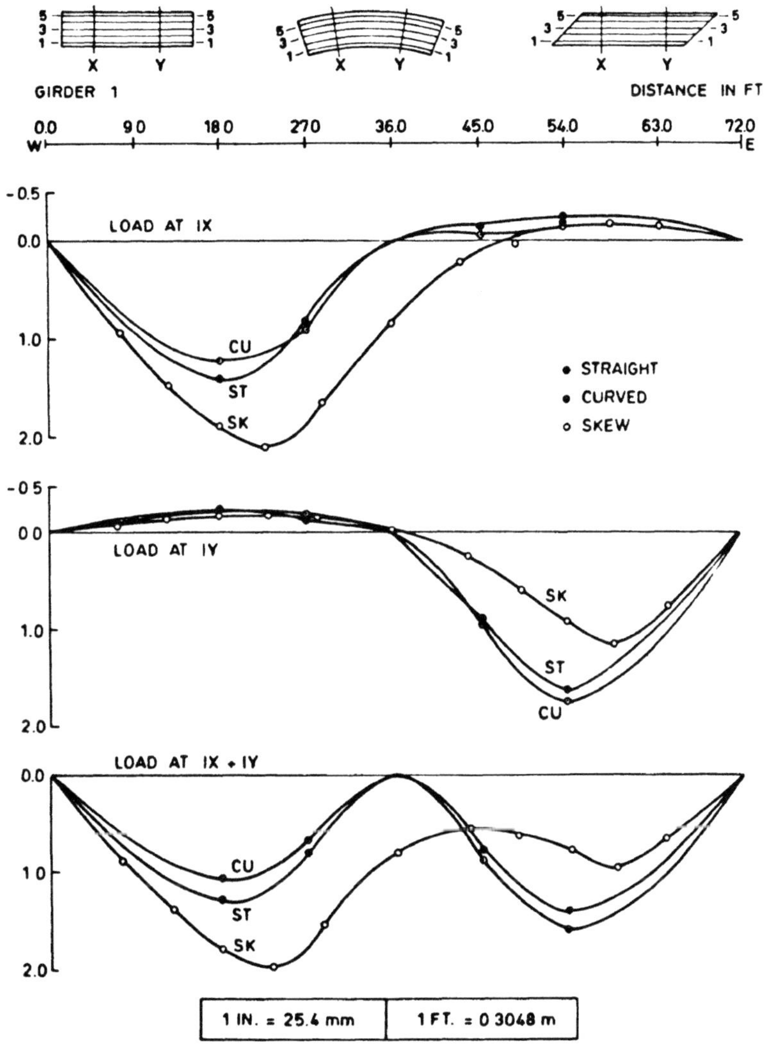

FIG 3 VERTICAL DEFLECTIONS (INCHES) ALONG GIRDER 1 FOR 100 KIP LOADS AT 1X, 1Y, 1X + 1Y

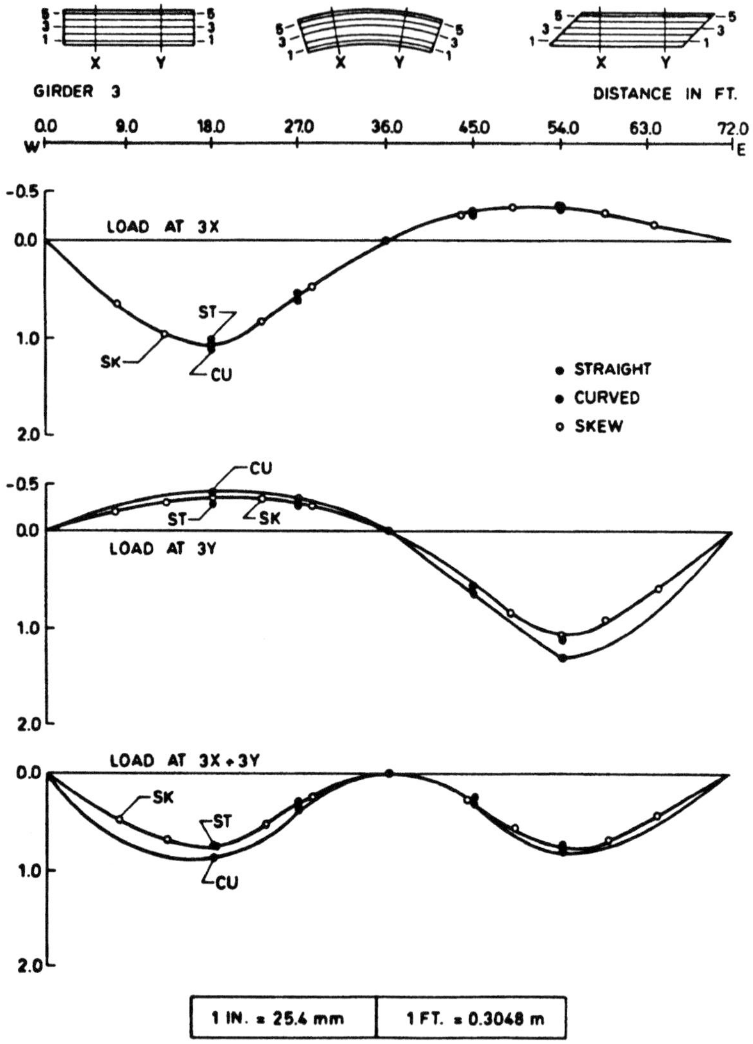

FIG. 4 VERTICAL DEFLECTIONS (INCHES) ALONG GIRDER 3 FOR 100 KIP LOADS AT 3X, 3Y, 3X + 3Y

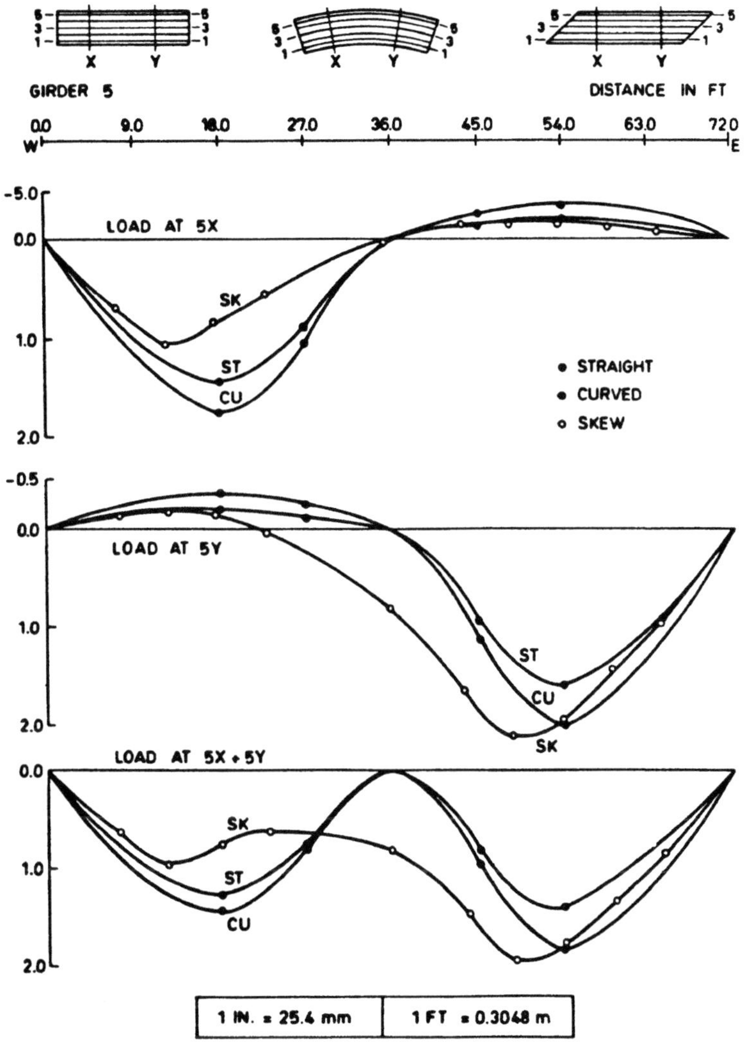

FIG. 5 VERTICAL DEFLECTIONS (INCHES) ALONG GIRDER 5 FOR 100 KIP LOADS AT 5X , 5Y , 5X + 5Y

1.5 times the analytically predicted values.

5.4 Internal Moments

A comparison of theoretical and experimental internal moments at Sections A and D for the six load cases 1X, 3X, 5X, 1X + 1Y, 3X + 3Y is presented in Table 2 for the straight, curved and skew bridge models, respectively. The agreement between theory and experiment, while not as close as in the case of reactions, is adequate.

While the total moments for the straight and curved bridge change only a little for point loads moving from girder 1 to girder 5, the skew bridge total midspan moments vary significantly. For loads on the acute side of the skew span, the total moments are much higher, and for loads on the obtuse side of the span much lower than for the straight or curved bridges. This is especially important for the positioning of high concentrated live loads such as heavy construction vehicles. As might be expected, the differences in moment values for the three different bridge types are least when the loads are placed over the central girder 3.

Two further points indicating the difference in the behavior of the skew bridge model compared to the straight and curved models are not featured in Table 2 but will be mentioned here. One is that for the important load cases of dead load and conditioning loads the skew bridge possesses substantially lower internal moments than the straight or curved bridges. Secondly, the negative moment over the center support is found to be smaller for the skew bridge than for the straight or curved bridges, independently of the load position on the bridge.

The transverse distribution of the maximum total positive moments for the straight, curved and skew bridge models to the individual girders provides an assessment of the load distribution properties of the box girders. While restrictions of space prevent a detailed presentation here of the theoretical and experimental percentages of the total moment taken by the individual girders for different load cases, it can be stated that reasonable agreement was found between theoretical and experimental values.

Analytical and experimental results indicate that the distribution of maximum total moments to individual girders for the straight vs. the curved bridge are almost identical at Section A for the diaphragmed Span I and only slightly different at Section D for the undiaphragmed Span II. However, they differ from those of the skew bridge which shows larger percentages especially of the experimental internal moments being taken by the exterior girders. This is especially true in the vicinity of location 1A near the staggered midspan diaphragm in the skew bridge.

TABLE 2

THEORETICAL AND EXPERIMENTAL TOTAL MOMENTS AT SECTIONS A AND D (LOADS NORMALIZED TO 100 KIPS)

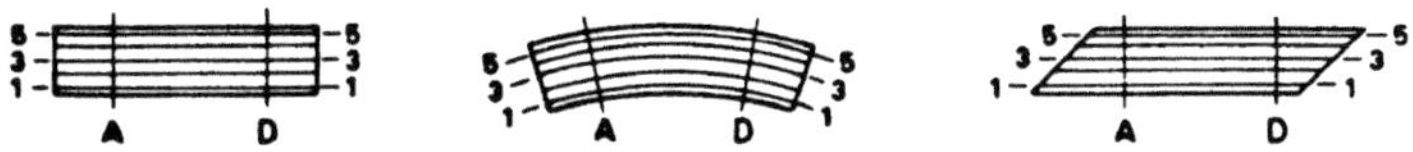

LOAD CASE	BRIDGE MODEL	INTERNAL MOMENTS (FT-KIPS)			
		SECTION A		SECTION D	
		THEORY	EXPER.	THEORY	EXPER
1X	STRAIGHT	565	524	-118	-125
	CURVED	557	515	- 97	- 75
	SKEW	693	657	-209	-183
3X	STRAIGHT	561	535	-116	-123
	CURVED	569	538	-116	-124
	SKEW	561	556	-131	-123
5X	STRAIGHT	565	542	-118	-127
	CURVED	582	586	-139	-148
	SKEW	420	407	- 58	- 62
1X + 1Y	STRAIGHT	448	404	443	461
	CURVED	458	427	457	431
	SKEW	632	597	212	268
3X + 3Y	STRAIGHT	446	406	448	443
	CURVED	453	467	452	399
	SKEW	430	433	430	437
5X + 5Y	STRAIGHT	448	429	443	475
	CURVED	444	440	445	467
	SKEW	211	273	634	576

1 KIP = 4.448 kN	1 FT-KIP = 1.356 kN-m

6. COMPARISONS OF THEORETICAL AND OVERLOAD STRESS EXPERIMENTAL RESULTS

6.1 Preliminary Remarks

In order to provide an indication of the modification of the structural response of the straight, curved and skew box girder bridge models with overload, selected comparisons of reactions, deflections and moments are presented in this section. Details are available in References [1-9]. The chosen results are from the undiaphragmed Span II with loads at Section Y so as to eliminate the effects of the midspan diaphragms in Span I.

6.2 Reactions and Deflections

In Fig. 6 the behavior of the total vertical east end reaction and deflection at the location 5Y for a point load applied at 5Y after the removal of the 30 ksi (207 MPa) and 60 ksi (414 MPa) conditioning load is depicted. As mentioned in Section 3, this point load was actually applied at several different locations to produce nominal stresses in the tensile reinforcement of the girder where it acted of 30 ksi (207 MPa). The value of this point load was about 20 kips (89 kN), and it has not been normalized so that it is retained at its actual value.

The reaction graphs show only a minor change in the values of the reactions after overload stresses of up to 60 ksi (414 MPa). The deflections at 5Y, however, increase when the same point load is reapplied after increased cracking and deterioration of the bridge models. As expected, the deflection at 5Y of the skew model is greater even than that of the curved model where girder 5 is the longest (outermost) girder and should show large deflection.

6.3 Moments

The point load of about 20 kips (89 kN) at location 5Y has been taken as a basis of comparison for the assessment of changes in the transverse distribution of the moments at Sections A and D after the successive application of conditioning loads resulting in nominal maximum tensile stresses in the steel reinforcement at Sections X and Y of 24 ksi to 60 ksi (166 MPa to 414 MPa). In Fig. 7a the theoretical distribution in percentage of the total section moment to each girder for a point load at 5Y is given for all three bridge models. This distribution does not change with an increase in stress level because of the assumption of linear elastic behavior in the analytical models. The experimentally measured percentages in Fig. 7b indicate fluctuations of up to 4% in the girder distributions, but comparisons with theoretical percentage predictions continue to be reasonable.

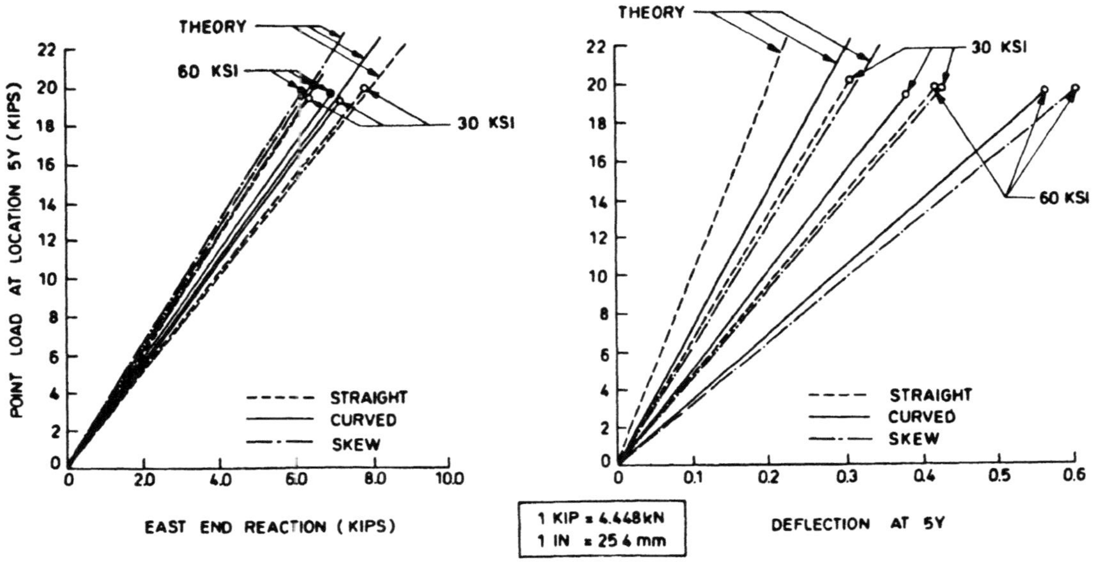

FIG. 6 COMPARISON OF TOTAL VERTICAL EAST REACTION (KIPS) AND DEFLECTION (INCHES) AT 5Y FOR A POINT LOAD AT 5Y AFTER 30 AND 60 KSI CONDITIONING LOADS

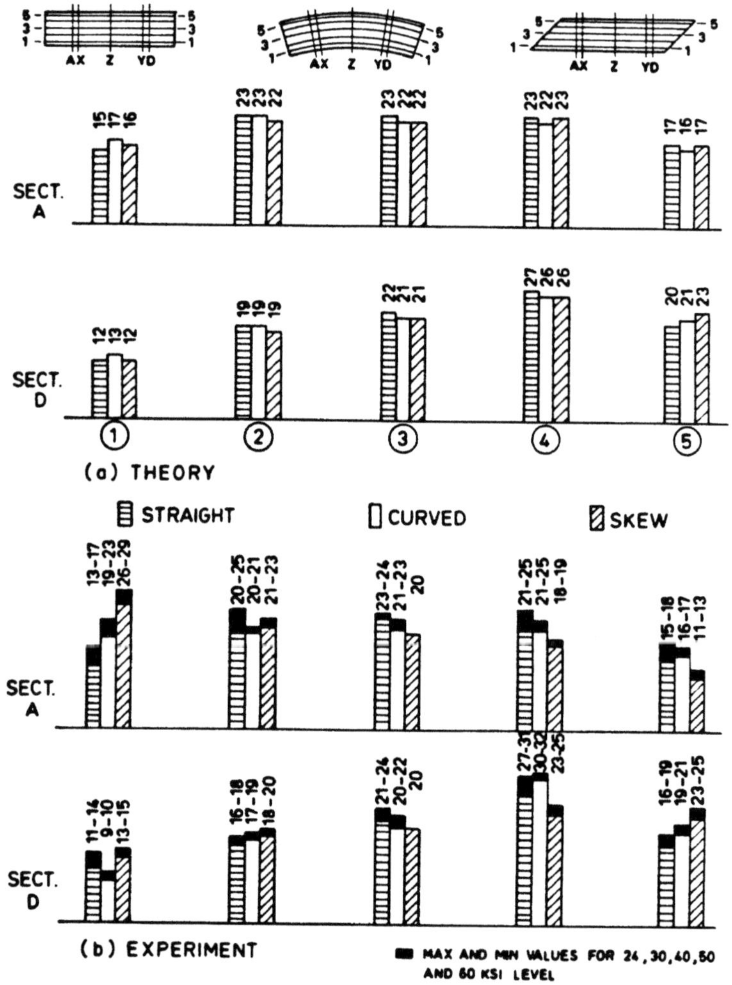

FIG. 7 COMPARISON OF PERCENTAGES OF TOTAL MOMENT AT SECTIONS A AND D EACH GIRDER FOR A POINT LOAD AT 5Y AFTER 24 TO 60 KSI CONDITIONING LOADS

7. CONCLUSIONS

The finite element computer programs based on linear elastic analysis of an uncracked homogeneous concrete structure are observed to predict the structural response of all three types of box girder bridges satisfactorily and accurately, especially with regard to reactions and moments at sections of interest. Experimental values of deflections at working stress levels lie between 1.0 and 1.5 times the theoretical values but the experimental trend is suitably predicted by the analytical model.

The straight, curved and skew box girder bridge models do indicate broad similarities of overall structural behavior. However, the structural behavior of the curved box bridge closely resembles that of the straight bridge, while the skew bridge features marked differences. All loadings produce end moments and torques only in the case of the skew bridge. Furthermore, the skew bridge is sensitive to the transverse location of the load on the top deck, showing higher deflections and moments for loads on the acute side of the span and lower moments for loads on the obtuse side of the span. Similarly the outermost girder registers the highest deflections in the case of the curved bridge model.

The general response of the skew bridge is favorable under conditioning loads (and also under dead load) in that smaller total section moments are produced compared to the straight and curved bridges. The skew center bent diaphragm also offers the advantage of participating in the longitudinal bending and thus attenuating the internal negative moments in the center bent region.

In all cases the midspan diaphragms in Span I contribute towards reducing the maximum deflections under eccentric loads. For the skew bridge the diaphragm has the effect of transferring a greater portion of the section moment to the exterior girders.

The level of deterioration in the bridge models as measured herein does not produce any major deviation in structural behavior over the range 30 ksi - 60 ksi (207 MPa - 414 MPa). The reactions can be adequately predicted and the transverse distribution of moments to each girder is substantially unaffected. The deflections given by theory need to be augmented by a suitable magnification factor.

Finally, it should be emphasized that all three bridge models exhibited excellent load carrying abilities and structural behavior even after high overloads. All three models had ultimate capacities well above DL + 4LL even though the skew bridge had substantially less longitudinal reinforcement than the straight or curved bridges.

8. REFERENCES

1. Scordelis, A. C., Bouwkamp, J.G. and Wasti, S.T., "Structural Behavior of a Two Span Reinforced Concrete Box Girder Bridge Model," Vols. I, II, and III, SESM Reports Nos. 71-5, 71-6 and 71-7, University of California, Berkeley, Calif., 1971.

2. Scordelis, A.C., Bouwkamp, J.G., and Wasti, S.T., "Structural Response of Concrete Box Girder Bridge," Journal of the Structural Division, Proceedings of the American Society of Civil Engineers, Volume 99, No. ST10, October 1973.

3. Bouwkamp, J.G., Scordelis, A.C., and Wasti, S.T., "Ultimate Strength of Concrete Box Girder Bridge," Journal of the Structural Division, Proceedings of the American Society of Civil Engineers, Vol. 100, No. ST1, January 1973.

4. Scordelis, A.C., Bouwkamp, J.G., and Larsen, P.K., "Structural Behavior of a Curved Two Span Reinforced Concrete Box Girder Bridge Model," Vols. I, II and III, SESM Reports Nos. 74-5, 74-6 and 74-7, University of California, Berkeley, Calif., 1974.

5. Scordelis, A.C., and Larsen, P.K., "Structural Response of Curved RC Box Girder Bridge," Journal of the Structural Division, Proceedings of the American Society of Civil Engineers, Vol. 103, No. ST8, August 1977.

6. Scordelis, A.C., Larsen, P.K., and Elfgren, L.G., "Ultimate Strength of Curved RC Box Girder Bridge," Journal of the Structural Division, Proceedings of the American Society of Civil Engineers, Vol. 103, No. ST8, August 1977.

7. Scordelis, A.C., Bouwkamp, J.G., Wasti, S.T., and Anicic, D., "Structural Behavior of a Skew Two Span Reinforced Concrete Box Girder Bridge Model," Vols. I, II, III and IV, SESM Reports Nos. 80-1, 80-2, 80-3, and 80-4, University of California, Berkeley, Calif., 1980.

8. Scordelis, A.C., Wasti, S.T., and Seible, F., "Structural Response of Skew RC Box Girder Bridge," Journal of the Structural Division, Proceedings of the American Society of Civil Engineers, Vol. 108, No. ST1, January 1982.

9. Scordelis, A.C., Bouwkamp, J.G., Wasti, S.T., and Seible, F., "Ultimate Strength of Skew RC Box Girder Bridge," Journal of the Structural Division, Proceedings of the American Society of Civil Engineers, Vol. 108, No. ST1, January 1982.

10. Meyer, C., and Scordelis, A.C., "Computer Program for Prismatic Folded Plates with Plate and Beam Elements," Structural Engineering and Structural Mechanics Report No. SESM 70-3, University of California, Berkeley, February 1970.

11. Willam, K.J., and Scordelis, A.C., "Computer Program for Cellular Structures of Arbitrary Plan Geometry," Structural Engineering and Structural Mechanics Report No. UC SESM 70-10, University of California, Berkeley, Calif., September 1970.

LATERAL LIVE LOAD DISTRIBUTION IN PRESTRESSED CONCRETE HIGHWAY BRIDGES

Celal N. Köstem, Professor of Civil Engineering

Fritz Engineering Laboratory, 13, Lehigh University
Bethlehem, Pennsylvania, 18015 USA

1 INTRODUCTION

The design of prestressed concrete highway bridges in the USA is governed by the provisions of the "Standard Specifications for Highway Bridges" of The American Association of State Highway and Transportation Officials (AASHTO) (7). These specifications, or slightly modified versions, have been extensively used in the design of bridges throughout the world (4). The AASHTO Specifications are known to be rather conservative, and in some areas ambiguous. Some, if not most, bridge engineers tend to implement the provisions faithfully, without due consideration to the first paragraph of the Specifications, which reads "In any case where the specifications provide an empirical formula as a design convenience, a rational analysis based on a theory accepted by the Committee on Bridge Superstructures of the American Association of State Highway and Transportation Officials, with stresses in accordance with the specifications, will be considered as compliance with the Specifications." The flexibility provided by these statements should be considered by bridge engineers in the review of generally accepted research findings and their implementation in the design of bridges.

This paper summarizes the results of extensive analytical research programs which were aimed at the identification of the design parameters which effect the lateral live load distribution in simple span highway bridges, with or without skew, and with prestressed concrete I-beams and spread box-beams.

1.1 Design Loading

The design loading employed in major bridges follow the HS

20-44 design vehicle (7). For simplified design purposes this loading vehicle is replaced by a lane loading and a concentrated loading. The research reported herein have investigated the practicality of this design vehicle. This vehicle is a pragmatic load. However, it is known that this loading should be subject to careful scrutiny in view of the changes in the frequently encountered vehicular axle combinations and weights (4). The research employed the design vehicle, rather than the simplified equivalent lane loadings. Thus, the results reflect the intent of the AASHTO Specifications more realistically (7).

1.2 Distribution Factor

In the design of bridges beams are individually designed through the use of the distribution factor, such that the fraction of the vehicular loading is applied to the single isolated simple beam. This fraction is called the lateral live load distribution factor, or simply the distribution factor, and has been defined as "s/k," (7), where "s" is center to center spacing of the beams and k is a constant. If the beam spacing is defined in feet, then for prestressed concrete I-beam bridges the value of "k" is 5.5 (s/1.676, if "s" is in meters). The implication of this formula is that regardless of the geometry of the bridge, that is, regardless of the span length, width, skew, etc., the load distribution does not vary so long as the beam spacing is constant. It has been found by previous research on the prestressed concrete spread box-beam bridges without skew that such a simplification can not be warranted for many bridge geometries (4). More complex, but highly realistic formulae, were developed and were adopted by the AASHTO Specifications (7). A similar work was conducted for I-beam bridges and the formulae were developed, pending the official adoption into the AASHTO Specifications (5,6).

Since the detailed formulae for the new distribution factor provisions are to appear in the future AASHTO Specifications, this paper will focus on the pertinent aspects of the distribution factors but not necessarily the formulae.

2 EFFECTS OF BRIDGE GEOMETRY AND DIAPHRAGMS

The research have indicated that for simple span right prestressed concrete I-beam bridges the design parameters that govern the load distribution are the beam spacing, as it is already recognized by the current Specifications, span length of the bridge, and width of the bridge. (Spacing/span length) ratio is critical in the definition of the effective participatory area of a given beam. This ratio also facilitates the consideration of the extent of the single or bi-directional bending of the deck slab. Furthermore, the width of the bridge also needs to be considered. The width can be taken into account either as is, or through the specification of the number of beams and the beam spacing. It is also noted that depending upon whether the load distribution formulae is to be used for old bridges as a rating tool, or for the design of new bridges, the number of loaded traffic lanes also needs to be considered as a secondary factor.

In view of the many independent design parameters that need to be investigated in assessing their full contribution to the load distribution, the conduct of a test program using full scale bridges became a highly improbable proposition. However, an analytical research program that can include the contribution of all design parameters was expeditiously executed. In the reported research it has been found that an analytical experiment can be designed and executed using a finite element software package. In excess of 300 right bridges were analyzed having span lengths from about 40 ft. (12.20 m) to 140 ft. (42.70 m), and with the number of beams varying from 3 to 15. Since the computation of the distribution factors require the definition of the mid-span transverse influence line for the definition of loaded lanes, each bridge had to be analyzed for ten different loading conditions. Through the full computerization of the input and output options of the finite element analysis, and automatic file transfers from program to program, it was possible to conduct the extensive parametric investigation with great ease, after the development of the required computer programs (1).

All the theoretical analysis results were benchmarked against available test results of full scale bridges and the tests conducted in the laboratories under controlled conditions for scaled down bridges (4). An acceptable agreement, even with the relatively coarse "production tool" version of the finite element mesh, was observed, the maximum discrepancy being less than 5% (1).

2.1 Curb and Parapet

The research have indicated that even though in the design of the superstructure the curb and parapet are not considered as the primary load carrying members, these structural components still participate in the load distribution process (1,2). If the curb and parapet follow the so-called "New Jersey Barrier" geometry, about one-half of the moment of inertia of this structural component contributes to the structural stiffness of the bridge. Thus, maintenance of the structural integrity and provision for full composite interaction between the curb and parapet, and the deck slab, will have positive effects on the load distribution. Figure 1 shows that regardless of how small the contribution might be, the curb and parapet still provide some stiffness.

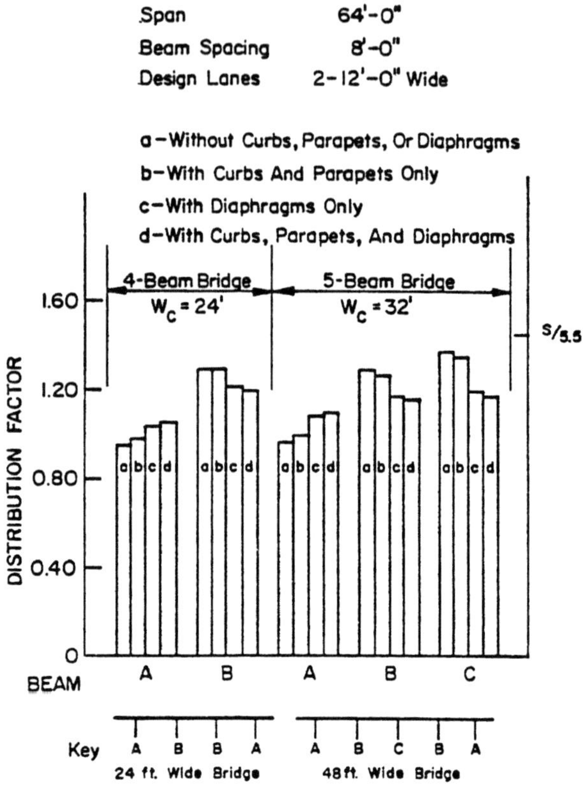

Fig. 1 Distribution Factors for the 4-Beam and 5-Beam Bridges

2.2 Diaphragms

The structural contribution of the diaphragms is still a misunderstood phenomenon. It has been found that if the diaphragm's stiffness is comparable to that of the beams, and also if the beams and the diaphragms are monolithically constructed (or post tensioned in the transverse direction after the construction) the beams and diaphragms will work as a grillage. However, if the diaphragms are cast in-situ with the bridge deck, and also if they have limited depth, the structural contribution of the diaphragms is not as critical as assumed, but never quantified by the bridge engineers (1,2). The role of the diaphragms can be noted in Figure 1.

Certain bridge specifications require that if the span length is less than some set length then mid-span diaphragms are to be used. However, if the span length is larger than this set value, instead of midspan diaphragm, the use of diaphragms at the third points is specified. The research have showed that for a real bridge, with an adverse loading condition as would be commonly encountered in the traffic stream, the diaphragms at midspan lessen the portion of the load carried by the beam immediately under the vehicle (1,6). The effects of the diaphragms are imperceptible for the beams sufficiently away from the vehicle. Furthermore, at the maximum moment section of the bridge, i.e. mid-span, more uniform load distribution can be achieved through the use of mid-span diaphragms. Third-point diaphragms do not have any justifiable contribution. Therefore, it is recommended that if any diaphragm is to be used it should be at the mid-span. Also, if more diaphragms are to be used due to any reason that may be put forth, then the diaphragms should be located at mid-span and at quarter points. These conclusions can easily be observed in Figure 2.

2.3 Skew

Most bridge specifications tend to leave the inclusion of the skew in the design computations to the designer. For the skew angle of 90°-75° the bridges can safely be designed as right bridges (noting that the skew angle of 90° corresponds to the right bridge). In the definition of the skew another parameter that might be looked into is the distance between the slab and the centroid of the beam, i.e. "e." Figure 3 shows the effect of the skew and the eccentricity of the beams on the distribution of the moments between various beams. In the figure Beams-A, -B, and -C correspond to exterior beams, interior beam, and the center beam in a five beam bridge. The span length was about 70 ft. (21.35 m) and the bridge carried four traffic lanes. In the figure e/d=0.0 corresponds to fully non-composite construction and e/d=0.5 corresponds to fully composite interaction between the girders and the deck. It can be noted that depending upon the extent of the composite interaction there is a substantial change in the torsional properties of the beam, thus the bending moment carried by each beam changes. From this figure additional observations can be made (13):

- The beam directly under the load carries a major portion of the total load as a composite structure. Increase in moment coefficients in exterior and center beams is balanced by the decrease in the moment coefficient of the exterior beam. The remaining differences are carried by the slab.
- The reduction and the rate of reduction in moment coefficients for the interior beams seems to be

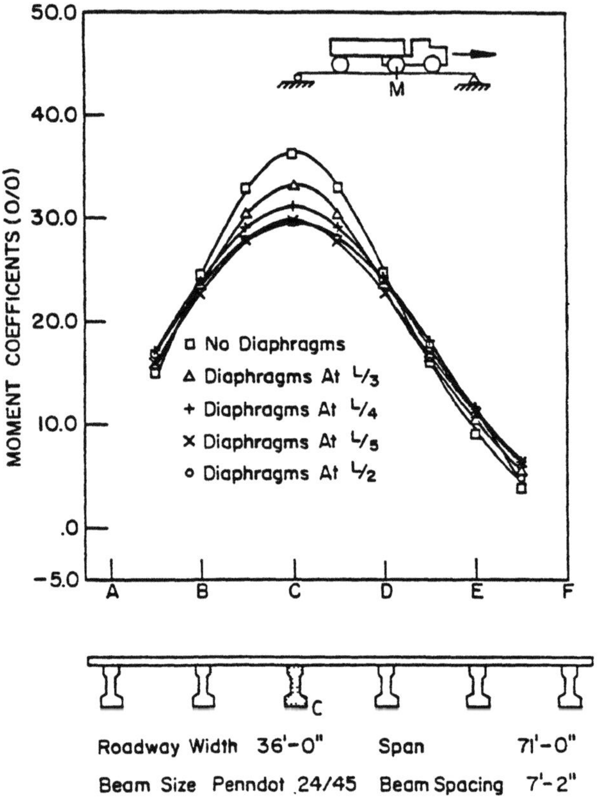

Fig. 2 Influence Lines for Moment, Beam C - With and Without Diaphragms

almost the same for both composite and noncomposite construction.

- The effect of the skew in bridges with skew angle greater than 75° can be neglected. The skew for bridges with skew angle 75°-60° may be neglected, with the understanding that this will lead to a more conservative design. At the same time, this results in adverse plate bending moments at the corners of the deck slab. For bridges with skew angle less than 60°, and especially less than 45°, the effect of the skew can not be neglected.
- The correction terms to be applied to the distribution factors for bridges with non-negligible skew are

presented in Reference 1, and a more refined version can be found in Reference 3.

For skewed prestressed concrete I-beam bridges the distribution factor for interior and exterior beams has the following general form (2):

$$DF_{\emptyset} = DF_{\emptyset=90^\circ}(1.0 - (C_1(s/L) - C_2)\cot\emptyset)$$

where C_1 and C_2 are constants, s and L are beam spacing and the span length respectively, and $\emptyset$ is the skew angle. The formulae for the load distribution for prestressed concrete spread box-beam bridges have a slightly more complex form and the formulae used for interior and the exterior beams are different (2).

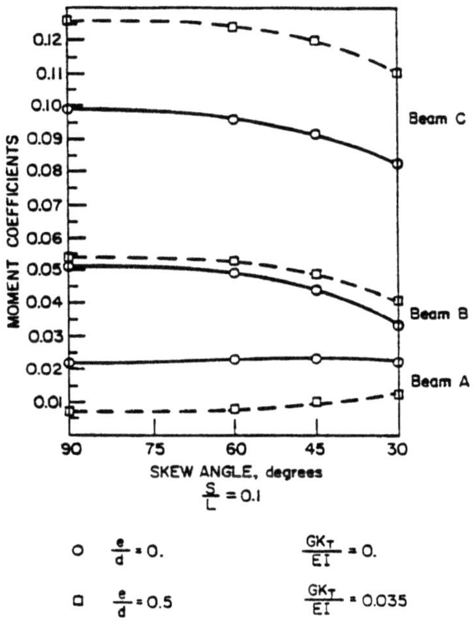

Fig. 3 Moment Coefficients in a 5-Beam Bridge Including the Beam Eccentricity and Torsional Rigidity

3 DISTRIBUTION FACTORS FOR RIGHT BRIDGES

The realistic distribution factor formulae for simple span prestressed concrete box-beam bridges without skew has already been developed by the researchers at Lehigh University and adopted by AASHTO (7).

In order to provide uniformity between the formulae for spread box-beam and the I-beam bridges, the following general format of formula has been employed:

$$DF = (2\ N_L/N_B) + k_1\ (s/L)^n + k_2$$

where N_L, N_B, s, and L correspond to the number of traffic lanes, number of beams, beam spacing and the span length respectively. k_1 and k_2 are the algebraic expressions in terms of bridge width, number of beams and number of traffic lanes; and n is a constant (5,6). It has been found that even though these formulae are slightly more complicated then the current s/5.5 expression given in the AASHTO Specifications (7), they are far more realistic.

The best illustration of these developed formulae, and the proposed specification, can be seen through inspection of Figs.

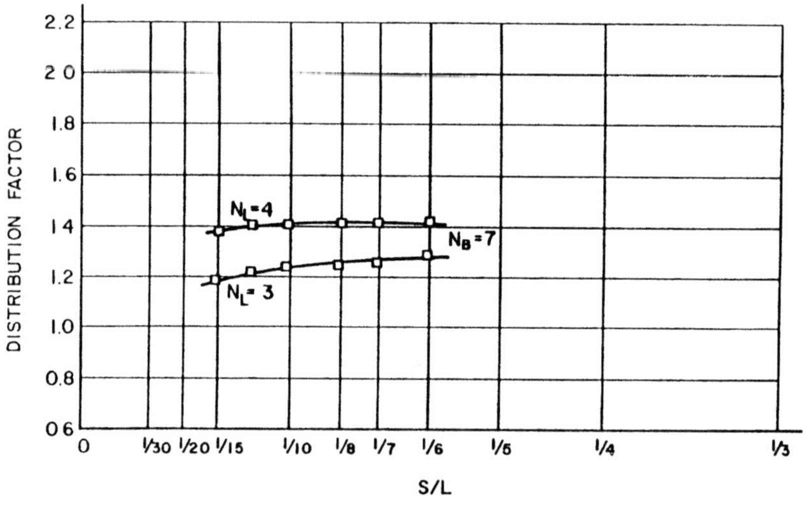

Fig. 4 Distribution Factors for Interior Beam
42 Ft. Wide, 7 Beam Bridges

4 and 5. These figures contain the finite element study results, as shown by discrete symbols.

For example, using Fig. 4 for a bridge with span length of 49 ft. (14.95 m), and beam spacing of 7 ft. (2.14 m), it can be found that the interior beam distribution factors for 3 and 4 traffic lanes will be 1.24 and 1.42, respectively. If AASHTO provisions are applied, then the distribution factor would be 7/5.5 = 1.27. The difference at first glance may not be large, however, due to the large number of bridges that are built of this type, even small differences will have a contributory effect on the overall cost and safety of the bridges in question.

A study of Fig. 5 shows that the distribution factors decrease more rapidly then the decrease in the beam spacing, i.e. to go from "n+1" to "n" beam bridge.

4 CONCLUSIONS

Simple span prestressed concrete bridges with in-situ reinforced concrete deck and with prestressed concrete I-beams or spread box-beams are one of the most commonly used types of bridge

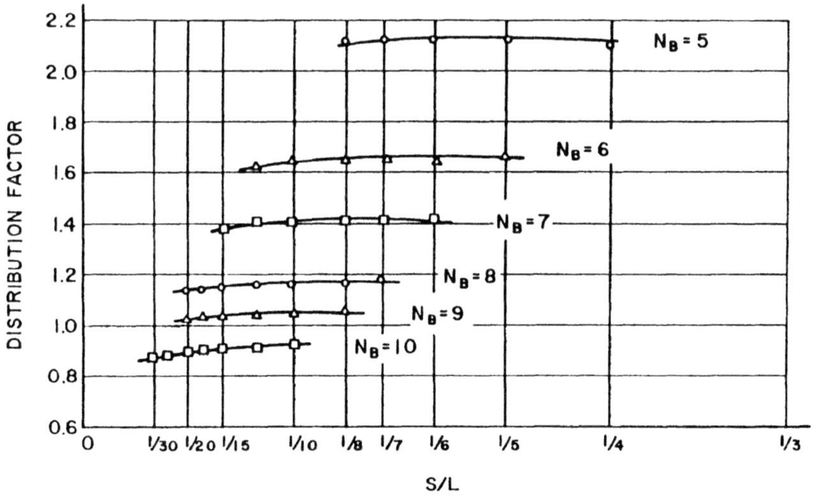

Fig. 5 Distribution Factors for Interior Beam
42 Ft. Wide Bridges (N_L = 4)

superstructures throughout the world. In the past the design of the beams has followed the basic distribution factor approach, using s/k as the distribution factor. The field studies and analytical research have indicated that a distribution factor of this simplicity may be good for the preliminary design, but is too crude for the final design of the bridges. Refined distribution factors should be used. Formulae for both right and skewed bridges have already been developed and verified, and are available in the literature (1,3,5,6).

The curb and parapet, depending upon the construction practices, do help in the distribution of the load. Their contribution should, preferably, not be taken into account in the design process, but for the maintenance inspection of the bridges attention should be paid to the structural integrity of these components.

The shallow diaphragms that are not built as a part of the total bridge structure contribute marginally in the lateral distribution of the live load. If diaphragms of any type are to be used, there should always be a midspan diaphragm.

To verify the completed design of a bridge "reverse design" process, i.e. the use of s/k, should not be employed. Instead, consideration should be given either to the new formulae or to the finite element analysis of the total superstructure, if the structure is critical enough and/or if a number of similar bridges are to be built.

1. DeCastro, E. S. and C. N. Kostem. Load Distribution in Skewed Beam-Slab Highway Bridges, Fritz Engineering Laboratory Report No. 378A.7 (Lehigh University, Bethlehem, Pennsylvania, USA, 1975).
2. DeCastro, E. S., C. N. Kostem, and D. A. VanHorn. A Pilot Study of Factors Affecting Live-Load Distribution in Pre-Stressed Concrete Beam-Slab Bridges, Fritz Engineering Laboratory Report No. 387.4 (Lehigh University, Bethlehem, Pennsylvania, USA, 1979).
3. DeCastro, E. S., C. N. Kostem, D. R. Mertz, and D. A. VanHorn. Live Load Distribution in Skewed Prestressed Concrete I-Beam and Spread Box-Beam Bridges, Fritz Engineering Laboratory Report No. 387.3 (Lehigh University, Bethlehem, Pennsylvania, USA, 1979).
4. Zellin, M. A., C. N. Kostem and D. A. VanHorn. Structural Behavior of Beam-Slab Highway Bridges-A Summary of Completed Research and Bibliography, Fritz Engineering Laboratory Report No. 387.1 (Lehigh University, Bethlehem, Pennsylvania, USA, 1973).

5. Zellin, M. A., C. N. Kostem, D. A. VanHorn, and J. M. Kulicki. Lateral Distribution of Live Load in Prestressed Concrete I-Beam Bridges, Fritz Engineering Laboratory Report No. 387.2A (Lehigh University, Bethlehem, Pennsylvania, USA, 1975).
6. Zellin, M. A., C. N. Kostem, D. A. VanHorn, and J. M. Kulicki. Live Load Distribution Factors for Prestressed Concrete I-Beam Bridges, Fritz Engineering Laboratory Report No. 387.2B (Lehigh University, Bethlehem, Pennsylvania, USA, 1976).
7. Standard Specifications for Highway Bridges, The American Association of State Highway and Transportation Officials (Washington, D.C., USA, 1977).

SKEW SLAB BRIDGES

Christian Menn

Professor fur Baustatik und Konstruktion
Eidgenossische Technische Hochschule
Zurich

SUMMARY

The applications, typical dimensions, structural behavior, analysis methods, and ultimate strength considerations of skew slab bridges are presented.

The effects of cracking and inelastic behavior on the deflections and servicability of skew slab bridges are reviewed. The need for prestressing in order to control behavior and provide servicability is discussed.

Theories of elasticity and plasticity are discussed with respect to their applicability to the design of skew slab bridges. It is shown that an inherent stress condition can be superimposed on the computed ultimate section forces with no effect on ultimate strength.

Finally, the relationships necessary for designing a skew slab bridge for ultimate strength are presented.

1. INTRODUCTION

Skew slab bridges are especially suitable for overpasses with limited construction depth. The material outlay for these structures is relatively high, but their construction is extremely simple. Slab bridges with a span of 20 m or less for simply supported systems, or 25 m for continuous systems are economical. For larger spans, the available construction depth generally determines whether the cross section is designed as a plate or a hollow box girder.

1.1 Dimensions and Cross Sections

The plan dimensions of a skew plate (Fig. 1) are labelled as follows: Span l, width b, angle of skew φ.

Various types of cross section are possible (Fig. 2). Both solid plates and voided plates are used, with and without lateral overhangs. Solid plates of not more than 80 cm thickness are suitable. Correspondingly the span over which a solid skew plate is appropriate is one of 25 to 40 m for simply supported bridges, depending on the relationship between l/b and φ (Fig. 3).

1.2 Load Carrying Behavior

The load carrying behavior of a skew plate is characterized by the fact that as the width b increases, the direction of the larger principal moment in the middle of the plate increasingly deviates from the bridge axis and approaches the perpendicular to the skew support lines (Fig. 4).

Stress concentrations in the obtuse corners result from the fixity of the free edges in the rigid support lines. This does not significantly influence the behavior of the plate. Through admissible cracking or limited plastic deformation in the obtuse corners, the stress condition in the plate is only slightly altered.

In addition to economic considerations and crack behavior, the anticipated deflection of the plate determines whether or not prestressing should be employed.

The cracking moment $M_R = (b \cdot h^2/6) \cdot f_{ct}$ for slab bridges is approximately five times smaller than the dead load moment in the middle of the plate. Therefore, normally reinforced slab bridges should be considered as thoroughly cracked, and they behave in principle like a normally reinforced girder.

The deflection of a girder is computed as follows:

$$\delta = \frac{q \cdot l^4}{\alpha \cdot EI} \quad \text{or} \quad \frac{q \cdot l^4}{\alpha \cdot h^3}$$

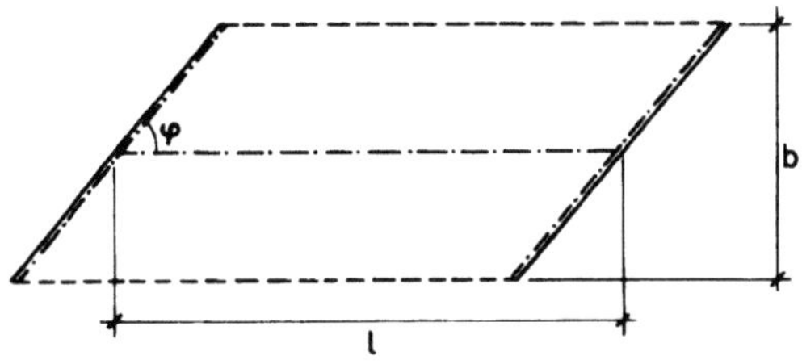

Fig. 1

FIG. 1: DIMENSIONS OF A SKEW PLATE

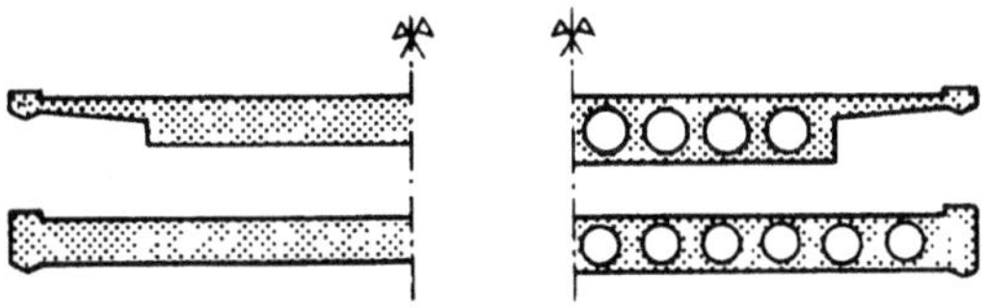

Fig. 2

FIG. 2: TYPES OF CROSS SECTIONS

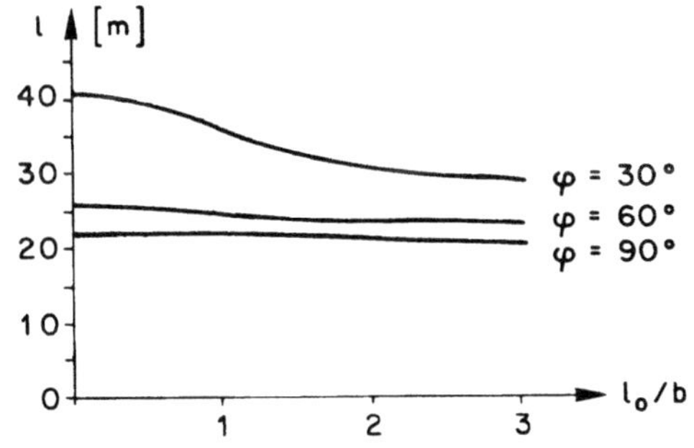

Fig. 3

FIG. 3: RECOMMENDED SPAN FOR SIMPLY SUPPORTED BRIDGES

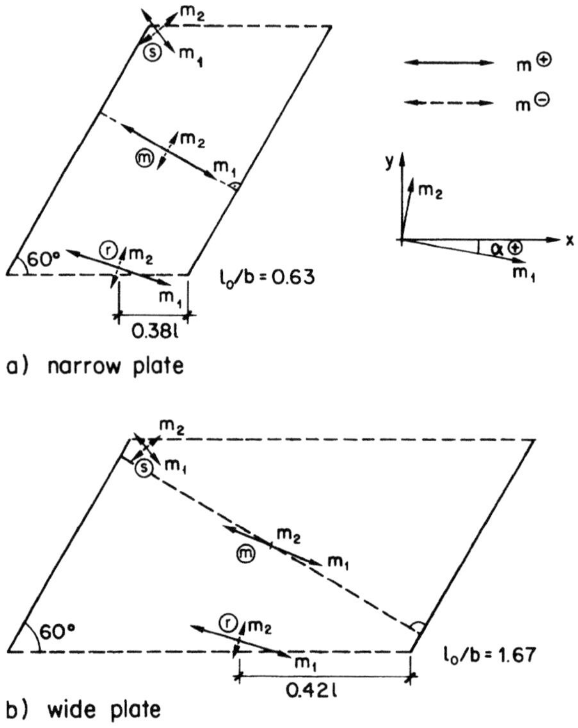

narrow plate	m_1	m_2	α
(m)	1.000	0.164	29.0
(r)	1.048	-0.104	18.3
(s)	0.448	-0.480	51.0
	$ql_0^2/8$	$ql_0^2/8$	[o]

wide plate	m_1	m_2	α
(m)	1.060	0.016	22.0
(r)	1.088	-0.096	16.8
(s)	0.272	-0.304	47.5
	$ql_0^2/8$	$ql_0^2/8$	[o]

c) points of design

Fig 4

FIG. 4: LOAD CARRYING BEHAVIOR UNDER UNIFORMLY DISTRIBUTED LOAD

Consequently, l^4/h^3 is a deflection index which gives an indication of the extent of the deflection. An 80 cm plate with a span of 20 m exhibits a deflection value of

$$\frac{l^4}{h^3} = \frac{20^4}{0.8^3} \doteq 300 \cdot 10^3$$

which is a very high value. A girder with a slenderness ratio l/h = 20 acquires such a high value only when the spans reach 40 m or more. Of course no one would design such a girder in normally reinforced concrete. Thus slab bridges of substantial span should always be prestressed. Normally reinforced plates frequently suffer inadmissibly large deflections.

2. CALCULATION OF SECTION FORCES IN SKEW PLATES

The controlling position of live load on a skew plate depends on the pattern of reinforcement and is different for each reinforcement direction. Simplifications are therefore indispensable. Generally the section forces at a few characteristic points in the plate (Fig. 5) are calculated for several specific load positions. These results are then used to obtain a simplified reinforcement pattern for the slab.

A plate element is subjected to the following section forces:

$$m_x,\ m_y,\ m_{xy},\ v_x,\ v_y$$

The shear stresses due to the shear forces v_x and v_y are usually small and generally require no special reinforcement. The determination of the necessary reinforcement is primarily dependent on the moments m_x, m_y, m_{xy}.

2.1 Theory of Plasticity

Any reliable analysis of the ultimate limit state must be based on the theory of plasticity, since an accurate non-linear theory, taking into account the real behavior of the structure past the quasi-elastic range, is too complicated for practice.

A check for ultimate strength is normally formulated as follows:

$$F_d(\gamma_g \cdot G,\ \Sigma\gamma_q \cdot Q) \leq \frac{1}{\gamma_m} \cdot R_d(f_{ck},\ f_{sk})$$

where | associated safety factors:

G	= dead load	γ_g
Q	= live load	γ_q
f_{ck}	= characteristic concrete strength	γ_m
f_{sk}	= characteristic steel strength	

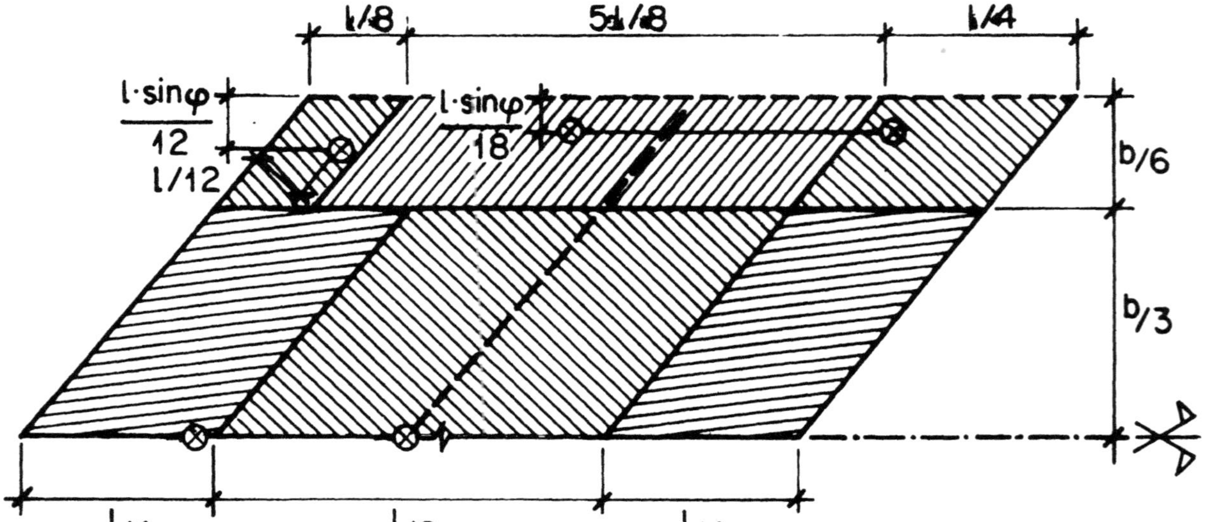

FIG. 5: CHARACTERISTIC POINTS OF DESIGN AND REINFORCEMENT ZONES

The ultimate strength check may be expressed as

design value of forces $\geq$ design value of capacity.

For the calculation of the design internal forces, the static method of the theory of plasticity must be used, since it yields a lower bound on the value of the ultimate load capacity of the system.

The kinematic (yield line) method of the theory of plasticity for slabs is unsuitable and should be avoided. Particularly in the case of skew slabs it is very difficult to prove that the chosen mechanism does not overestimate the ultimate load, in other words that the chosen mechanism fulfills the conditions of plasticity. This method is both obsolete and dangerous since it yields an upper bound on the ultimate load capacity of the system.

According to the preferred static method, the determination of the reinforcement may be based on any distribution of internal forces which satisfies the conditions of equilibrium.

Today the internal forces of a relatively difficult system such as a skew slab are almost always calculated with a finite element computer program. The programs are based on the theory of elasticity in that they assume linear elastic material behavior. Under high stresses this assumption is no longer valid. As a result of cracking, the rigidity of the structure is altered, resulting in a redistribution of internal forces. An inherent stress condition is therefore superimposed on the elastic stress conditions. This reduces high local stresses and exploits the structure's available load carrying reserves.

2.2 Inherent Stress Conditions

The internal forces under an inherent stress condition are not a consequence of a load. The number of independent inherent stress conditions of a system corresponds to the degree of statical indeterminacy. A plate, as an infinitely statically indeterminate system, exhibits an infinite number of independent inherent stress conditions.

Inherent stress conditions are defined in principle as follows:

for beams: $\frac{dV}{dx} = 0$ or $\frac{d^2M}{dx^2} = 0$

for plates: $\frac{\partial v_x}{\partial x} + \frac{\partial v_y}{\partial y} = 0$ or $\frac{\partial^2 m_x}{\partial x^2} + 2\cdot\frac{\partial^2 m_{xy}}{\partial x\cdot\partial y} + \frac{\partial^2 m_y}{\partial y^2}$

Such an inherent stress condition can result from support

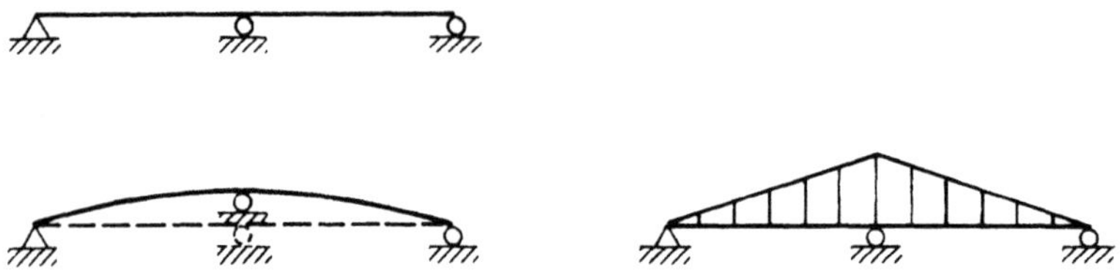

a) support displacement

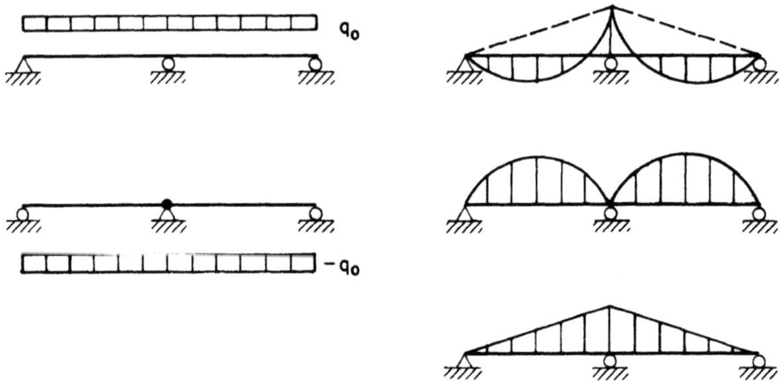

b) q_0 , $-q_0$ acting on different static systems

Fig 6

FIG 6: INHERENT STRESS CONDITIONS

displacements (Fig. 6a) or curvature of elements, due for example to variations in temperature. An inherent stress condition also results when two different statical models of the same system are loaded with q_0 and $-q_0$ respectively (Fig. 6b), and the results are superimposed.

According to the theory of plasticity, an inherent stress condition does not change the load carrying capacity of a system. Thus, in the calculation of the ultimate capacity of a system, any state of inherent stresses may be superimposed on the stresses resulting from an elastic analysis of the structure under the applied loads (Fig. 7).

Therefore the design forces may be formulated as follows:

$$F_d = F_{d,E} + F_0$$

where $F_{d,E}$ represents the design force according to the theory of elasticity.

F_0 represents an inherent stress condition.

or:

$$F_d = \Sigma k_i \cdot F_{d,i}$$

where $F_{d,i}$ represents the design force according to statical model number 'i'.

$$\Sigma k_i = 1.0$$

2.3 Prestressing and Servicability

In principle, the design forces must be calculated in such a way that the corresponding reinforcement provides optimum structural behavior with respect to servicability.

For normally reinforced structures this is customarily the case when the design forces are computed on the basis of the theory of elasticity and the stiffness of the homogeneous concrete cross section. Nonetheless it is possible to superimpose an inherent stress condition which takes into consideration the arrangement of the reinforcement and approximates the stiffness contribution of the cracked concrete section.

In prestressed structures the redundant moments due to prestress generally create a favorable inherent stress condition which should be taken into account in the design of the reinforcement.

The redundant moments represent a part of the entire

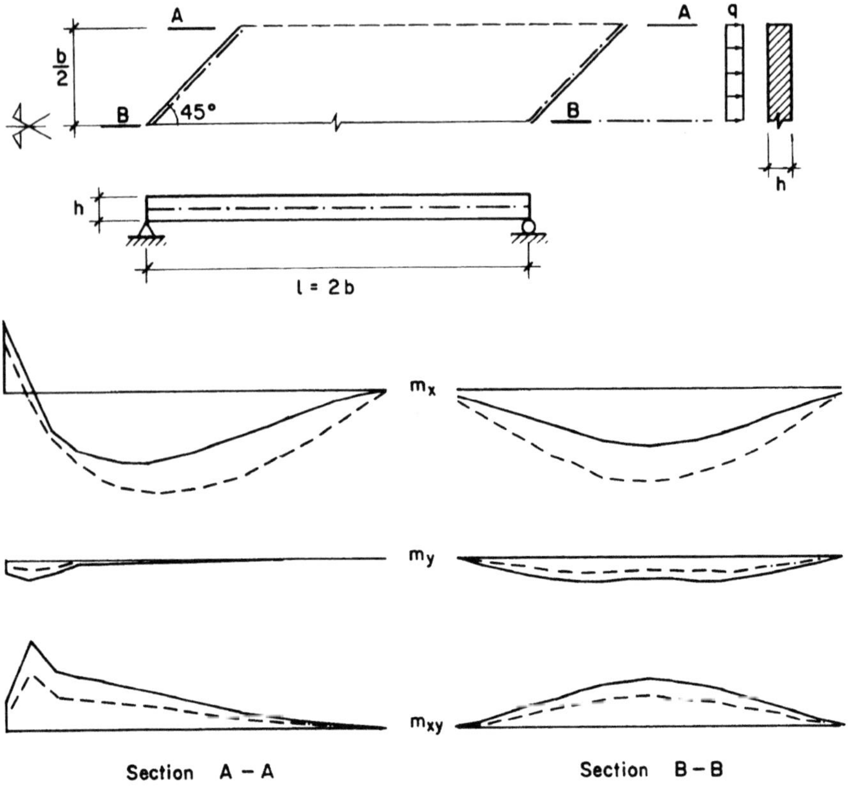

Fig 7

——— F_{d_E}

-·-·-·-· F_d

FIG 7: DESIGN INTERNAL FORCE DISTRIBUTIONS

prestressing moment m_p:

$$m_p = m_{op} + m_{zp}$$

where $m_{op} = - p \cdot e$

m_{zp} represents the redundant moment due to prestress.

For plates the total prestressing moments m_p are generally computed with the same computer program as used for the applied loads, with the prestressing represented by anchorage forces and the deviation forces of the tendons. This yields the redundant moment

$$m_{zp} = m_p - m_{op}$$

The value chosen for the width over which a prestressing tendon is considered effective is not of great importance. More important is the necessity for using the same width (for example $b_p \simeq 2 \cdot h$) throughout the design calculation.

In the formula for the design ultimate forces, the redundant moments due to prestressing (an inherent stress condition) are generally included with a factor of 1.0 (Fig. 8). Taking into account a possible moment redistribution, however, this factor may actually vary from about 0.8 to 1.2.

$$m_d = m_{d,E} + (0.8 \div 1.2) \cdot m_{zp}.$$

This inherent stress need not necessarily correspond exactly to the redundant moments from the effective prestress. The inherent stress can also be determined from a fictitious prestress. This fictitious prestress must not, however, differ substantially from the effective prestress, particularly with respect to the reinforced zone and the section of the tendons.

A proposed prestressing arrangement for a narrow simply supported skew slab is shown in Fig. 9.

3. DESIGN CHECKING FOR ULTIMATE STRENGTH

The reinforcement in a skew slab bridge normally consists of prestressing tendons and a grid of normal mild steel reinforcement.

Proportioning of the reinforcement for strength generally requires several trials in order to satisfy the design requirements. Each trial proceeds as follows.

First, the mechanical reinforcement content in the x and y directions is estimated based on the computed internal forces and the

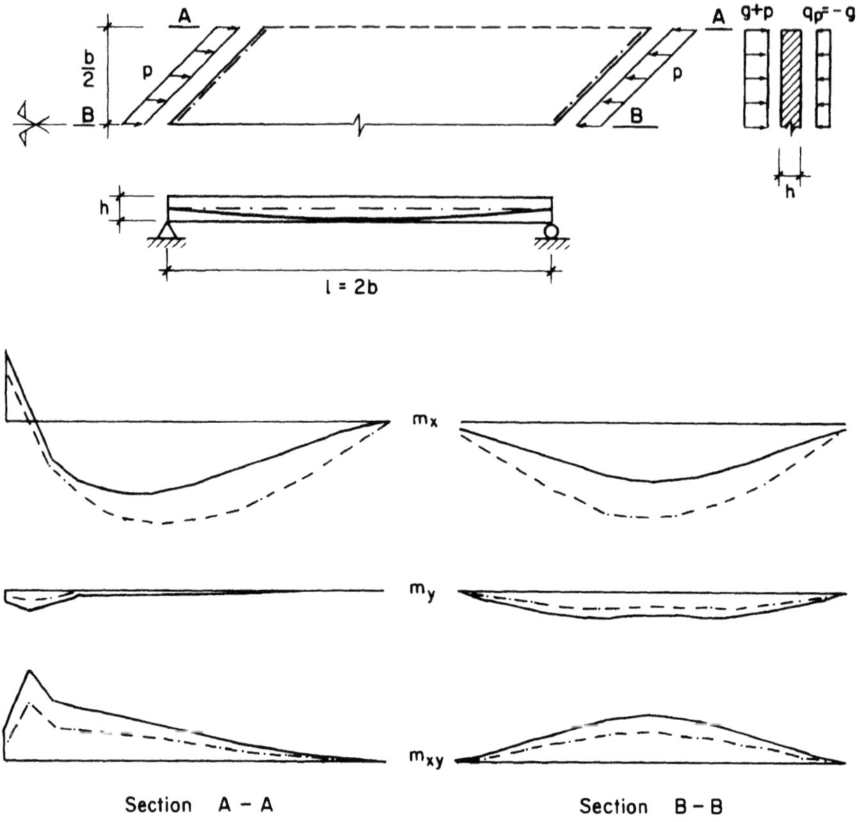

Fig 8

——— $1.8\ m_{g+p}$

-•-•-•-• $1.8\ m_{g+p} + 1.0\ m_{zp}$

FIG. 8: DESIGN INTERNAL FORCE DISTRIBUTIONS CONSIDERING m_{zp}

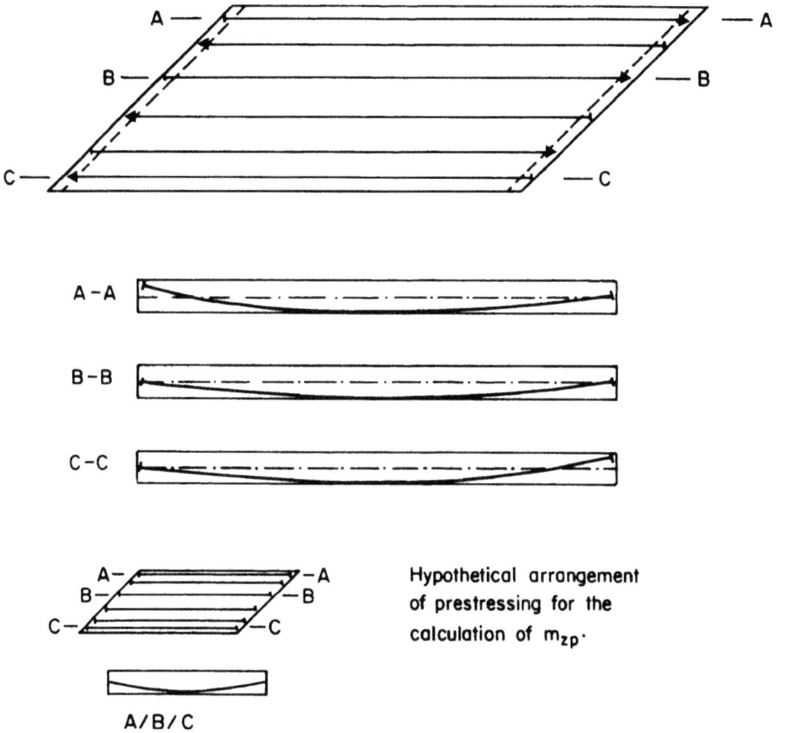

Fig 9

FIG. 9: PRESTRESSING ARRANGEMENT FOR NARROW SIMPLY SUPPORTED SLABS chosen reinforcement directions.

$$\omega_x,\ \omega_y,\ \omega_{xy}$$

A fictitious reinforcement in the principal directions is then calculated (Fig. 10).

$$\omega_1 = \omega_\xi = \omega_x \cdot \cos^2\varphi_1 + \omega_y \cdot \sin^2\varphi_1 + \omega_{xy} \cdot \sin 2 \cdot \varphi_1$$

$$\omega_2 = \omega_\eta = \omega_x \cdot \sin^2\varphi_1 + w_y \cdot \cos^2\varphi_1 - \omega_{xy} \cdot \sin 2 \cdot \varphi_1$$

where $\tan 2 \cdot \varphi_1 = (2 \cdot \omega_{xy})/(\omega_x - \omega_y)$

From this fictitious reinforcement, the ultimate internal resisting moments may be calculated.

$$m_{u\xi} = d^2 \cdot f_{cu} \cdot \omega_\xi \cdot (1 - \frac{\omega_\xi}{2})$$

$$m_{u\eta} = d^2 \cdot f_{cu} \cdot \omega_\eta \cdot (1 - \frac{\omega_\eta}{2})$$

The design ultimate applied moments, expressed as

$$m^* = 1.8 \cdot m_{g+q} + (1.0 \div 1.3) \cdot m_{zp}$$

must now be transformed to the principal directions for comparison with the provided ultimate internal resisting moments.

$$m^*_\xi = m^*_x \cdot \cos^2\varphi_1 + m^*_y \cdot \sin^2\varphi_1 + m^*_{xy} \cdot \sin 2 \cdot \varphi_1$$

$$m^*_\eta = m^*_x \cdot \sin^2\varphi_1 + m^*_y \cdot \cos^2\varphi_1 - m^*_{xy} \cdot \sin 2 \cdot \varphi_1$$

$$m^*_{\xi\eta} = \frac{1}{2} \cdot (m^*_y - m^*_x) \cdot \sin 2 \cdot \varphi_1 + m^*_{xy} \cdot \cos 2 \cdot \varphi_1$$

According to the theory of plasticity, safety against collapse is guaranteed by satisfying the following inequalities:

Positive moments:

$$\underbrace{(m_{u\xi} - m^*_\xi)}_{\geq 0} \cdot \underbrace{(m_{u\eta} - m^*_\eta)}_{\geq 0} - (m^*_{\xi\eta})^2 \geq 0$$

Negative moments:

$$\underbrace{(m'_{u\xi'} + m^*_{\xi'})}_{\geq 0} \cdot \underbrace{(m'_{u\eta'} + m^*_{\eta'})}_{\geq 0} - (m^*_{\xi'\eta'})^2 \geq 0$$

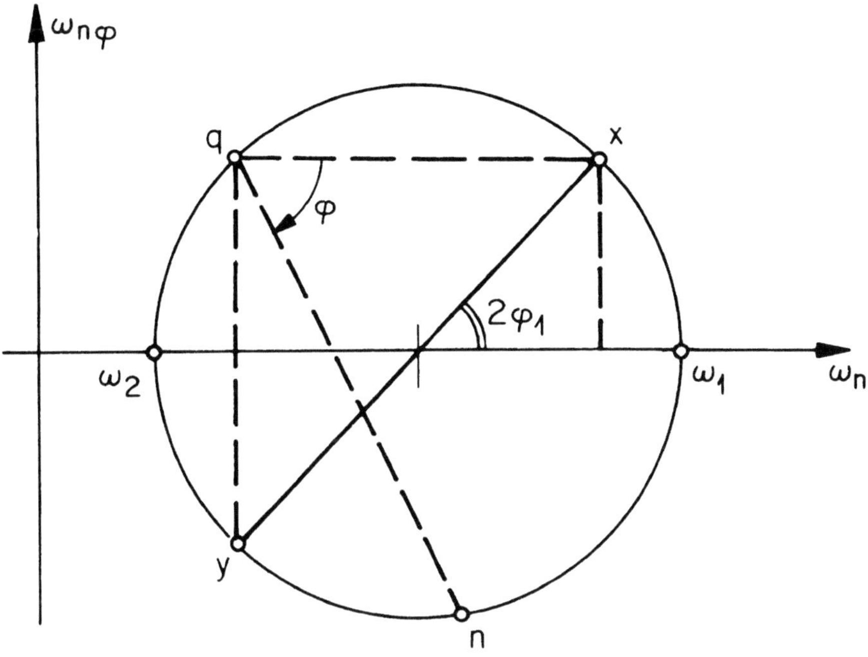

FIG 10: MOHR'S CIRCLE FOR THE TRANSFORMATION OF THE MECHANICAL REINFORCEMENT

IV. DYNAMICS AND VIBRATION

FLUID-STRUCTURE INTERACTION IN ANALYSIS OF BRIDGE PIERS. PART I : THEORETICAL PRELIMINARIES

Nuri Akkaş and Çetin Yılmaz

Department of Civil Engineering
Middle East Technical University , Ankara, Turkey

ABSTRACT. The theoretical preliminaries needed for the analysis of fluid-structure interaction are presented. After the derivation of the wave equation which governs the response of an acoustic fluid, the effect of the surrounding fluid on the response of a spring-mass system is studied. A submerged rigid cylinder, which may be considered as the model of a bridge pier, is subjected to a sinusoidal excitation. The hydrodynamic force acting on the cylinder is calculated analytically.

1. INTRODUCTION

The dynamic analysis of structural systems immersed in fluid media is of significant interest in many fields of engineering. Fluid-structure interaction occurs in the earthquake response of bridge piers. The dynamic response of such immersed structural systems is coupled to the fluid response through dependence of the effective mass, damping, and stiffness of these systems on the acceleration, velocity, and displacement of the fluid field, respectively. In the range of frequencies characteristic of the free vibration of these systems, the primary effect of the fluid is to increase the effective mass of the structural system. In the analysis of the forced or transient response of these systems, the effects of the fluid on the effective damping and stiffness of the system may be of considerable importance also.

In the present work we are concerned with the analysis of fluid-structure systems in which the physical behavior of the fluid, to a first approximation, is characteristic of an acoustic medium. The constraining assumptions are that the fluid is inviscid, irrotational, compressible and subject only to small displacements from or oscillations about some equilibrium position. The behavior of the structural systems is assumed to be linearly elastic. In the presence of fluid free surfaces, it is generally considered consistent with the above assumptions to use the simplified, high frequency approximation to the free surface condition which corresponds to the absence of gravity waves. Within these constraints, the analytical methods described in the present work may be routinely applied to approximate the vibratory response of submerged bridge piers.

As piers respond to earthquake ground motions, they cause corresponding motions in the water in contact with the pier surfaces. This results in changes of water pressure acting on the piers which, in turn, affect the dynamic response of the piers. This coupled hydroelastic phenomenon is termed fluid-structure interaction.

In the following discussion, the equations of motion governing the response of an acoustic fluid medium are derived from the principles of conservation of mass and momentum. Similar derivations can be found in many textbooks (Lamb, 1945). Following the derivation of the governing equations, the analytical solutions for two simple problems are presented. The applicability of analytical techniques to realistic fluid-structure interaction problems is limited to simple geometries. The solution of more realistic problems is feasible generally through numerical techniques which will be the topic of Part II of this presentation. Analytical solutions, on the other hand, are not completely useless. They provide one with an insight to the interaction phenomenon. The experience and insight gained from analytical solutions prove to be extremely useful in the interpretation of the results obtained from numerical techniques.

2. EQUATIONS OF FLUID MOTION

The equations governing the motion of a homogeneous, isentropic, Newtonian fluid medium are given by (Temkin, 1981)

$$\frac{1}{\rho}\frac{D\rho}{Dt} + \vec{\nabla}.\vec{u} = 0, \tag{1}$$

$$\frac{\partial}{\partial t}(\rho u_i) + \frac{\partial}{\partial X_j}(\rho u_i u_j) = \frac{\partial \sigma_{ij}}{\partial X_j} + F_i \tag{2}$$

where

$\rho = \rho(t)$ = density of the fluid
$u = u(X, Y, Z) = (u_1, u_2, u_3)$
= velocity of the fluid at a point (X, Y, Z)
F_i= body force at a point (X, Y, Z) of the fluid
σ_{ij}= dynamic stress in the fluid at a point (X, Y, Z).

Finally,

$$\frac{D(\ )}{Dt} = \frac{\partial(\ )}{\partial t} + u_j \frac{\partial(\ )}{\partial X_j} \tag{3}$$

is the material derivative. As usual, repeated indices imply summation.

In characterizing the behavior of an acoustic fluid, we assume tnat the fluid is initially at rest with uniform density, ρ. The fluid is then disturbed by some or all material elements being displaced small amounts with corresponding changes in density. The fluid is subsequently allowed to return to some equilibrium position and oscillate about it. The gradients of both the fluid velocity and density are assumed to be negligibly small. Thus a consistent approximation to equations (2) is seen to be

$$\rho \frac{\partial u_i}{\partial t} = \frac{\partial \sigma_{ij}}{\partial X_j} + F_i \quad . \tag{4}$$

The dynamic stress, σ_{ij}, in a Newtonian fluid can be related to rate of strain in the fluid by the relationship

$$\sigma_{ij} = -p\delta_{ij} + 2\mu(e_{ij} - \frac{1}{3} e_{kk}\delta_{ij}) \tag{5}$$

where δ_{ij} is the Kroenecker delta function, μ is the viscous shear coefficient, and e_{ij} are rates of fluid strain.The pressure is p.

For a homogeneous, isentropic, elastic medium the stress, T_{ij} is related to the material strain by

$$T_{ij} = K\varepsilon_{kk}\delta_{ij} + 2G(\varepsilon_{ij} - \frac{1}{3} \varepsilon_{kk}\delta_{ij}) \tag{6}$$

where K is the bulk modulus and G is the shear modulus of the material. Following the assumptions which characterize the acoustic fluid medium, we may further approximate the behavior of the fluid as being analogous to that of an elastic medium with zero shear modulus. We write

$$\sigma_{ij} = K\varepsilon_{kk}\delta_{ij} \quad . \tag{7}$$

Stress is now defined as positive in compression and ε_{kk} is the divergence of a material fluid volume given by

$$\varepsilon_{kk} = \frac{\partial h_k}{\partial X_k} = \frac{\partial h_1}{\partial X_1} + \frac{\partial h_2}{\partial X_2} + \frac{\partial h_3}{\partial X_3} = \vec{\nabla} \cdot \vec{h} \tag{8}$$

where $\vec{h} = \vec{h}(X,Y,Z,t) = (h_1 h_2 h_3)$ = small displacement of the fluid from equilibrium at a point (X,Y,Z).

In order to make use of the above relations we must return to equation (4) and express the fluid velocity in terms of the displacement field $\vec{h}$. We write

$$\frac{\partial^2 \vec{h}}{\partial t^2} = \vec{\nabla} p \tag{9}$$

for which we have assumed the body forces, $\vec{F}$, to be uniform throughout the fluid domain and have set them equal to zero.

On combining equations (7), (8), and (9) to eliminate either displacement or stress we obtain, respectively,

$$c_f^2 \nabla^2 p = \frac{\partial^2 p}{\partial t^2} \tag{10}$$

and

$$c_f^2 (\vec{\nabla}^2 \vec{h} + \text{curlcurl } \vec{h}) = \frac{\partial^2 \vec{h}}{\partial t^2} \tag{11}$$

where

$$c_f = \sqrt{K/\rho} \tag{12}$$

is the speed of sound in the fluid.

Equations (10) and (11) are both forms of the wave equation for which various methods of solution are well established. Equation (10) is a scalar equation in one unknown, p, whose solution yields the dynamic pressure field in the fluid domain. Equation (11) is a vector equation in one, two, or three unknowns governing the time-varying displacements of the fluid medium.

Recall that equations (10) and (11) have been derived under the assumption that the acoustic fluid is compressible. This implies that the speed of sound in the fluid, c_f, is finite. If the fluid is assumed to be incompressible, which is a reasonable assumption for water, the speed of sound becomes infinitely large. In this case, the response of the fluid is governed by either one of the following equations :

$$\nabla^2 p = 0 \tag{13}$$

$$\bar{\nabla}^2 \bar{h} + \text{curlcurl } \bar{h} = 0 \tag{14}$$

In the following sections, the pressure dependent formulation of the wave equation, equation (10), will be used to show the effect of the interacting fluid on the response of a structure.

3. SINGLE-DOF MASS-SPRING SYSTEM

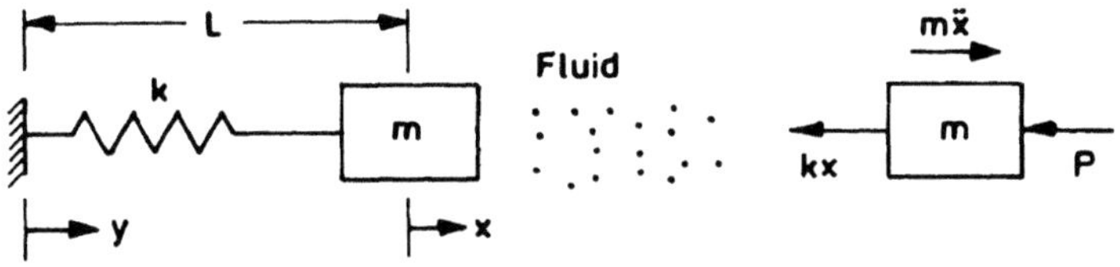

Fig. 1 A submerged single degree-of-freedom mass-spring system

Consider a single-degree-of-freedom mass-spring system interacting with an acoustic fluid medium as shown in Fig. 1. The hydrodynamic pressure acting on the face of the mass is denoted by p. Assuming that this face has a unit area p is also equal to the hydrodynamic force acting on the mass. The displacement of the mass m from its position at rest is denoted by x. The spring constant is k. The mass is assumed to be at a distance L from the origin of the y axis fixed in space.

The equation of motion governing the displacement x(t) of this idealized structure is

$$m\ddot{x} + kx = -p \qquad (15)$$

in which dots denote differentiation with respect to time. The response of the acoustic fluid, under the assumptions stated in the previous section, is governed by the classical one-dimensional wave equation :

$$\frac{\partial^2 \Phi}{\partial y^2} = \frac{1}{c_f^2} \ddot{\Phi} \qquad (16)$$

in which $\Phi(y,t)$ is the velocity potential. The fluid is assumed to be compressible. The hydrodynamic pressure on the mass is related to the velocity potential through the following relation :

$$p = -\rho \dot{\Phi} \qquad \text{at } y = L, \qquad (17)$$

in which ρ is the mass density of the fluid. The kinematic (or interface) boundary condition between the mass and the fluid states that the velocities of the mass and the fluid particles at the interface must be equal so that no gap can occur between the two. This condition is stated mathematically as follows :

$$\dot{x} = \frac{\partial \Phi}{\partial y} \quad \text{at } y = L . \tag{18}$$

The assumption that displacements are small has been used in equation (18). From equation (16) it is seen that we need two boundary conditions on Φ in y. Equation (18) provides one of them. The second one is obtained from the fact that, for propagating waves, the radiation condition must be satisfied at $y = \infty$. This implies that as $y \to \infty$, Φ must go to zero. The definition of the problem will be completed by assigning the initial conditions which are taken as

$$x = 0, \quad \dot{x} = V, \quad \Phi = 0, \quad \dot{\Phi} = 0 \quad \text{at } t = 0, \tag{19}$$

in which V is a constant.

The problem is now well-defined and the solution can be obtained by various means. The Laplace Transform method used in the present work yields the following :

$$x = \frac{V}{\Omega} e^{-\alpha t} \operatorname{Sin}\Omega t \tag{20}$$

in which

$$\Omega = \omega \sqrt{1 - \frac{\alpha^2}{\omega^2}} , \tag{21a}$$

$$\omega = \sqrt{\frac{k}{m}} , \tag{21b}$$

$$\alpha = \frac{\rho c_f}{2m} . \tag{21c}$$

Note that ω is the natural circular frequency of vibration (rads/sec) of the single-degree-of-freedom mass-spring system in vacuo. The frequency of the same system interacting with the fluid medium shall be called the modified frequency and denoted by Ω.

It is obvious from a study of equation (20) that the solution is similar to that for a single-degree-of-freedom mass-spring-damper system in vacuo. In other words, the fluid acts as a damper as expected. The term α/ω corresponds to the well-known damping ratio $\xi = c/2m\omega$ of structural dynamics (Chopra, 1981). Here c is the damping coefficient for the structure. The presence of the fluid has the effect of decreasing the natural circular frequency of vibration and, thus, increasing the natural period of vibration.

One significant difference between the response characteristics of a submerged mass-spring system and a mass-spring-damper system *in vacuo* is the following : According to Chopra (1981), damping ratios less than or equal to 0.2 constitute a range which includes most structures *in vacuo*. Thus, damping effects on the frequency and period of such structures are negligible. On the other hand, the same can not be said for a mass-spring system in a fluid medium. The α/ω for the interacting system is not necessarily less than 0.2. It is commonly larger than 0.2 and it may even be larger than 1.0. Recall that, for ρ=1000 kg/m³, c_f=1500 m/sec and ω=200 rad/sec,

$$\frac{\alpha}{\omega} = \frac{\rho c_f}{2m\omega} = \frac{3750}{m}.$$

Thus, for a mass m > 3750 kg the α/ω ratio will be less than 1.0. For smaller masses it will be even larger than 1.0. The generalization stated by Chopra (1981) is, therefore, not applicable for damping caused by the presence of the fluid. The difference between structural damping and fluid damping can be explained as follows :

Recall that

$$\xi = \frac{c}{2m\omega} \quad \text{and} \quad \frac{\alpha}{\omega} = \frac{\rho c_f}{2m\omega} \tag{22}$$

which are the damping ratios for a damped structure *in vacuo* and a structurally undamped structure in fluid. ξ depends *on the ratio* between c and m which are both related to the properties of *the same structure*. On the other hand, α/ω depends on the ratio between the quantities which are related to the structure *and* the surrounding fluid. Thus, it is, in principle, possible to come *up* with α/ω ratios which are much different from corresponding ξ ratios. In case α/ω turns out to be greater than 1.0, the solution becomes

$$x = \frac{V}{\Omega} e^{-\alpha t} \text{ Sinh } \Omega t \tag{23}$$

with

$$\Omega = \omega \sqrt{\frac{\alpha^2}{\omega^2} - 1}\,.$$

This solution is not even sinusoidal.

As a second example, let us now study the response of a single degree-of-freedom mass-spring system interacting with a fluid subjected to base motion. The governing equation is

$$m\ddot{x} + kx = -p - m\ddot{x}_B \tag{24}$$

in which x_B is the displacement of the support shown in Fig.1. The fluid response is governed by the wave equation (16) as before but, now, it is more advantageous to use it in its plane wave form :

$$\frac{\partial \Phi}{\partial y} + \frac{1}{c_f}\dot{\Phi} = 0. \tag{25}$$

The kinematic boundary condition becomes

$$\frac{\partial \Phi}{\partial y} = \dot{x} + \dot{x}_B \quad \text{at} \quad y = L, \tag{26}$$

because the fluid particle velocity is equal to the absolute velocity of the mass. Using equation (17) we obtain the hydrodynamic force as

$$p = \rho c_f(\dot{x} + \dot{x}_B) \quad \text{at} \quad y = L. \tag{27}$$

Thus, the final form of the governing equation is

$$m\ddot{x} + \rho c_f \dot{x} + kx = -m\ddot{x}_B - \rho c_f \dot{x}_B \ . \tag{28}$$

Let us assume that the base displacement is

$$x_B = B\ e^{i\Omega t} \ . \tag{29}$$

The solution will be of the form :

$$x = \bar{x}\ e^{i\Omega t} \ . \tag{30}$$

Substituting equations (29) and (30) into (28) we obtain the real part of x and it is

$$\text{Real } x = \frac{B\{[(\omega^2 - \Omega^2) - \frac{\rho^2 c_f^2}{m^2}] \text{Cos}\Omega t + \frac{\rho c_f}{m\Omega}\omega^2 \ \text{Sin}\Omega t\}}{[\ \frac{(\omega^2 - \Omega^2)^2}{\Omega^2} - \frac{\rho^2 c_f^2}{m^2}\ ]} \ . \tag{31}$$

Here, $\omega = \sqrt{k/m}$ as before.

Now consider the response of a single-degree-of-freedom mass-spring-damper system in vacuo subjected to a base motion. The governing equation is (Chopra, 1981)

$$m\ddot{x} + c\dot{x} + kx = -m\ddot{x}_B \ . \tag{32}$$

A comparison of equations (28) and (32) shows that, in fluid-structure interaction problems with base motion, base velocity $\dot{x}_B$ enters the equation in addition to the base acceleration $\ddot{x}_B$. This is another important difference between the responses of a structure in vacuo and of that in a fluid medium. The damping force proportional to the base velocity is added to the damping force proportional to the relative velocity of the mass. Having discussed some significant aspects of fluid-structure interaction using a single-degree-of-freedom system, we can now consider a slightly more complicated and more realistic model which is studied below.

4. SUBMERGED LONG RIGID CYLINDER

A long rigid cylinder, surrounded by an infinite fluid medium, is subjected to a rigid body motion in the rz-plane as shown in Fig. 2.The motion of the cylinder is described by the velocity V_g. The fluid is compressible. Small vibrations of the fluid are governed by the wave equation in cylindrical coordinates :

$$\frac{\partial^2 \Phi}{\partial r^2} + \frac{1}{r}\frac{\partial \Phi}{\partial r} + \frac{1}{r^2}\frac{\partial^2 \Phi}{\partial \Theta^2} = \frac{1}{c_f^2}\frac{\partial^2 \Phi}{\partial t^2} \tag{33}$$

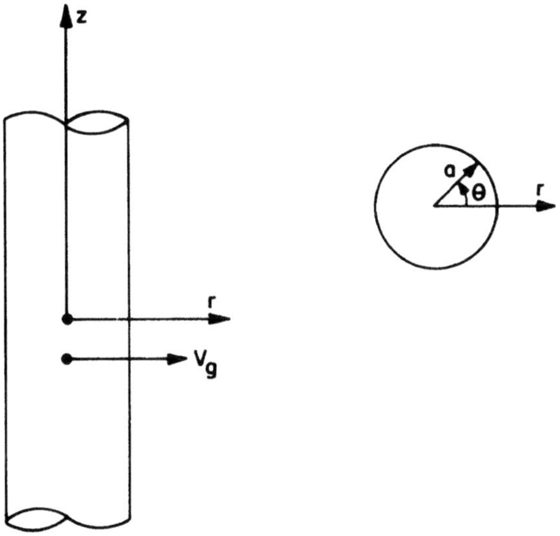

Fig. 2 A submerged long rigid cylinder

in which Φ is the velocity potential and c_f is the speed of sound in the fluid. Fluid particle velocity in radial direction is $\partial\Phi/\partial r$ and the hydrodynamic excess pressure is $p = -\rho\partial\Phi/\partial t$ where ρ is the fluid mass density.

Boundary conditions are

$$\frac{\partial \Phi}{\partial r} = V_g \quad \text{at} \quad r = a, \tag{34a}$$

$$\frac{\partial \Phi}{\partial r} = 0 \quad \text{as} \quad r \to \infty. \tag{34b}$$

Here, a is the radius of the cylinder. Note that V_g corresponds to the rigid body motion and for a steady state analysis we have :

$$V_g = v_g\, e^{-i\omega t} \tag{35}$$

From (34a) :

$$\frac{\partial \Phi}{\partial r} = v_g \, \mathrm{Cos}\Theta \, e^{-i\omega t} \quad \text{at} \quad r = a . \tag{36}$$

Accordingly,

$$\Phi(r,\Theta,t) = \Phi_1(r) \, \mathrm{Cos}\Theta \, e^{-i\omega t} . \tag{37}$$

Substituting (37) into (33) :

$$\Phi_1'' + \frac{1}{r}\Phi_1' - \frac{1}{r^2}\Phi_1 + \frac{\omega^2}{c_f^2}\Phi_1 = 0 , \tag{38}$$

in which the derivatives are with respect to r. The solution of (38) for outgoing waves is (Redwood, 1960).

$$\Phi_1(r) = A\, H_1^{(1)}\left(\frac{\omega r}{c_f}\right) \tag{39}$$

in which A is a constant to be determined from the boundary condition at r = a and $H_1^{(1)}$ is the Hankel function of the first kind and first order. The general solution of (33) is, therefore,

$$\Phi(r,\Theta,t) = A\, H_1^{(1)}\left(\frac{\omega r}{c_f}\right) \mathrm{Cos}\Theta \; e^{-i\omega t} \tag{40}$$

The boundary condition (36) yields

$$A = v_g / \left\{\frac{d}{dr}\, H_1^{(1)}\left(\frac{\omega r}{c_f}\right)\right\}_{r = a} . \tag{41}$$

The hydrodynamic pressure at r = a is

$$p_a = -\rho\left\{\frac{\partial \Phi}{\partial t}\right\}_{r=a} = i\rho\omega\{\Phi\}_{r=a} . \tag{42}$$

Substituting (40) and (41) into (42), one obtains

$$p_a = i\rho c_f\, V_g \; \frac{\{H_1^{(1)}(\xi)\}_{\xi=\omega a/c_f}}{\{\frac{d}{d\xi}\, H_1^{(1)}(\xi)\}_{\xi=\omega a/c_f}} \, \mathrm{Cos}\theta , \tag{43}$$

in which $\xi = \omega r/c_f$

Recall that (Redwood, 1960)

$$\frac{d}{d\xi} H_1^{(1)}(\xi) = H_o^{(1)}(\xi) - \frac{1}{\xi} H_1^{(1)}(\xi) \ . \qquad (44)$$

Substituting (44) into (43), one obtains

$$p_a = i\rho c_f V_g \frac{H_1^{(1)}(\omega a/c_f)}{\{H_o^{(1)}(\omega a/c_f) - \frac{c_f}{\omega a} H_1^{(1)}(\omega a/c_f)\}} \operatorname{Cos}\theta \ . \qquad (45)$$

Call the terms on the right hand side of (45) β, except for the $\operatorname{Cos}\theta$ term. Thus,

$$p_a = \beta \operatorname{Cos}\theta \ . \qquad (46)$$

The total hydrodynamic force, P_T, acting on a cylinder of height H is

$$P_T = 4aH\beta \int_o^{\pi/2} \operatorname{Cos}^2\theta \, d\theta = \pi a H \beta \ . \qquad (47)$$

Finally,

$$P_T = i\pi a H \rho c_f V_g \frac{H_1^{(1)}(\omega a/c_f)}{\{H_o^{(1)}(\omega a/c_f) - \frac{c_f}{\omega a} H_1^{(1)}(\omega a/c_f)\}} \ . \qquad (48)$$

The following two approximations bring the above equation into a simpler form to handle.

a) High frequency approximation :

It is assumed that $\omega a/c_f >> 1$. In this case, the second term in the denominator in (48) can be neglected. Thus,

$$P_T = i\pi a H \rho c_f V_g \frac{H_1^{(1)}(\frac{\omega a}{c_f})}{H_o^{(1)}(\frac{\omega a}{c_f})} \ . \qquad (49)$$

For $\frac{\omega a}{c_f} >> 1$, from Abramowitz and Stegun (1965),

$$\frac{H_1^{(1)}(\frac{\omega a}{c_f})}{H_o^{(1)}(\frac{\omega a}{c_f})} = \frac{\exp\{i(\frac{\omega a}{c_f} - \frac{\pi}{4} - \frac{\pi}{2})\}}{\exp\{i(\frac{\omega a}{c_f} - \frac{\pi}{4})\}} ,$$

$$= e^{-i\pi/2} = -1 \ . \qquad (50)$$

Substituting (50) into (49) one obtains

$$P_T = \pi a H \rho c_f V_g \ . \tag{51}$$

Assuming that the highest frequency that is encountered in earthquakes is about 10 cps; thus, for (19) to be valid, one must have

$$2\pi f a / c_f >> 1 \ .$$

This yields,

$$a >> \frac{c_f}{2\pi f} \qquad \text{or} \qquad a >> 24 \text{ m.}$$

This is very unrealistic. Therefore, high frequency approximation is not a reasonable one.

b) Low frequency approximation :

It is assumed that $\omega a/c_f << 1$. In this case, (48) can be brought to the following form :

$$P_T = - i\pi a^2 H \rho \omega V_g \ . \tag{52}$$

This yields the added mass results because $(\pi a^2 H\rho)$ is the mass of the fluid displaced by the cylinder. For this assumption,

$$2\pi f a << c_f$$

and this shows that, for reasonable frequencies, realistic dimensions will always satisfy the above equation. Therefore, the low frequency approximation, which is equivalent to the added mass approach, is satisfactory for many practical cases.

After noting the limitations of the high and low frequency approximation as given above, now let us go back to the exact solution (48). Let

$$P_T = i\pi a H \rho c_f V_g \Lambda \tag{53}$$

in which

$$\Lambda = \frac{H_1^{(1)}(\omega a/c_f)}{\{H_o^{(1)}(\omega a/c_f) - \frac{c_f}{\omega a} H_1^{(1)}(\omega a/c_f)\}} \tag{54}$$

Using the following identities (Abramowitz and Stegun, 1965)

$$H_\nu^{(1)}(z) = J_\nu(z) + i\, Y_\nu(z) \ , \tag{55}$$

we obtain

$$\Lambda = \frac{J_1 + i\, Y_1}{(J_o + iY_o) - \frac{1}{z}(J_1 + iY_1)} , \tag{56}$$

in which $z = \omega a/c_f$ and the argument of the Bessel functions has been dropped. Equation (56) can be brought to the following form

$$\Lambda = \frac{J_1 + i\, Y_1}{(J_o - \frac{1}{z} J_1) + i(Y_o - \frac{1}{z} Y_1)} . \tag{57}$$

By multiplying and dividing (57) with complex conjugate of the denominator one obtains

$$\Lambda = \frac{\alpha + i\beta}{\gamma} , \tag{58}$$

in which

$$\alpha = J_1(J_o - \frac{1}{z} J_1) + Y_1(Y_o - \frac{1}{z} Y_1) , \tag{59a}$$

$$\beta = Y_1(J_o - \frac{1}{z} J_1) - J_1(Y_o - \frac{1}{z} Y_1) , \tag{59b}$$

$$\gamma = (J_o - \frac{1}{z} J_1)^2 + (Y_o - \frac{1}{z} Y_1)^2 . \tag{59c}$$

Accordingly, (53) becomes

$$P_T = i\pi a H\rho c_f\, V_g \frac{(\alpha + i\beta)}{\gamma} . \tag{60}$$

Using equation (35) one obtains

$$P_T = \pi a H\rho c_f\, v_g(\mathrm{Sin}\omega t + i\mathrm{Cos}\omega t) \frac{(\alpha + i\beta)}{\gamma} . \tag{61}$$

Let $\Delta = \pi a H\rho c_f\, v_g$. Thus

$$P_T = \Delta\{(\frac{\alpha}{\gamma} \mathrm{Sin}\omega t - \frac{\beta}{\gamma} \mathrm{Cos}\omega t) + i(\frac{\alpha}{\gamma} \mathrm{Cos}\omega t + \frac{\beta}{\gamma} \mathrm{Sin}\omega t) . \tag{62}$$

Finally,

$$P_T = \mathrm{Re}(P_T) + i\mathrm{Im}(P_T) , \tag{63}$$

in which

$$\mathrm{Re}(P_T) = \frac{\Delta}{\gamma}(\alpha\mathrm{Sin}\omega t - \beta\mathrm{Cos}\omega t), \tag{64a}$$

$$\mathrm{Im}(P_T) = \frac{\Delta}{\gamma}(\alpha\mathrm{Cos}\omega t + \beta\mathrm{Sin}\omega t). \tag{64b}$$

Nondimensionalize P_T as follows :

$$\bar{P}_T = \frac{P_T}{\pi a^2 H\rho g} , \tag{65}$$

in which g is the gravitational acceleration. Let $\chi = \pi a^2 H\rho g$ which is the weight of the fluid displaced by the cylinder of height H. Then,

$$\bar{P}_T = \mathrm{Re}(\bar{P}_T) + i\mathrm{Im}(\bar{P}_T) \tag{66}$$

in which

$$\mathrm{Re}(\bar{P}_T) = \frac{\Delta}{\chi\Upsilon} (\alpha \mathrm{Sin}\omega t - \beta \mathrm{Cos}\omega t), \tag{67a}$$

$$\mathrm{Im}(\bar{P}_T) = \frac{\Delta}{\chi\Upsilon} (\alpha \mathrm{Cos}\omega t + \beta \mathrm{Sin}\omega t) . \tag{67b}$$

The amplitude of $\mathrm{Re}(\bar{P}_T)$ is obtained by setting $\omega t = \mathrm{Arctan}(-\alpha/\beta)$ and that of $\mathrm{Im}(\bar{P}_T)$ by setting $\omega t = \mathrm{Arctan}(\beta/\alpha)$. The results are

$$|\mathrm{Re}(\bar{P}_T)| = \frac{\Delta}{\chi\Upsilon}\{\alpha \mathrm{Sin}(\mathrm{Arctan}\ (-\tfrac{\alpha}{\beta})) - \beta \mathrm{Cos}(\mathrm{Arctan}(-\tfrac{\alpha}{\beta}))\} , \tag{68}$$

$$|\mathrm{Im}(\bar{P}_T)| = \frac{\Delta}{\chi\Upsilon}\{\alpha \mathrm{Cos}(\mathrm{Arctan}(\tfrac{\beta}{\alpha})) + \beta \mathrm{Sin}(\mathrm{Arctan}(\tfrac{\beta}{\alpha}))\} . \tag{69}$$

A computer program has been coded for the calculation of the above results. The numerical results obtained from the approximate solution (52) and the exact solution (68) are compared in Fig. 3. It is seen that, for the range of frequencies realistic for earthquakes, the low frequency (or the added mass) approximation yields results very close to the exact solution.

5. SUMMARY AND CONCLUSIONS

The effect of the surrounding fluid on the response of a single-degree-of-freedom spring-mass system is studied analytically. An analytical method of solution is used also for the steady state response analysis of a submerged rigid cylinder. The flexibility of the cylinder has not been considered to keep the analytical solution feasible. The solution of more realistic problems, in which, for instance, flexibility of the pier is considered, generally requires numerical techniques which will be the topic of Part II of the present paper.

Acknowledgement : This work has been supported by Yüksel Proje A.Ş., architect-engineers for the Karakaya Dam Bridge project. The authors are indebted to Mr. Erdal Erkakan, project manager.

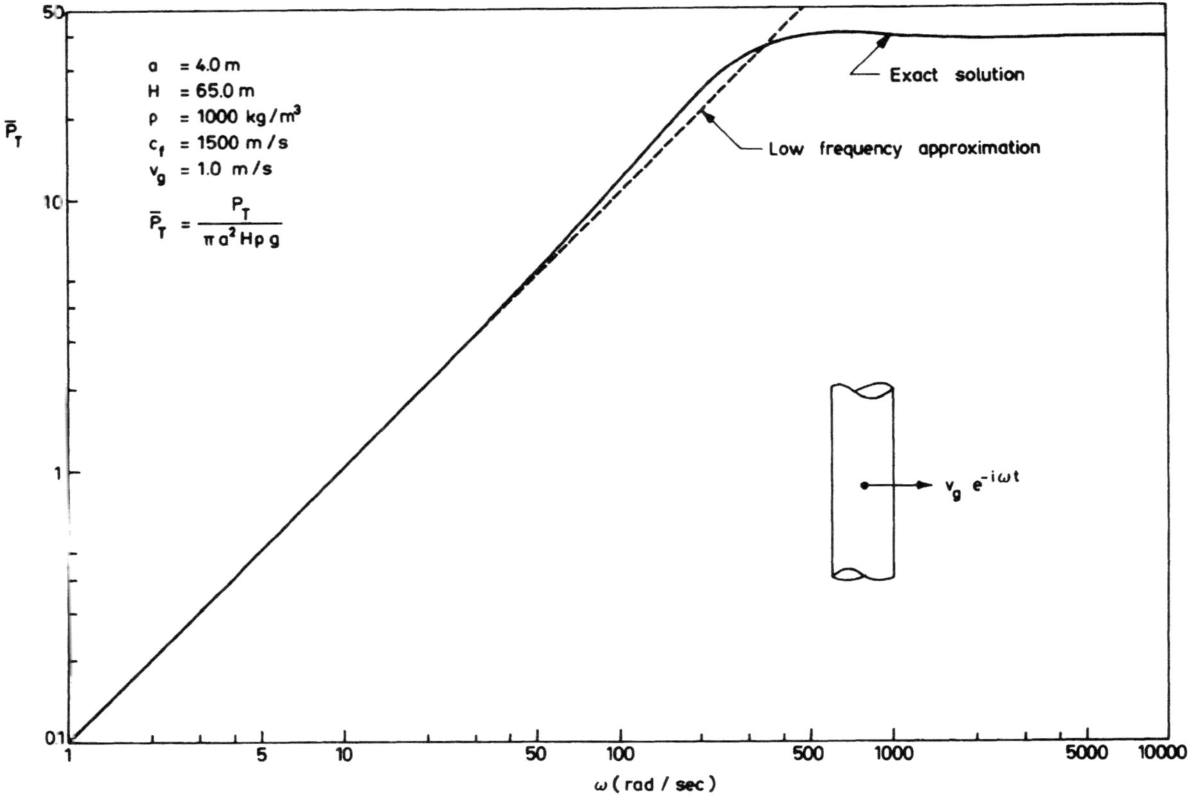

Fig. 3 Comparison of low frequency approximation with exact solution

REFERENCES

1. Abramowitz, M., and Stegun, I.A., (editors), Handbook of Mathematical Functions, Dover Publ., New York, 1965.

2. Chopra, A., Dynamics of Structures, A Primer, EERI, California, 1981.

3. Lamb, H., Hydrodynamics, Dover Publ., 6th edn., New York, 1945.

4. Redwood, M., Mechanical Waveguides, Pergamon Press, Oxford, 1960.

5. Temkin, S., Elements of Acoustics, John Wiley and Sons, New York, 1981.

FLUID-STRUCTURE INTERACTION IN ANALYSIS OF BRIDGE PIERS. PART II: NUMERICAL APPROACH

Çetin Yılmaz and Nuri Akkaş

Department of Civil Engineering
Middle East Technical University, Ankara, Turkey

ABSTRACT. The development of fluid finite element is presented. Earthquake response analysis of a submerged thick cylindrical shell is studied. The response of the same shell in vacuo is also investigated. The effect of the fluid-shell interaction on the displacements, stresses and the base moment of the shell is discussed. It was noted that the presence of the fluid may increase the base moment by quite a significant percentage.

1. INTRODUCTION

In Part I of the present work the theoretical preliminaries needed for the analysis of fluid-structure interaction have been presented. In Part I it was stated that the solution of realistic problems is feasible generally through numerical techniques. The finite element formulation for the numerical analysis of the fluid-structure interaction problems will be presented in this part of the work.

Numerical solution methods which have been applied to interaction problems were initially dependent on approximate decoupling of the structural response from the fluid motion. Solutions for the fluid response were based on potential theory (Geers, 1971 ; Clark, 1973) or, in the case of shock response, on plane and spherical wave approximations (Bedrossian et al., 1972, 1972). These approximations have the advantage of not requiring explicit or discrete modeling of the fluid volume. They are, however, limited to applications in which the structure is deeply submerged and the effects of the free surface are negligible. Zienkiewicz and Newton (1969) developed acoustic-fluid finite elements based on a Galerkin approximation to the wave equation using pressure as the dependent field variable. This method was used to compute the natural frequencies of submerged structures in finite fluid regions and in unbounded fluid regions in which the fluid-structure interaction effects are negligible beyond finite distances from the structure. This method was extended by Hunt (1974) and Zarda (1976) to applications in which the far field effects are important by a procedure which matches finite element solutions at finite external fluid boundaries with known analytical solutions for infinite fluid cases. A similar formulation is presented in (Nastran, 1976) and implemented in the finite element, structural analysis computer program, NASTRAN. In this case the finite element equations were derived from variational methods using a form of the wave equation in which the fluid pressure is the dependent field variable. Pressure analog methods have also been developed which use either "lumped" (Everstine et al., 1975) or consistent (Schroeder and Marcus, 1974, 1976) formulations equivalent to the previous finite element pressure methods. The pressure analog methods are derived from an analogy between fluid pressure and one of the Cartesian displacement components in finite element formulation of classic linear elasticity. These methods have been implemented using readily available finite element, structural analysis programs such as Nastran.

Further investigations have produced structurally compatible, acoustic fluid finite elements for linear two-dimensional analysis based on pressure as the dependent variable (Nahavandi et al. 1975; Marcus, 1978). Bathe and Hahn (1979) have implemented two-and three-dimensional, acoustic fluid elements in the structural analysis program ADINA which is capable of nonlinear and large deformation analyses. These methods require the addition of separate acoustic fluid finite elements to the element libraries of existing finite element, structural analysis programs.

In all of the preceding finite element methods, the boundary conditions at the fluid-structure interfaces are satisfied such that the acoustic-fluid finite elements are compatible with structural elements. The coupling of the fluid-structure interaction is achieved directly through the assembly of the system mass, damping, and stiffness matrices. The generality and versatility of the finite element method in purely structural applications is thus extended to the acoustic fluid-structure application with relative ease.

This present work represents an extension of techniques developed in (Akay et al., 1977; Akkaş et al., 1979) for the implementation of two-dimensional and axisymmetric, acoustic fluid finite elements in the structural analysis, finite element program SAPIV. The finite element formulation is derived from the application of the method of weighted residuals using Galerkins criterion to a form of the wave equation based on fluid displacement as the independent field variable. Implementation of acoustic fluid finite elements is achieved with a minimum of computer programming and without addition of separate elements to the existing SAPIV library by recognition of the equivalence between the above displacement formulation and the classical linear elasticity formulation used by SAPIV.

In the derivation of acoustic fluid finite element equations, several crucial assumptions are made which approximate the physical properties and behavior of the fluid medium. The most important of these assumptions in regard to the structural compatibility of acoustic fluid elements, is that fluid displacements are small and oscillatory about some equilibrium position. This assumption allows the fluid stress tensor to be derived based on an analogy between the stress state in a linearly elastic medium and that of an acoustic fluid medium. In order to further simplify the governing equations of fluid motion, the acoustic fluid medium is assumed to be inviscid, irrotational, and isentropic. Changes in fluid density due to compression of the fluid are assumed to be negligibly small.

In the following discussion, three different finite element formulations are derived using either variational or weighted residual methods. Two of these formulations use pressure as the dependent field variable and the third uses displacement. The displacement formulation is shown to be equivalent to the classic linear elasticity formulation commonly used in finite element, structural analysis programs such as SAPIV. The present method for implementation of acoustic fluid, finite elements in SAPIV is then discussed.

2. PRESSURE - DEPENDENT FINITE ELEMENT FORMULATION

Finite element formulations for acoustic-fluid elements may be readily obtained from equation (10) of Part I by either variational methods or the method of weighted residuals (Huebner, 1975). In the classical variational formulation, the problem is to find the unknown function or functions which extremize (maximize, minimize) or make stationary a functional or system of functionals subject to given boundary conditions. In order to apply the method we must obtain a variational statement for the continuum problem described by equation (10) of Part I. We begin by rewriting this equation in the form

$$\rho\ddot{p} - K\vec{\nabla} \cdot \vec{\nabla}p = 0 \quad (1)$$

We define the sum of the terms on the left hand side of equation (1) to be the generalized force,

$$J = \rho\ddot{p} - K\vec{\nabla} \cdot \vec{\nabla}p \quad (2)$$

For the case of shock or pressure wave propagation in an inviscid fluid initially at rest, the flow field may be described as homentropic. Hence the volume integral of the product of the variation of pressure, δp, times the generalized force will produce a variation of the generalized energy :

$$\delta H = \int_\Omega J\delta p dV = \int_\Omega (\rho\ddot{p} - K\vec{\nabla} \cdot \vec{\nabla}p)\delta p \, dxdydz \tag{3}$$

We may substitute the identity

$$\vec{\nabla} \cdot \{\delta p(\vec{\nabla}p)\} = \delta p\vec{\nabla} \cdot \vec{\nabla}p + \vec{\nabla}(\delta p) \cdot \vec{\nabla}p \tag{4}$$

into equation (3) to obtain

$$\delta H = \int_\Omega \rho\ddot{p}\delta p dxdydz + \int_\Omega K\vec{\nabla}(\delta p) \cdot \vec{\nabla}p \, dxdydz - \int_\Omega K\vec{\nabla} \cdot \{\delta p(\vec{\nabla}p)\} \, dxdydz \tag{5}$$

Applying the divergence theorem to the third integral in equation (5) produces an integral over the surface s,

$$\int_\Omega K\vec{\nabla} \cdot \{\delta p(\vec{\nabla}p)\} \, dxdydz = \int_{\partial s} K\delta p\vec{\nabla}p \cdot \vec{n} \, ds \tag{6}$$

where the vector $\vec{n}$ is the outward unit normal to the surface s. The first and second integrals in equation (5) may be rewritten using the identities

$$\delta(\frac{\partial p}{\partial t})^2 = 2 \frac{d}{dt} (\frac{\partial p}{\partial t})\delta p = 2\ddot{p}\delta p, \delta(\vec{\nabla}p \cdot \vec{\nabla}p) = 2\vec{\nabla}p \cdot \vec{\nabla}(\delta p) \tag{7}$$

Applying equations (6) and (7) to equation (5) results in

$$\delta H = \delta\{\int_\Omega (\frac{1}{2} \rho \, (\dot{p})^2 + \frac{1}{2} K\vec{\nabla}p \cdot \vec{\nabla}p) \, dxdydz\} - \int_{\partial s} K\delta p(\vec{\nabla}p) \cdot \vec{n} \, ds \tag{8}$$

The generalized energy internal to the fluid is the value of the first integral of equation (5) in which we can identify a term representing kinetic energy and a term representing potential energy. The energy flow through the boundary is given by the second integral.

In the finite element method of solution, the solution domain Ω is now divided into m elements of r nodes. We then express the behavior of the unknown function p within each element as

$$p^{(e)}(x,y) = \sum_1^r N_i(x,y)p_i = |N|\{p\}^{(e)} \tag{9}$$

where p_i is the nodal value of p at node i. A stiffness matrix $|k_{ij}|$ is formed from the internal potential energy

$$V = \int_\Omega \frac{1}{2} K\vec{\nabla}p.\ \vec{\nabla}p\ dxdydz \qquad (10)$$

by the equation

$$k_{ij} = \frac{\partial^2 V}{\partial p_i \partial p_j} = \int_\Omega K(\frac{\partial N_i}{\partial x}\frac{\partial N_j}{\partial x} + \frac{\partial N_i}{\partial y}\frac{\partial N_j}{\partial y} + \frac{\partial N_i}{\partial z}\frac{\partial N_j}{\partial z})dxdydz. \qquad (11)$$

A mass matrix $|m_{ij}|$ is formed from the internal kinetic energy

$$T = \int_\Omega \frac{1}{2}\rho(\dot{p})^2\ dxdydz \qquad (12)$$

by the equation

$$m_{ij} = \frac{\partial^2 T}{\partial p_i \partial p_j} = \int_\Omega \rho\ N_i N_j\ dxdydz \qquad (13)$$

The set of simultaneous equations describing the fluid pressure at discrete points may then be written in matrix form as

$$|m_{ij}|\{\ddot{p}_j\} + |k_{ij}|\{p_j\} = \{f_i\} \qquad (14)$$

where $\{p_j\}$ is the vector of fluid pressures at the nodal points. The matrix $\{f_i\}$ is the vector of generalized forces transmitted to the fluid and is given by the equation

$$f_i = \int_{\partial s} KN_i(\frac{\partial N_i}{\partial x}N_x + \frac{\partial N_i}{\partial y}N_y + \frac{\partial N_i}{\partial z}N_z)ds \qquad (15)$$

An alternative to the preceeding formulation is given by the method of weighted residuals. The resulting equations governing the fluid pressure field are identical to those previously derived. The mathematics of the weighted residual formulation are, however, a bit more direct and we do not have to rely on the abstract identification of energy terms as was previously necessary.

Applying the method of weighted residuals involves basically two steps. The first step is to assume the general functional behavior of the dependent field variable so as to satisfy the boundary conditions and to approximately satisfy the governing differential equation. Substitution of this approximation into the governing differential equation and boundary conditions then results in some error called a residual. This residual is required to vanish in some average sense over the entire solution domain.

The second step is to solve the equation or equations resulting from the first step and thereby specialize the general functional form to a particular function, which then becomes the solution to the field problem.

Thus we want to find an approximate functional representation for the field variable p governed by the differential equation

$$K\nabla^2 p - \rho\frac{\partial^2 p}{\partial t^2} = L(p) = 0 \tag{16}$$

in the domain Ω bounded by the surface s and subject to the boundary conditions

$$\frac{\partial p}{\partial n} = \vec{\nabla} p \cdot \vec{n} = -\rho \ddot{\vec{s}}_n \tag{17}$$

on fluid-structure interfaces, and

$$\frac{\partial p}{\partial n} = 0$$

on the fluid free surfaces. In equation (17), $\vec{n}$ is the outward unit normal from the fluid-structure interface and $\ddot{\vec{s}}_n$ is the normal acceleration of the structure surface.

The unknown exact solution p is approximated by

$$\tilde{p} \simeq \sum_{i=1}^{m} N_i p_i \tag{18}$$

where the m functions N_i are chosen to satisfy the boundary conditions (17) and the m parameters p_i are unknown. Substitution of $\tilde{p}$ into equation (16) produces

$$L(\tilde{p}) = R \tag{19}$$

where R is the error or residual that results from approximating p by $\tilde{p}$. Minimization of R over the domain Ω is accomplished by forming a weighted average of the error which is made to vanish over the solution domain. The error distribution principle most often used to derive finite elements is Galerkin's method which requires that

$$\int_\Omega |L(\tilde{p})| N_i \, dxdydz = 0, \quad i = 1,2,\ldots,m \tag{20}$$

Since the equation (16) is valid at any point in the solution domain, it also holds for any collection of points defining an arbitrary element of the whole domain. Thus the functions N_i may be recognized as interpolation functions $N_i(e)$ defined over the domain of the element and the p_i are approximate nodal pressures.

Thus from Galerkin's method we can write the equations governing the behavior of an acoustic-fluid finite element as

$$\int_{\Omega^{(e)}} N_i^{(e)} \left| K\left(\frac{\partial^2 p^{(e)}}{\partial x^2} + \frac{\partial^2 p^{(e)}}{\partial y^2} + \frac{\partial^2 p^{(e)}}{\partial z^2}\right) - \rho \frac{\partial^2 p^{(e)}}{\partial t^2} \right| \, dxdydz^{(e)} = 0, \quad i = 1,2,\ldots,r \qquad (21)$$

where r is the number of nodes for the given element. We now write

$$p^{(e)} = |N| \{p\}^{(e)} \qquad (22)$$

Upon substitution of equation (22) into equation (21) and integrating the Laplacian terms by parts we obtain (dropping the (e) superscripts)

$$\int_{\partial s} K |N|^t \vec{\nabla} p . \vec{n} ds - \int_{\Omega} K\left(\frac{\partial N_i}{\partial x} \left|\frac{\partial N}{\partial x}\right| + \frac{\partial N_i}{\partial y} \left|\frac{\partial N}{\partial y}\right| + \frac{\partial N_i}{\partial z} \left|\frac{\partial N}{\partial z}\right|\right) \{p\} \, dxdydz - \int_{\Omega} \rho N_i |N| \{\ddot{p}\} \, dxdydz = 0 \qquad (23)$$

or, in matrix form,

$$|m_{ij}| \{\ddot{p}\} + |k_{ij}| \{p\} = \{f_i\} \qquad (24)$$

In writing equation (24) we have implied the following definitions.

$$m_{ij} = \int_{\Omega} \rho \, N_i N_j \, dxdydz \qquad (25)$$

$$k_{ij} = \int_{\Omega} K\left(\frac{\partial N_i}{\partial x} \frac{\partial N_j}{\partial x} + \frac{\partial N_i}{\partial y} \frac{\partial N_j}{\partial y} + \frac{\partial N_i}{\partial z} \frac{\partial N_j}{\partial z}\right) dxdydz \qquad (26)$$

$$f_i = \int_{\partial s} K N_i \left(\frac{\partial N_i}{\partial x} n_x + \frac{\partial N_i}{\partial y} n_y + \frac{\partial N_i}{\partial z} n_z\right) ds \qquad (27)$$

We observe that this is the same result as obtained previously in equations (11), (13), and (15).

3. DISPLACEMENT-DEPENDENT FINITE ELEMENT FORMULATION

We now proceed to derive the finite element equations of fluid motion for acoustic fluid elements based on the displacement formulation, equations (11) of Part I. The method of weighted residuals using Galerkin's criterion is applied to the boundary value problem defined by these equations, the kinematic boundary condition

$$\dot{\vec{h}} = \dot{\vec{s}}_n \quad \text{on } s \tag{28}$$

and the dynamic free surface boundary condition

$$\vec{\nabla}.\vec{h} = 0 \quad \text{on fluid free surfaces} \tag{29}$$

where $\dot{\vec{s}}_n$ is the normal velocity of the structure surface. We begin by writing the field displacements as

$$\{h\}^{(e)} = \begin{Bmatrix} h_x(x,y,z,t) \\ h_y(x,y,z,t) \\ h_z(x,y,z,t) \end{Bmatrix}^{(e)} = \begin{Bmatrix} \sum_1^r N_i(x,y,z)h_{xi}(t) \\ \sum_1^r N_i(x,y,z)h_{yi}(t) \\ \sum_1^r N_i(x,y,z)h_{zi}(t) \end{Bmatrix}^{(e)}$$

$$= \begin{Bmatrix} |N|\{h_x\}^{(e)} \\ |N|\{h_y\}^{(e)} \\ |N|\{h_z\}^{(e)} \end{Bmatrix} = \{|N|\{h^{(e)}\}\} \tag{30}$$

Equations (11) of Part I may be rewritten as

$$K\left(\frac{\partial^2 h_x}{\partial x^2} + \frac{\partial^2 h_x}{\partial x \partial y} + \frac{\partial^2 h_x}{\partial x \partial z}\right) - \rho\frac{\partial^2 h_x}{\partial t^2} = 0 \tag{31}$$

$$K\left(\frac{\partial^2 h_y}{\partial x \partial y} + \frac{\partial^2 h_y}{\partial y^2} + \frac{\partial^2 h_y}{\partial y \partial z}\right) - \rho\frac{\partial^2 h_y}{\partial t^2} = 0 \tag{32}$$

$$K\left(\frac{\partial^2 h_z}{\partial x \partial z} + \frac{\partial^2 h_z}{\partial y \partial z} + \frac{\partial^2 h_z}{\partial z^2}\right) - \rho\frac{\partial^2 h_z}{\partial t^2} = 0 \tag{33}$$

Applying the method of weighted residuals using Galerkin's criterion to equation (31) yields (dropping the (e) superscripts)

$$\int_\Omega N_i \left| K\left(\frac{\partial^2 h_x}{\partial x^2} + \frac{\partial^2 h_x}{\partial x \partial y} + \frac{\partial^2 h_x}{\partial x \partial z}\right) - \rho \frac{\partial^2 h_x}{\partial t^2} \right| dxdydz = 0 \tag{34}$$

Consider, for a moment, the integral of spacial derivatives in (34),

$$\int_\Omega N_i \left(\frac{\partial^2 h_x}{\partial x^2} + \frac{\partial^2 h_x}{\partial x \partial y} + \frac{\partial^2 h_x}{\partial x \partial z}\right) dxdydz$$

$$= \int_\Omega N_i \frac{\partial}{\partial x}\left(\frac{\partial h_x}{\partial x} + \frac{\partial h_x}{\partial y} + \frac{\partial h_x}{\partial z}\right) dxdydz \tag{35}$$

We now substitute for h_x from equations (30) and integrate (35) by parts to obtain

$$\int_{\partial s} N_i \left(\left|\frac{\partial N}{\partial x}\right| + \left|\frac{\partial N}{\partial y}\right| + \left|\frac{\partial N}{\partial z}\right|\right)\{h_x\} n_x \, ds$$

$$- \int_\Omega \frac{\partial N_i}{\partial x}\left(\left|\frac{\partial N}{\partial x}\right| + \left|\frac{\partial N}{\partial y}\right| + \left|\frac{\partial N}{\partial z}\right|\right)\{h_x\} \, dxdydz \tag{36}$$

In a similar manner, after applying the method of weighted residuals with Galerkin's criterion to equations (32) and (33), we obtain the results

$$\int_{\partial s} N_i \left(\left|\frac{\partial N}{\partial x}\right| + \left|\frac{\partial N}{\partial y}\right| + \left|\frac{\partial N}{\partial z}\right|\right)\{h_y\} n_y \, ds$$

$$- \int_\Omega \frac{\partial N_i}{\partial y}\left(\left|\frac{\partial N}{\partial x}\right| + \left|\frac{\partial N}{\partial y}\right| + \left|\frac{\partial N}{\partial z}\right|\right)\{h_y\} dxdydz \tag{37}$$

and

$$\int_{\partial s} N_i \left(\left|\frac{\partial N}{\partial x}\right| + \left|\frac{\partial N}{\partial y}\right| + \left|\frac{\partial N}{\partial z}\right|\right)\{h_z\} n_z \, ds$$

$$- \int_\Omega \frac{\partial N_i}{\partial z}\left(\left|\frac{\partial N}{\partial x}\right| + \left|\frac{\partial N}{\partial y}\right| + \left|\frac{\partial N}{\partial z}\right|\right)\{h_z\} dxdydz \tag{38}$$

Substituting the results (36), (37), and (38) into equation (34) and the corresponding expressions derived from equations (32) and (33), respectively, yields the following result.

$$\int_{\partial s} K N_i \left(\left|\frac{\partial N}{\partial x}\right| + \left|\frac{\partial N}{\partial y}\right| + \left|\frac{\partial N}{\partial z}\right| \right)\{h\}.n \, ds$$

$$- \int_{\Omega} K\left(\frac{\partial N_i}{\partial x} + \frac{\partial N_i}{\partial y} + \frac{\partial N_i}{\partial z}\right)\left(\left|\frac{\partial N}{\partial x}\right| + \left|\frac{\partial N}{\partial y}\right| + \left|\frac{\partial N}{\partial z}\right| \right)\{h\}$$

$$dxdydz - \int_{\Omega} \rho \, N_i |N| \{\ddot{h}\} \, dxdydz = 0 \tag{39}$$

The first integral in equation (39) may be identified as the vector of generalized forces, $\{f_i\}$, upon substitution of the relation

$$\{p\}^{(e)} = K\vec{\nabla} \, . \, \{h\}^{(e)} = K|\vec{\nabla}N|\{h^{(e)}\} \tag{40}$$

We write

$$\{f_i\} = \int_{\partial s} N_i \, \{p_n\} ds \tag{41}$$

where $\{p_n\}$ are the generalized normal pressures on the surface of the element.

Furthermore, it can be shown that the second integral in equation (39) may be written as

$$|K_{ij}| = \int_{\Omega} |B_i|^t \, |C| \, |B_j| \, dxdydz \tag{42}$$

where

$$|B_i| = \begin{bmatrix} \frac{\partial N_i}{\partial x} & 0 & 0 \\ 0 & \frac{\partial N_i}{\partial y} & 0 \\ 0 & 0 & \frac{\partial N_i}{\partial z} \\ \frac{\partial N_i}{\partial y} & \frac{\partial N_i}{\partial x} & 0 \\ 0 & \frac{\partial N_i}{\partial z} & \frac{\partial N_i}{\partial y} \\ \frac{\partial N_i}{\partial z} & 0 & \frac{\partial N_i}{\partial x} \end{bmatrix} \tag{43}$$

and

$$|C| = K \begin{bmatrix} 1 & 1 & 1 & 0 & 0 & 0 \\ 1 & 1 & 1 & 0 & 0 & 0 \\ 1 & 1 & 1 & 0 & 0 & 0 \\ 0 & 0 & 0 & 0 & 0 & 0 \\ 0 & 0 & 0 & 0 & 0 & 0 \\ 0 & 0 & 0 & 0 & 0 & 0 \end{bmatrix} \tag{44}$$

Defining

$$|m_{ij}| = \int_\Omega \rho \, N_i N_j \; dxdydz \tag{45}$$

allows us, finally, to write the simultaneous equations (31) to (33) in the form

$$|m_{ij}| \{\ddot{h}\} + |k_{ij}| \{h\} = \{f_i\} \tag{46}$$

Cast in this form we can see that if we replace the constitutive matrix

$$|C| = \frac{E}{(1+\nu)(1-2\nu)} \; \cdot$$

$$\cdot \begin{bmatrix} 1-\nu & \nu & \nu & 0 & 0 & 0 \\ & 1-\nu & \nu & 0 & 0 & 0 \\ & & 1-\nu & 0 & 0 & 0 \\ & & & (1-2\nu)/2 & 0 & 0 \\ & & & & (1-2\nu)/2 & 0 \\ \text{Sym.} & & & & & (1-2\nu)/2 \end{bmatrix} \tag{47}$$

of classic linear elasticity with that of equation (44), the classic linear elastic finite element formulation becomes equivalent to the displacement dependent, acoustic fluid finite element formulation.

4. AN APPLICATION AND CONCLUSIONS

The numerical method described in the previous sections is now applied for the analysis of a bridge pier. With the impoundment of water in the Karakaya Dam, currently under construction in Turkey, a part of railway linking the cities of Malatya and Elazığ will be inundated. A new route with an approximately 2.1 km long crossing over the reservoir has been planned. The earthquake analysis of this multispan railway bridge on the Euphrates River has already been reported (Gülkan et al., 1982). Determination of the design earthquake ground motion from seismic hazard and geotechnical studies and derivation of the equivalent static force spectrum for design purposes have been presented in the mentioned reference . The emphasis in this work will be on the dynamic response analysis of the bridge piers including hydrodynamic effects. In the numerical analysis of the problem a computer program coded for earthquake response analysis of axisymmetric tower structures surrounded by water is used (Liaw and Chopra, 1973). The program is based on a pressure-dependent finite element formulation and the fluid is assumed to be incompressible.

A typical tall pier elevation and the pier cross section are illustrated in Fig.1. It is obvious that under operational conditions water will seep into the center of the box section, so partholes will be left in the walls to facilitate this action. Thus, it becomes necessary to consider the water in the box section in the finite element modelling.

As seen in Fig. 1 the pier cross section is of a box type. On the other hand, the computer program utilized is for axisymmetric structures only. It is necessary, therefore, to transform the actual cross section to an equivalent circular section (Fig.2). This can be achieved by requiring that the equivalent circular section have cross sectional area and moment of inertia equal to those of the actual section. It should be noted that the actual section has two different moments of inertia about its principal axes. We will have, indeed, two different equivalent circular sections. However, in the present work, the results for one of the equivalent sections only will be given. The actual pier, its equivalent model and the corresponding finite element model are shown in Fig. 3.

Figure 4 illustrates the first three mode shapes for the pier with and without water. The corresponding frequencies are also given in the same figure. It is noted that each one of the frequencies decreases when the effect of the water is taken into account.

A probability study of past seismicity, the known fault structures in the vicinity of the site and the attenuation characteristics of earthquakes in Turkey led to the establishment of 0.16g as the effective peak ground acceleration on firm ground for the present problem. It was also decided that the Taft record of 1952 and the El-Centro record of 1940 modified with respect to the appropriate peak acceleration should be employed in the studies of the superstructure and its interaction with the fluid medium. The use of two different earthquake records allows one to study the effect of the frequency content of an earthquake on the response of the system. In Fig. 5 the north-south component of the El-Centro record of 1940 which is modified for 0.16g is given. The Taft record is not shown but it is also modified for 0.16g.

The maximum responses of the pier model subjected to the modified Taft earthquake and to the modified El-Centro earthquake are illustrated in Figs. 6a and 6b, respectively. In these figures, maximum displacement, maximum bending stress and maximum shear stress are plotted along the length of the pier for both the *in vacuo* and submerged cases. The effect of the fluid interaction is clearly seen in these figures. For all the cases considered, the presence of the fluid amplifies the response of the structure. The maximum responses resulting from the modified El-Centro record are generally higher than those resulting from the modified Taft record.

The displacement at the top and the moment at the base of the pier subjected to the modified El-Centro record are given in Figs. 7 and 8, respectively, as functions of time. The corresponding responses for the modified Taft record are given in Fig. 9 and 10. The amplifying effect of the presence of water is clearly noted in these figures. Respective comparisons of Figs. 7 and 9 and Figs. 8 and 10 reveal the effect of the frequency content of the earthquake record on the structural response.

The numerical results given in this section are for a specific pier model only. The results indicate that the presence of water interacting with the structure affects the response of the latter significantly. The maximum responses are, in some cases, almost doubled. However, it should be emphasized that these conclusions are valid for the specific pier considered only. For another pier with different cross sectional dimensions and/or height the effect of water may turn out to be not so much significant. Thus it is suggested that each case be studied on its own and the conclusions thereof should not be generalized.

ACKNOWLEDGMENTS

The work described in this paper was performed in conjunction with consultancy services provided to Yüksel Proje A.Ş., architect-engineers for the project. The encouragement of Mr. Erdal Erk'akan, Project Manager, is appreciated. The sections on the finite element formulation are essentially based on a report by Mr. J.G. Bennett of the University of Michigan submitted to N.Akkaş as part of the requirements for a graduate course.

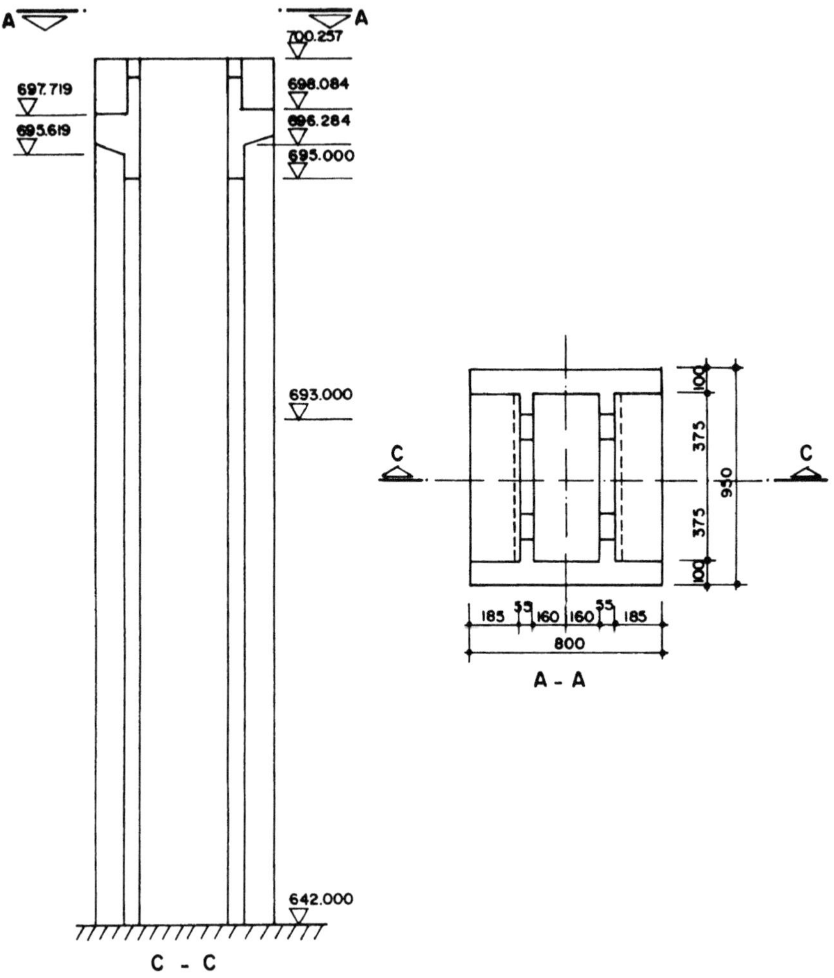

FIG. 1 ELEVATION AND CROSS SECTION OF THE PIER

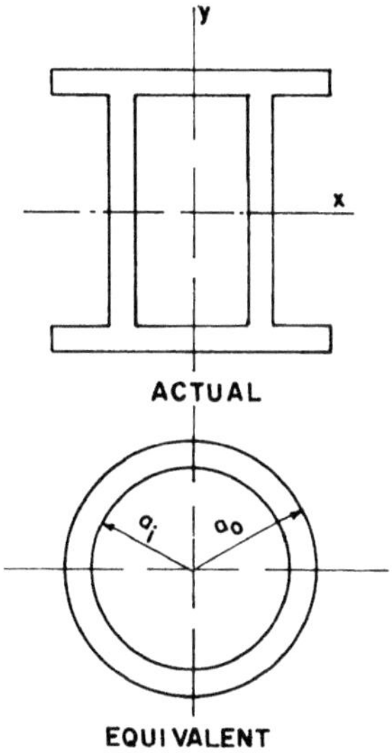

FIG. 2 ACTUAL AND EQUIVALENT CROSS SECTIONS

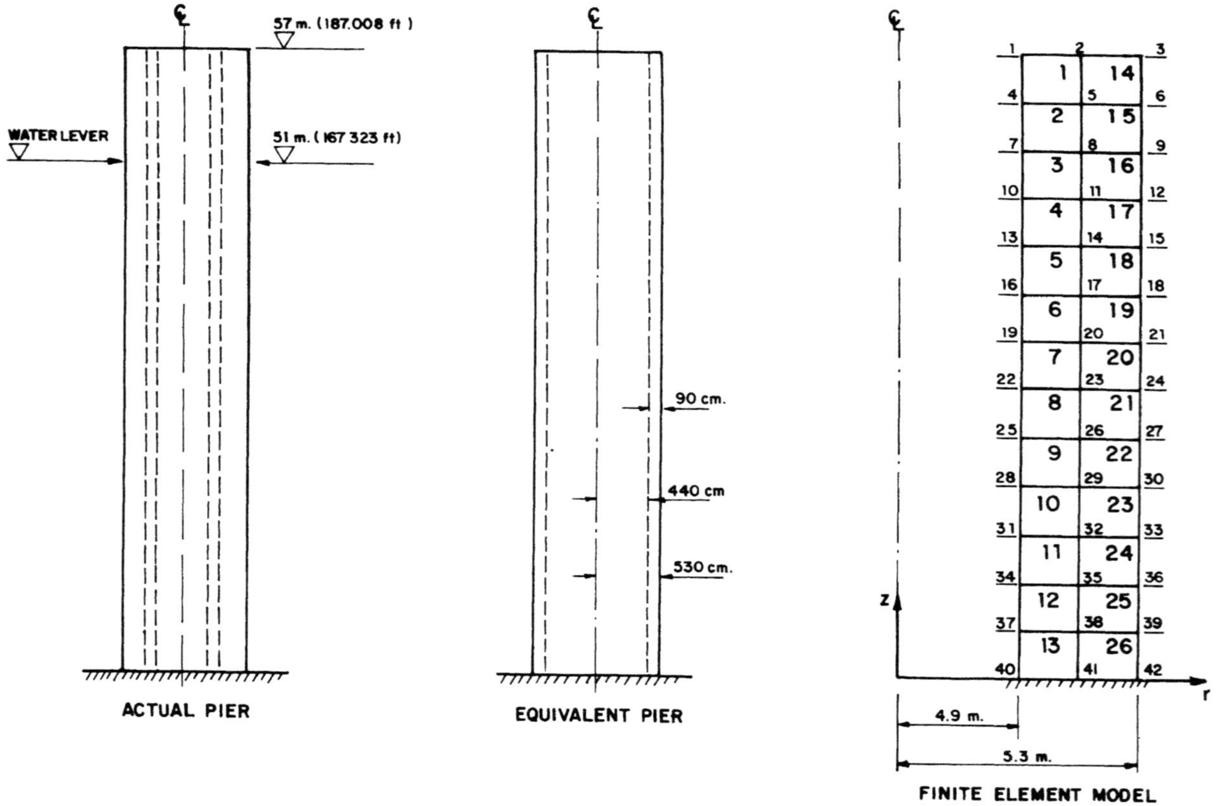

FIG. 3 ACTUAL AND EQUIVALENT PIERS AND THE FINITE ELEMENT MODEL

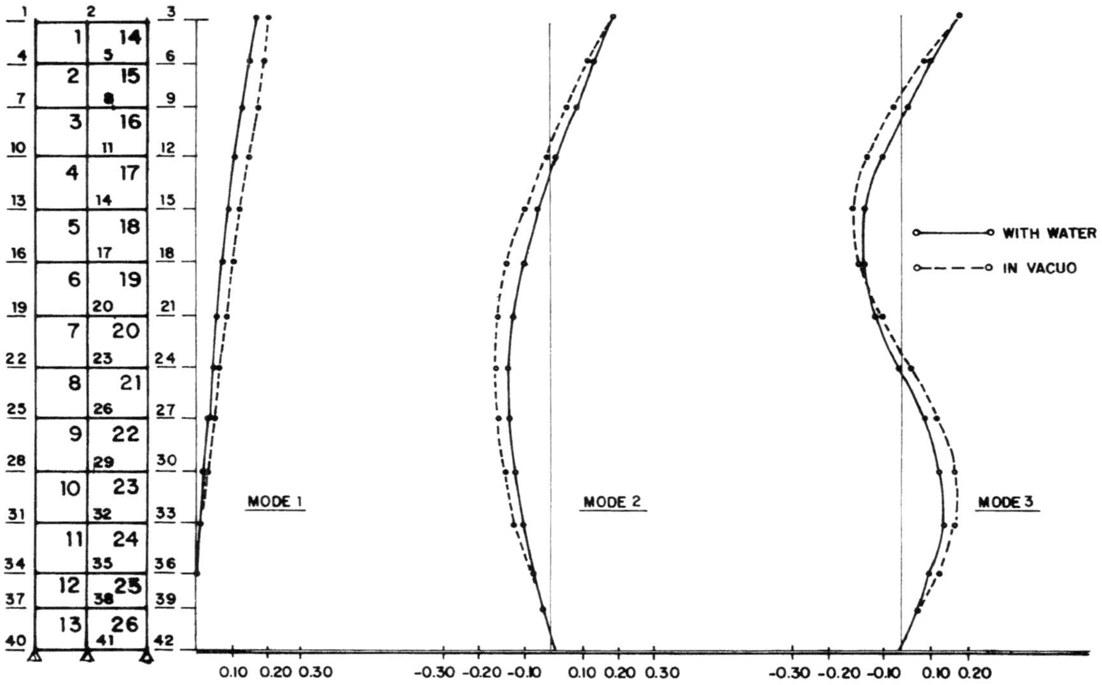

CASE	WATER LEVEL		FREQUENCY (CPS)		
	OUTSIDE	INSIDE	MODE 1	MODE 2	MODE 3
1	ZERO	ZERO	10.03	51.40	118.40
2	NORMAL	NORMAL	7.98	40.28	91.54

FIG. 4 MODE SHAPES AND FREQUENCIES

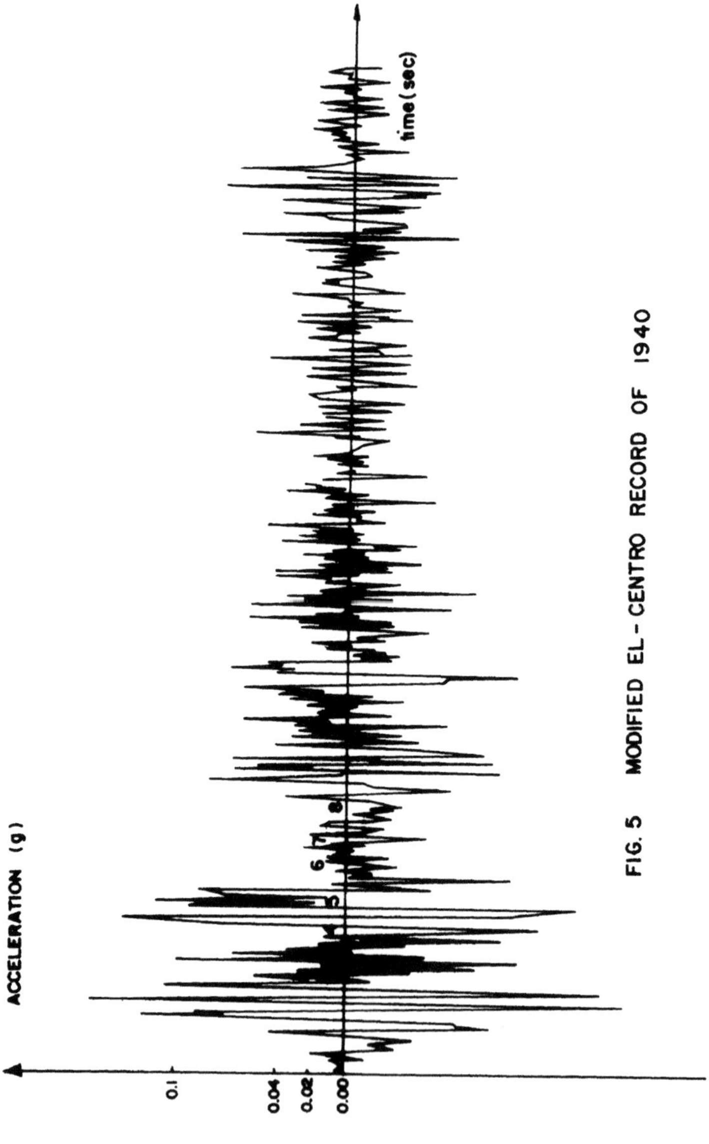

FIG. 5 MODIFIED EL - CENTRO RECORD OF 1940

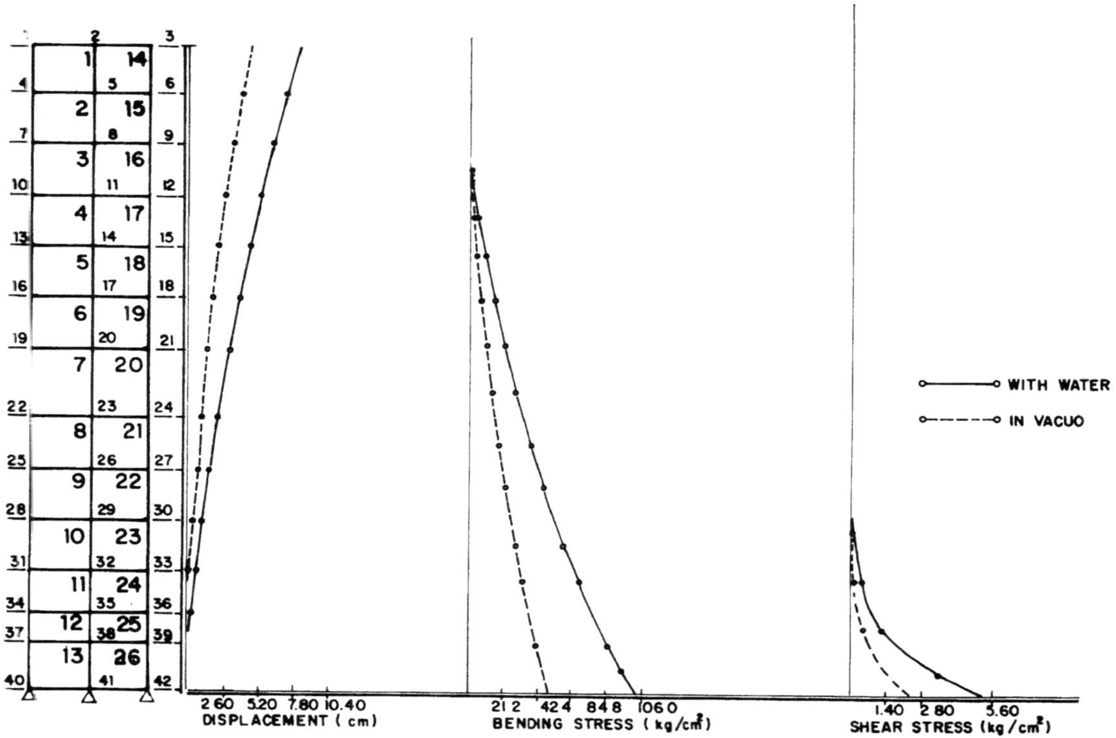

FIG. 6a MAXIMUM RESPONSES OF THE PIER TO THE MODIFIED TAFT RECORD

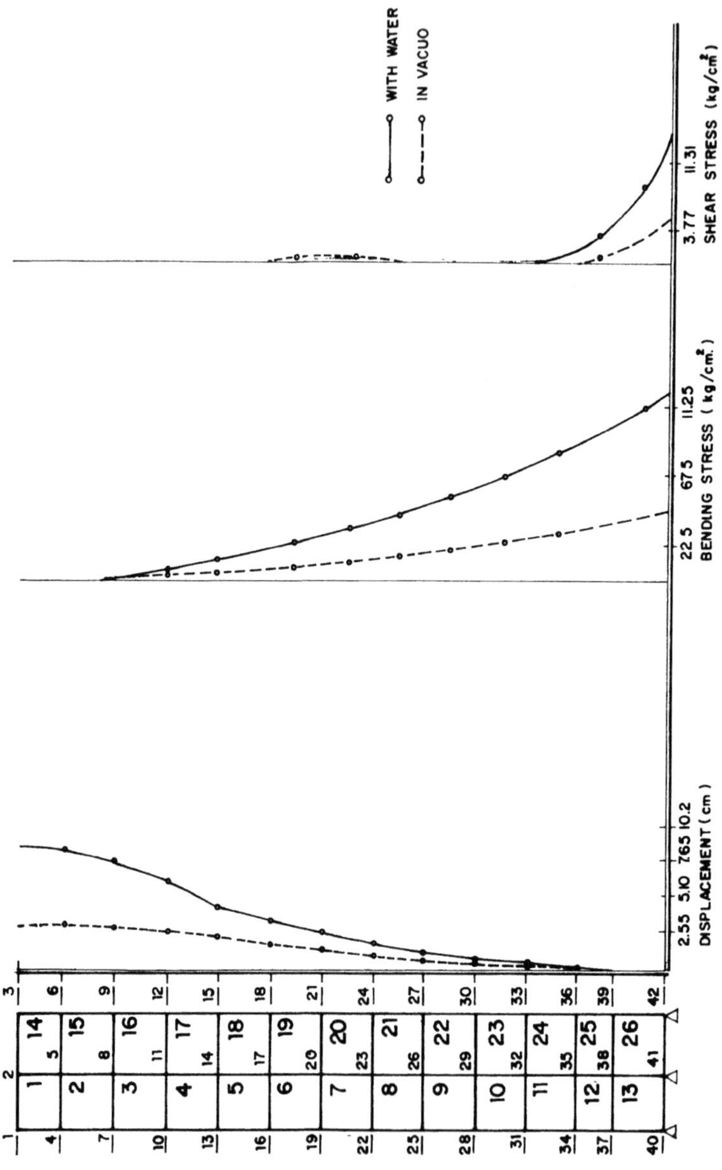

FIG. 6b MAXIMUM RESPONSES OF THE PIER TO THE MODIFIED EL-CENTRO RECORD

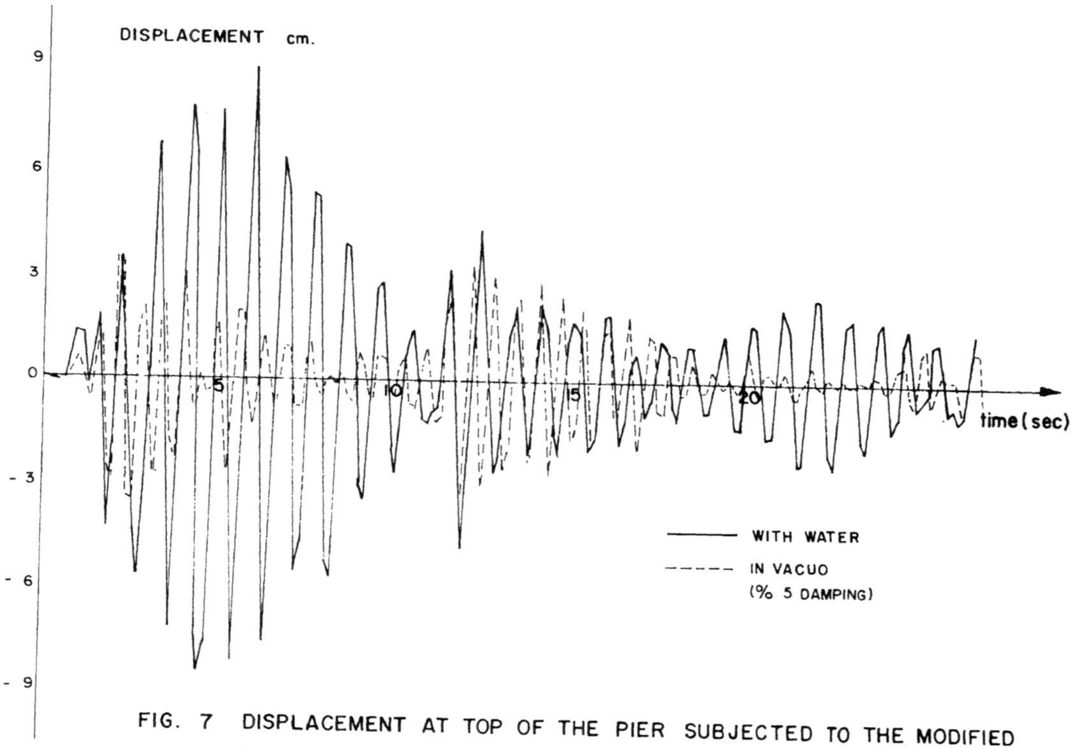

FIG. 7 DISPLACEMENT AT TOP OF THE PIER SUBJECTED TO THE MODIFIED EL-CENTRO RECORD

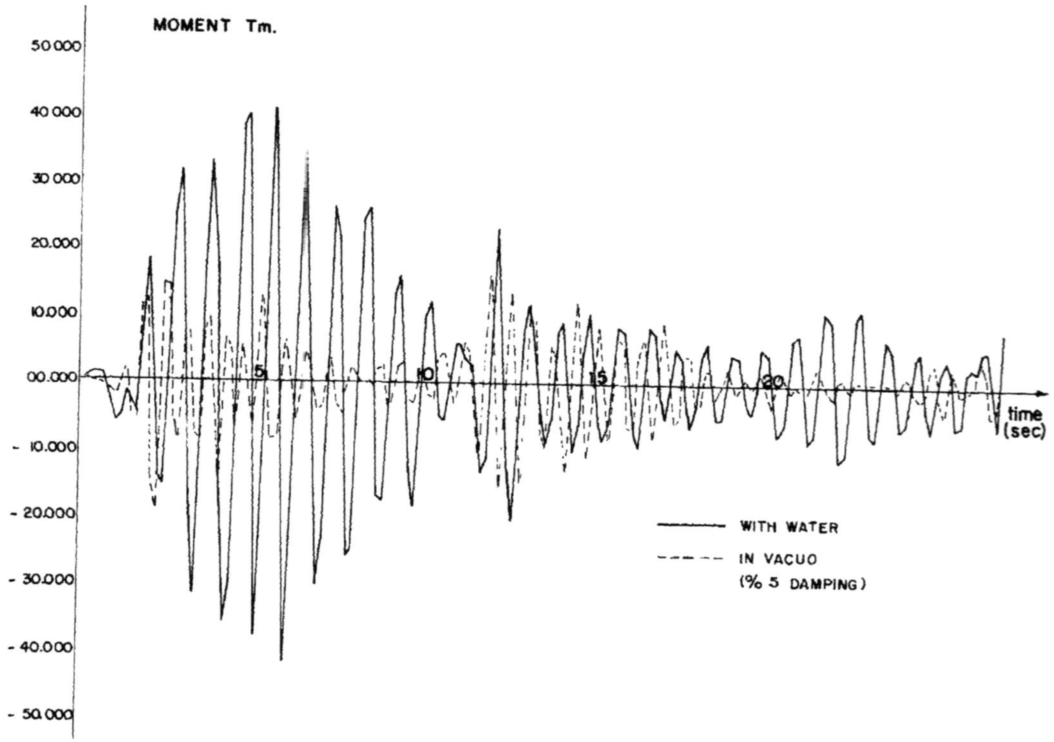

FIG. 8 MOMENT AT BASE OF THE PIER SUBJECTED TO THE MODIFIED EL-CENTRO RECORD

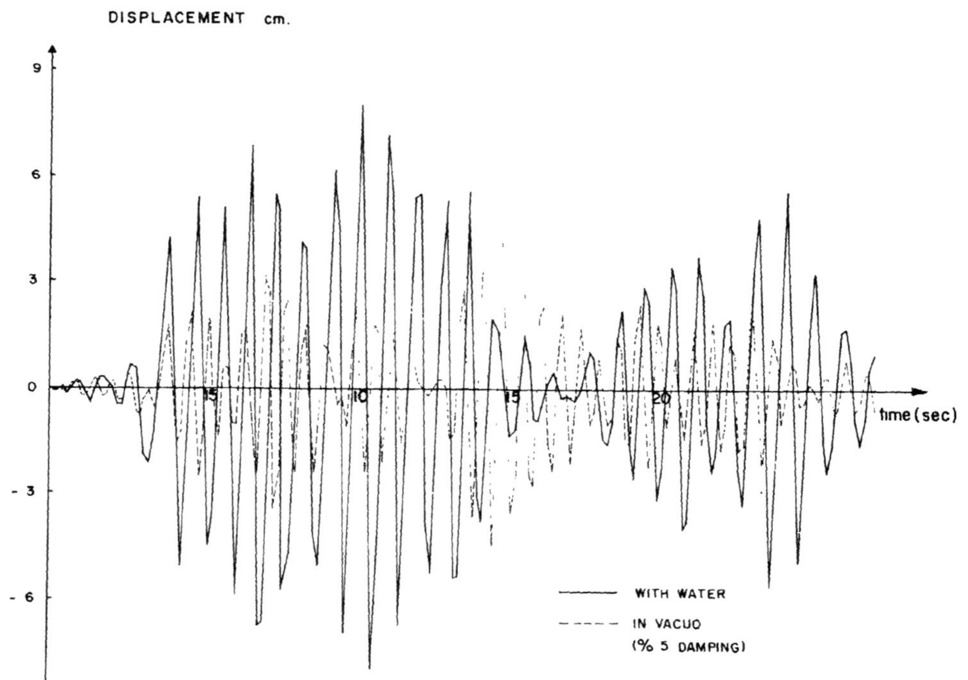

FIG. 9 DISPLACEMENT AT TOP OF THE PIER SUBJECTED TO THE MODIFIED TAFT RECORD

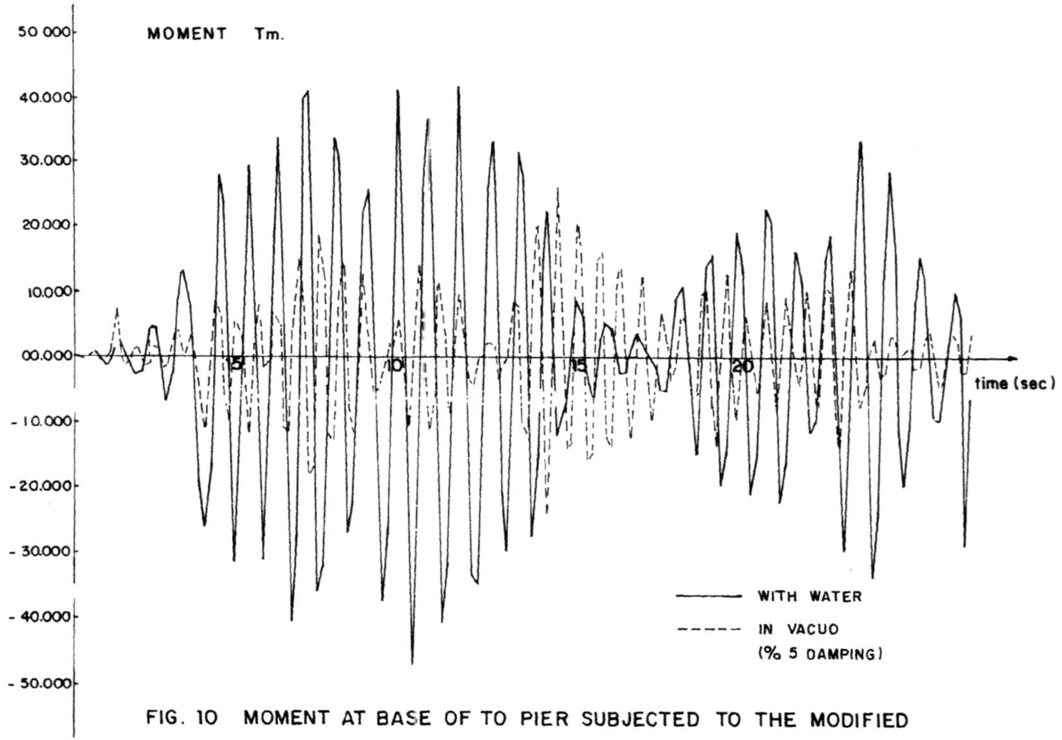

FIG. 10 MOMENT AT BASE OF TO PIER SUBJECTED TO THE MODIFIED TAFT RECORD

REFERENCES

1. Akay, H.U., Akkaş, N., and Yılmaz, Ç., "Analysis of Solid-Fluid Interaction Problems with SAPIV," Proc. Second Sap Users Conf., Paper No. 18, University of Southern California, 1977.

2. Akkaş, N., Akay, H.U., and Yılmaz, Ç., "Applicability of General-Purpose Finite Element Programs in Solid-Fluid Interaction Problems," Computers and Structures, Vol. 10, 1979, pp. 773-783.

3. Bathe, K.J., and Hahn, W.F., "On Transient Analysis of Fluid-Structure Systems," Computers and Structures, Vol. 10, 1979, pp. 383-391.

4. Bedrossian, Bedros and Di Maggio, Frank L., "Transient Response of Submerged Spheroidal Shells," Int. Journal Solids Structures, Vol. 8, 1972, pp. 111-129.

5. Bedrossian, Bedros and Di Maggio, Frank L., "Acoustic Approximations in Fluid-Shell Interactions," Journal Engr. Mech. Div., A.S.C.E., Vol. 98, No. EM3, June, 1972, pp. 731-743.

6. Clark, A.V., Jr., "A Study of Fluid-Structure Interaction and Decoupling Approximations," Naval Research Laboratory Report 7590, 1973.

7. Everstine, G.C., Schroeder, E.A., and Marcus, M.S., "The Dynamic Analysis of Submerged Structures," Nastran : Users' Experiences, NASA TM X-3278, National Space and Aeronautics Administration, 1975, pp. 419-429.

8. Geers, T.L., "Residual Potential and Approximate Methods for Three-Dimensional Fluid-Structure Interaction Problems," Journal of the Acoustical Society of America, Vol. 49, 1971, pp. 1505-1510.

9. Gülkan, P., Erdik, M., Akkaş, N., Yılmaz, Ç., and Öner, M., "Seismic Analysis of a Multispan Railway Bridge : A Case Study," Proc. 7th European Conference on Earthquake Engineering, September 20-25, 1982, Athens, Greece. Vol. 6, pp. 163-170.

10. Huebner, K.H., "The Finite Element Method for Engineers," John Wiley and Sons, Inc., 1975.

11. Hunt, J.T., Knittel, M.R., and Barach, D., "Finite Element Approach to Acoustic Radiation from Elastic Structures," Journal of the Acoustical Society of America, Vol. 55, 1974, pp. 269-280.

12. Liaw, C.Y., and Chopra, A.K., "Earthquake Response of Axisymmetric Tower Structures Surrounded by Water," EERC 73-25, University of California, Berkeley, 1973.

13. Marcus, M.S., "A Finite Element Method Applied to the Vibration of Submerged Plates," Journal of Ship Research, Vol. 22, No. 2, June 1978, pp. 94-99.

14. Nahavandi, A.N., Bomm, G.J., and Pedrido, R.R., "Structurally Compatible Fluid Finite Element for Solid-Fluid Interaction Studies," Nucl. Engng. Design, Vol. 35, 1975, pp. 335-347.

15. "The NASTRAN Theoretical Manual," NASA SP-222(03), National Aeronautics and Space Administration, Washington, D.C., 1976.

16. Schroeder, E.A., and Marcus, M.S., "Natural Frequencies of a Submerged Ring," CMD Report 27-74, David W. Taylor Naval Ship Research and Development Center, 1974.

17. Schroeder, E.A., and Marcus, M.S., "Finite Element Solution of Fluid-Structure Interaction Problems," David W. Taylor Naval Ship Research and Development Center Report 76-0145, 1976.

18. Zarda, P.R., "A Finite Element-Analytical Method for Modeling a Structure in an Infinite Fluid," Nastran : Users' Experiences, NASA TM X-3278, National Aeronautics and Space Administration, 1976, pp. 251-272.

19. Zienkiewicz, O.C., and Newton, R.E., "Coupled Vibrations of a Structure Submerged in a Compressible Fluid," Proceedings, International Symposium on Finite Element Techniques, Stuttgart, 1969, pp. 359-379.

VIBRATIONS AND DYNAMIC BEHAVIOUR OF ACTUAL BRIDGES

E. ZELLER

Associate Professor
Department of Civil Engineering - Institut Universitaire de Technologie - Université STRASBOURG-III (FRANCE)

INTRODUCTION

Bridges have been built since ancient times. As crossing and linking needs increased and developed, bridges have multiplied and improved.

In the beginning, the know-how was closely related to previous experience. From the point-of-view of materials used, as well as the knowledge of structural behaviour and an understanding of building techniques, bridges sum up the science and know-how of an era.

If we observe the evolution of the most important structures in history, we will be able to follow up the series of progressive innovations and improvements. Moreover, it can clearly be seen that those bridges nowadays "old", were modern and even ahead of their time since the most advanced scientific and technical elements had to be worked out and then assembled.

Nowadays, industry regularly provides high specification materials, and present technology supplies powerful means for a quick setting up and construction. But undoubtedly the most outstanding progress in the last twenty years is due to the elaboration of mathematical modelling of bridges, thus allowing better analysis.

If the hand solution procedures have permitted predicting fairly accurately the deflection and the stresses of structures under static or quite static loads, when it comes to deal with complex structures or carry out a dynamic analysis, the long numerical calculations required are only feasible by computer.

While the Finite Element Method and the processing power of computers were being developed, giving way to a real structural language, new measuring techniques have appeared which, together, by means of data acquisition and processing, enable us to closely determine the real structural behaviour of bridges and other structures.

If a bridge is subjected to a dynamic excitation load, the response varies with the time ; the practical vibration analysis must take into account all mechanical properties of the system.

Generally the designer is most interested in obtaining the maxima of the response data : stress and deflection, in order to chose the appropriate section and rigidity. Nevertheless it is also interesting to note that the dynamic problem does not have a single solution which can be approximated in the unique term of the theoretical maxima but all the elements of the response history are of interest.

Even if the objectives of a designer differ from those of a mechanical engineer, many aspects of the dynamic mechanism of a bridge are of direct interest to him. In fact the assessment of vibrational behavior and durability of bridges is the result of 2 types of investigation :

- observation and measuring of dynamic characteristics of bridges in service or subjected to dynamic tests.
- in laboratory simulated tests on bridge elements or material.

Due to the extent of the subject of investigation - large variety of structures, duration of a measuring period, cost of material... - the present results go beyond the possibilities of an engineer of the Civil Engineering Department of the University of Strasbourg. Thus the author could count on the experience and cooperation of some researchers, responsible for tne advanced investigation carried out in the different Laboratories and Institutes.

MM.De Back J, Kolstein, M.H, Van Leeuwen J.L.
University of Technologie, DELFT (Netherlands),
Lewin Institut.

MM.Caracilli J., Theuillout
Laboratoire Central des Ponts et Chaussées, LCPC,
PARIS (FRANCE)

Mr Cochet D. and Guillard Y.
Laboratóire Régional des Ponts et Chaussées,
LCPC, STRASBOURG (FRANCE)

Mr Gregoire R.
Centre d'Etudes Techniques de l'Industrie Mécanique,
CETIM, SENLIS (FRANCE)

Mr Lieurade H.P.
Institut de Recherche sur la Sidérurgie,
IRSID, St-Germain/Paris (FRANCE)

Mr Matsumoto Y.
University of Tokyo, Japan

Mr Pfeiffer M.R.
Frauenhofer Institut für Betriebsfestigkeit
LBF, DARMSTADT (GERMANY)

Mr Van Koten H.
Lab. for dynamics of structures
TNO, DELFT (Netherland).

CONTENTS

V. Simulation and modelling
   1. Choice of mechanical properties of bearings in order to improve the safety of a boxgirder bridge subjected to seismic vibrations
      a) mechanical characteristics of bearings
      b) choice of elastomer bearings in order to improve the mechanical behaviour
   2. Effect of profile imperfection of the bridge subjected to moving loads

CONCLUSIONS

I. METHOD OF MEASUREMENT AND ANALYSIS PROCESS

1. Dynamic test on actual bridges

A vertical force sufficient to produce meaningful amplitudes in bending can be applied in the central parts and in the symmetrical plan. According to the needs, the excitation force must be perfectly well-known, so the measurement of the response characteristics produces the transfer function of the mechanical system.

Otherwise, it will be sufficient to generate a movement by any means available, and then analyse the natural evolution (amplitudes, modes, frequencies, damping). In case of a random excitation due to road or railway traffic or wind, the analysis can only be carried out in statistical terms. The range of dynamic tests for the acquisition of the mechanical characteristics is shown in figure 1.

2. Analysis process of measuring data

For all measurements on actual structures, the apparatus and techniques for the acquisition and processing of data have to be made as compatible as possible. The complete equipment must be able to operate precisely all along the acquisition and treatment process. Gauges and transducers, data recording (including analog digital conversion) data analyses (including editing or automatic plotting) equipment provided with numerous options are now available.

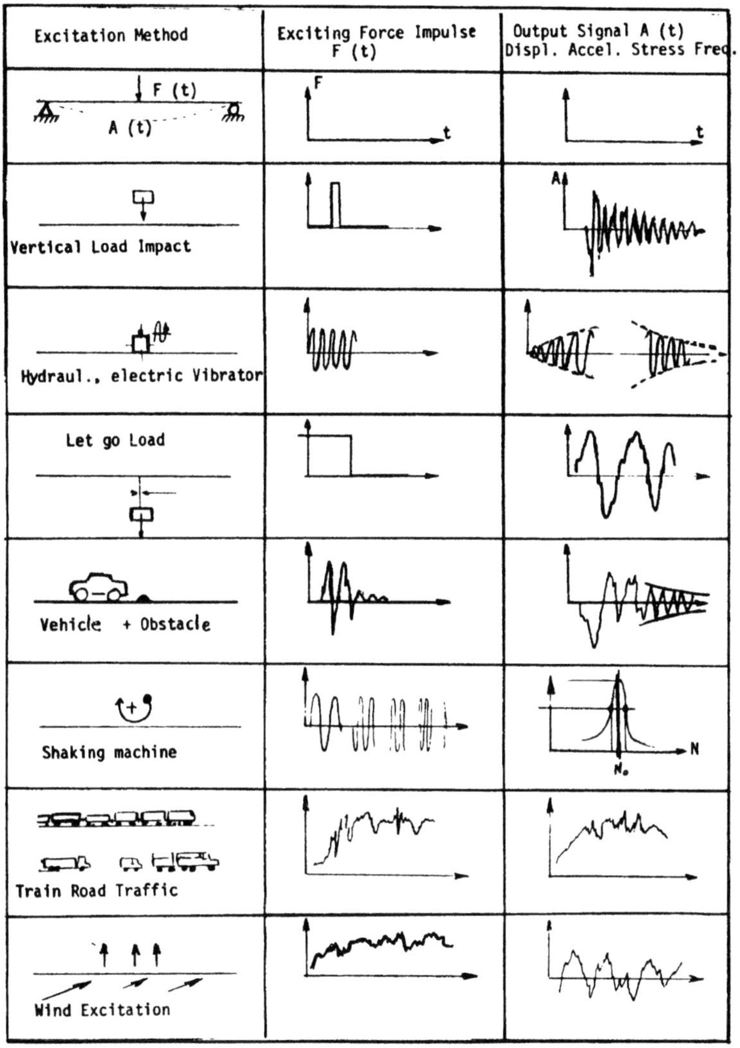

Excitation Method
Exciting Force Impulse F (t)
Output Signal A (t) Displ. Accel. Stress Freq.
F (t)
A (t)
F
t
t
Vertical Load Impact
A
Hydraul., electric Vibrator
Let go Load
Vehicle + Obstacle
Shaking machine
N
N₀
Train Road Traffic
Wind Excitation

Fig. 1

Position of the point of application of the excitation force

The position of the exciting force is a priori independent of the obtained characteristics, as long as it remains in the vertical symmetrical plan. Nevertheless, the further the spacing from the vibration knots of the corresponding mode, the bigger the amplitude response. Moreover, if we are dealing with a fixed frequency sinusoidal force, the initial excited mode generally degenerates quite fast to give way to all the other modes. However, when the system is excited in its central parts with a large energy the fundamental mode appears firstly with minute precision.

Position of the measuring gauges

In order to detect with precision the vibrating characteristics, one has to keep away from the vibration knot of the modes to be analysed. A systematic precaution consists in moving the gauges during the successive measurements.

Main types of gauges

Different types of measuring gauges, which provide enough precision, are presently available and it is easy to express the signal in millivolts with transducers. Numerical transcription of all data guarantees a perfect storage of the recorded data. All analytical studies can be carried out later.

Torsional behaviour

Though generally less significant, torsional behaviour combines in fact with behaviour in longitudinal vibration. In order to reveal the torsional characteristics of a bridge, the excitation force must be applied off the axis of the bridge; otherwise, the simple presence of a relatively important load as an integral part of the bridge during the excitation process is sufficient to degenerate the pure vertical movement and will reveal its torsional components.

An interesting technique for torsional analysis consists in placing 2 movement gauges in A and B and then assembling them in opposition.

Method of measurements	Measuring apparatus position	Out put signal measured datas
	accelerometer seismometer	acceleration frequencies displacement
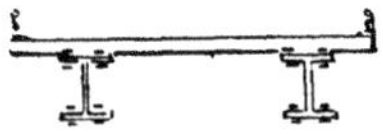	strain gages	stress strain frequencies moment
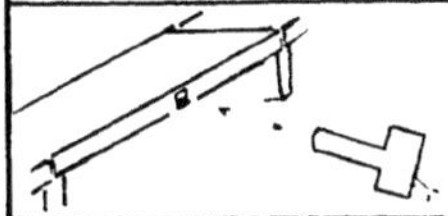	electronic theodolite "tracker"	displacement static + dynamic frequencies
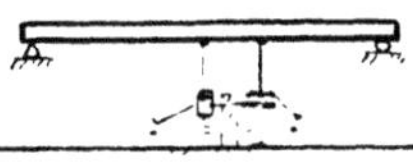	magnetic coil or capacity induction	displacement, frequencies static, dynamic
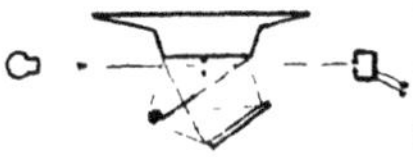	optical measure light laser... hologram	displacement frequencies deformation
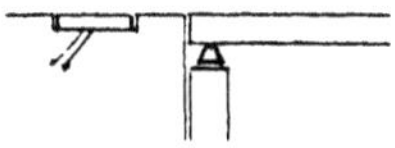	weighbridge strain gage quartz	axle load traffic statistics
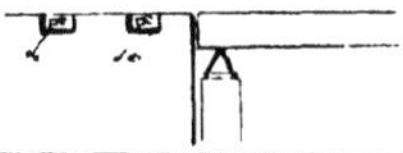	magnetic coils	Number of vehicles speed, intervals

Fig. 2

Fig. 3

Frequency analysis

For the treatment of vibrational characteristics different types of spectrum analysis (Narrow Band) are now popular. The common mathematical basis is the Fourier Transform. The discrete equivalent of Fourier's integral, called Discrete Fourier Transform D.F.T. can be written in two equations :

$$F(k) = \frac{1}{N} \sum_{n=0}^{n-1} f(n) \exp\left(-j \frac{2\pi nk}{N}\right) \qquad \text{forward transform}$$

$$f(x) = \sum_{k=0}^{N-1} F(k) \exp\left(j \frac{2\pi nk}{N}\right) \qquad \text{inverse transform}$$

The Fast Fourier Transform F.F.T. is more efficient in digital computing ; this algorithm reduces the number of complex multiplication from N^2 to $N \log_2 N$.

Analysis bandwidth T

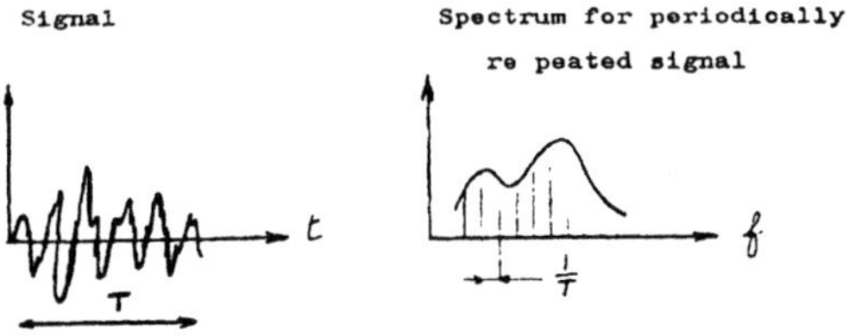

The analysis bandwidth must be greater than the Fourier line spacing $B > \frac{1}{T_{rep}}$

(T_{rep} = repetition time).

Fig. 4

A typical bandwidth choice would be approximately 1/5 of the bandwidth $B_{eff} = \frac{1}{T_E}$; T_E is the length of the pulse.

Nevertheless it is possible to increase T_{rep} to obtain a better likelihood of encountering problems with dynamic range and crest factor. The value of bandwidth B_{eff} can in most cases only be fixed by analysis of the data (If we gradually vary the bandwidth, a stage is reached where the results remain unchanged. Only if we want to express the results as an "energy spectral density" the bandwidth must be less than the value of the transient itself $B \leq B_{eff}$.

Window length and incremental step

It is always desirable to make the step somewhat shorter than T so that there is a certain amount of redundant information in the succession of analysis

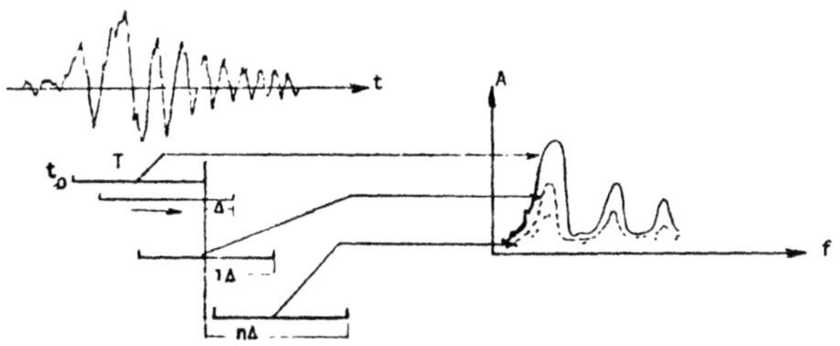

Fig. 5

Window type

The choice of the window type is also important. The Hanning function as well as the rectangular function are easily generated and give good results.

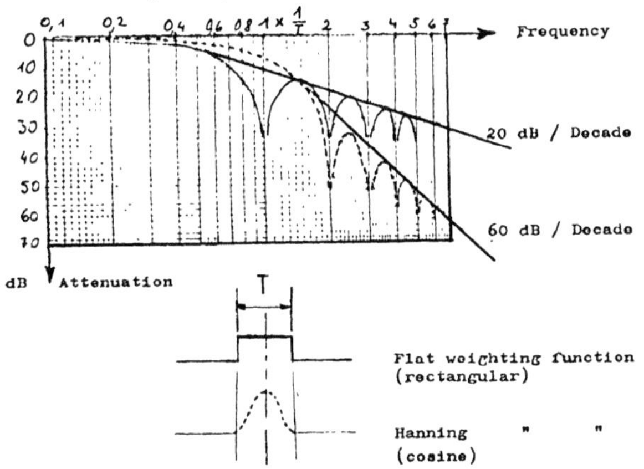

Comparison of the spectra of two common window functions (rectangular and Hanning)

Fig. 6

Advanced analysis and "Zoom FFT"

It has been a long time since the F F T became well known. The normal analysis is made from 0 to the Nyquist frequency F and the frequency lines up to F give an acceptable resolution. Howewer in some special cases of analysis it is necessary to have a finer resolution over a limited range of the spectrum. This fine resolution of the "Zoom F F T" is particulary useful for sharp resonances which are too closely spaced to be identified by normal analysis : Multiple assembled spans showing similar mechanical properties, deck elements presenting numerous harmonics, superposition of vibration modes at very close frequencies, flexion, torsion...

Signal and noise

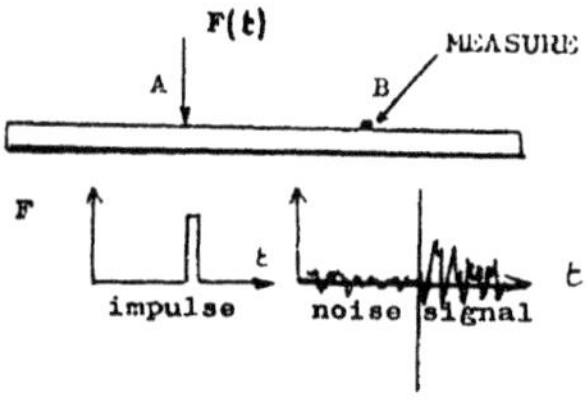

Fig. 7

A vertical impulse applied at a point A on a bridge structure gives at B : the dynamic characteristics of the structure (periodic) and the "noisiness" (random). Measurement procedures used to render the analysis clearer, and which guarantee that a mode, though not widely represented, will not be forgotten, consist in summing up the n sample data successively recorded.

If for 1 measure $\frac{\text{Signal}}{\text{Noise}} = a$, by addition of n response data (acceleration, stress...)

1° the periodical expected signal becomes more apparent

2° the average effect of the noise produces an attenuation

Consequently the above mentioned report becomes much more favorable.

$\frac{\text{Signal}}{\text{Noise}} = a\sqrt{n}$ and the accuracy of the parameter of interest is greatly improved.

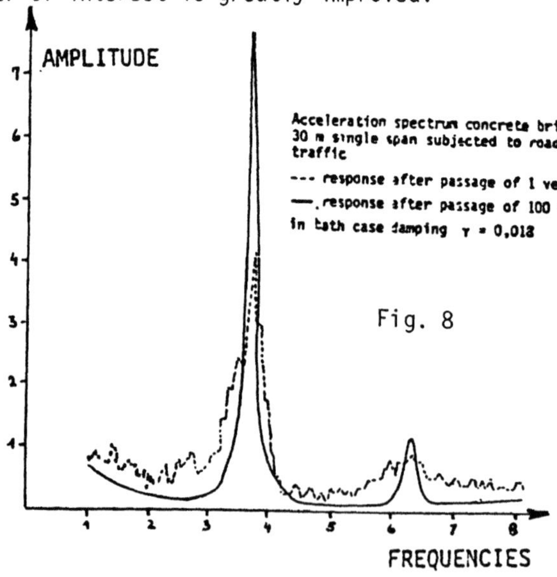

Fig. 8

3. Dynamic Response Components

Design specialists as well as mechanical engineers tend to interpret as clearly as possible the components of dynamic bridge behaviour. Different pictorial interpretations of Fourier Analysis greatly facilitate the comprehension of the relationship dynamic excitation response.

Spectral Density

The Forward Fourier Transform of a periodical signal becomes a continuous function of the frequency and in this case the spectral components have the dimension of spectral density. If we square the amplitudes we then represent the unit of energy per frequency unit ; the finite energy in a frequency range is obtained by integration of the corresponding power spectral surface. The power spectral density is specially adapted to stationary random function with a continuous spectrum and which has a statically constant and finite power. This is the case of the traffic loads. Various spectrum representations can be made.

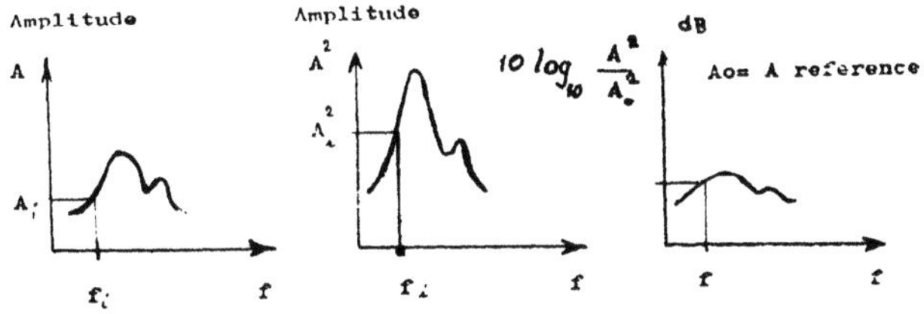

Fig. 9

Autocorrelation function, Convolution, Transfer function

This function is obtained by multiplying a function fx(t) by the function fx(t + τ) as follows :

$$Rxx\ (\tau) = \lim_{T \to \infty} \frac{1}{T} \int_{-T/2}^{T/2} f_x(t)\ f_x(t+\tau)dt$$

So Rxx (τ) gives a precise measure of the extent to which a signal fx (t) and a displaced version of itself fx (t + τ) correlate. If τ = 0 ; Rxx (0) gives us the mean square value of the function fx (t). The power spectrum is the Forward Fourier Transform of the autocorrelation function.

- Let us consider 2 time-functions f(t) and h(t). The convolution is defined as:

$$g\ (t) = \int_{-\infty}^{+\infty} f\ (\tau)\ h\ (t - \tau)\ d\tau$$

written in a symbolic form ($*$ ⟶ convolved with)

$$g\ (t) = f(t) * h\ (t)$$

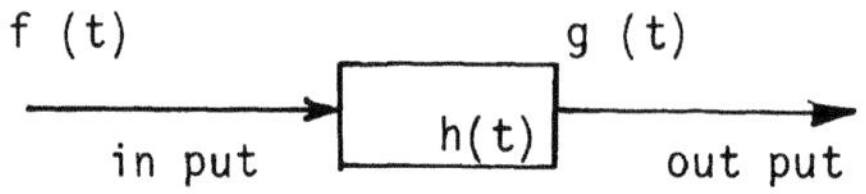

h (t) impulse response of the system f (t) can be considered as a succession of impulses (delta function) and each of them generates an impulse response with two caracteristics :

- the scaling is proportional to the level of f (t)
- the delay is constant, the time interval excitation-response

The complete response at time (t) is the sum of these impulse responses (integration over a dummy time variable τ).

- There is a parallelism between convolution in time and convolution in frequency domain .

If we call G (f) the Forward Fourier Transform of g (t) and respectively

F (f) of f (t) and H (f) of h (t) then

$$G\ (f) = F\ (f)\ .\ H\ (f)$$

The frequency spectrum of the output is obtained by multiplying the input spectrum by the transfer function.

Now let us consider :

. Fx (f) Fourier spectrum of the entry signal fx (t) applied to point x

. Fy (f) Fourier spectrum of the response signal fy (t) obtained at the point y.

The transform function from x to y is the complex ratio

$$Hxy\ (f) = \frac{Fy\ (f)}{Fx\ (f)}$$

by multiplying by Fx^* (f), the complex conjugate of Fx (t), numerator and denominator of the second member

$$Hxy\ (f) = \frac{Fy\ (f)}{Fx\ (f)} \quad x \quad \frac{Fx\ (f)}{Fx\ (f)} = \frac{Fxy\ (f)}{Fxx\ (f)}$$

$$Hxy\ (f) = \frac{\text{Cross spectrum between x and y}}{\text{Power spectrum at point x}}$$

Therefore the sample records fx (t) and fy (t) give us Fx (f) and Fy (f).

From these last two functions we can easily obtain the cross and power spectrum and also the transfer function (or frequency response).

In the case of traffic loads, wind changes, earthquake, vibrations, excitations are random ; the transfer function will be determining and this means that between input and output for a given frequency the phase angle is always the same. It should also be noticed that the transfer function applied to a given exciting random force can only provide the envelope of maximums and not all the intermediate steps so that the transfer function cannot be used directly for fatigue or prediction of the life-time of a structure.

II. DYNAMIC BEHAVIOUR OF BRIDGES SUBJECTED TO TRAFFIC

1. Simplified approach :

Until recently, bridges were heavier and had more limited spans, and at the same time the dynamic requirements to be taken into consideration were relatively less important.

The design methods were meant to deal with the structural problem in statistical terms, by using the so-called dynamic increase coefficient $\delta > 1$ or dynamic factor.

According to different design codes, simplified approaches were suggested, based on actual experience.These, at any rate, give bridges sufficient safety.

Considering 20 countries whose experience in bridge design is recognized, the dynamic coefficient diagrams according to the bridge span are located within a wide band shown in figure 10.

For example, for a road bridge (class B), in France, we have the following coefficient of dynamic action

$$\delta = 1 + \alpha + \beta$$

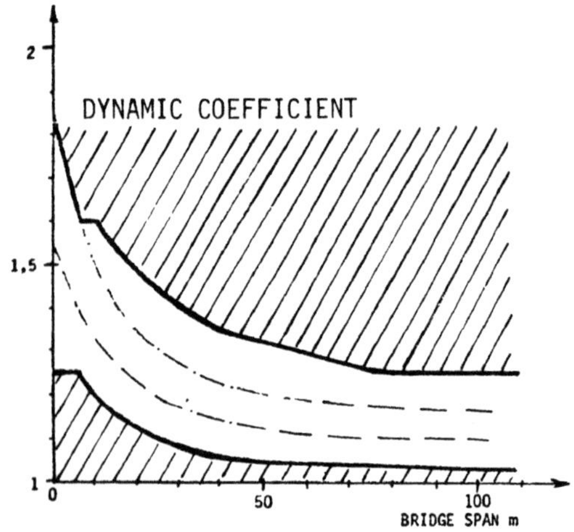

Fig. 10

α = a factor of elastic sensibility which depends on the length L of the element

$$\alpha = \frac{0,4}{1 + 0,2\ L}$$

β = factor similar to a mechanical admittance ;

S = dynamic load (vehicles) ;

G = dead load of the bridge

$$\beta = \frac{0,6}{1 + 4\ \frac{G}{S}}$$

2. Current research

Traffic conditions as well as higher stresses in material as lighter bridge decks lead to greater stress changes. Therefore the vibrational behavior and the risk of fatigue damage must be taken into account.

In order to study the relation between traffic load and stress in the different bridges components, the European Coal and Steel Community sponsored a large investigation program. Some interesting research on existing bridges has already been undertaken by various laboratories.

a) Road-traffic loads on a steel bridge

A single carriage way bridge with only two lanes, built with a simple structural system has been chosen and studied by Mr. J. DE BACK and MR. M.H. KOLSTEIN, Delft University, The Netherlands.

The bridge

A movable bascule bridge crossing a river has been chosen : "the Haagsche Schouw Bridge" on Highway 44 Den Haag -

Amsterdam.

The bridge structure is as follows :

A : two main steel girders
B : cross girders
C : longitudinal stringers
D : deck made of wooden boards
E : asphalt top layer

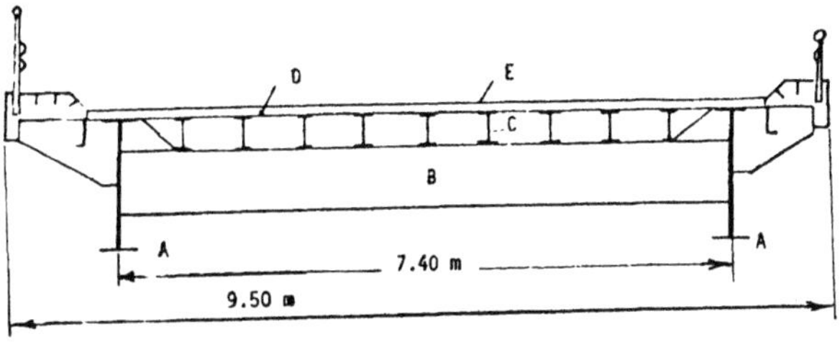

Fig. 11

Measuring experiment

Strain gages were applied to a cross girder (bottom flange) and to the stringer located under the traffic lanes.

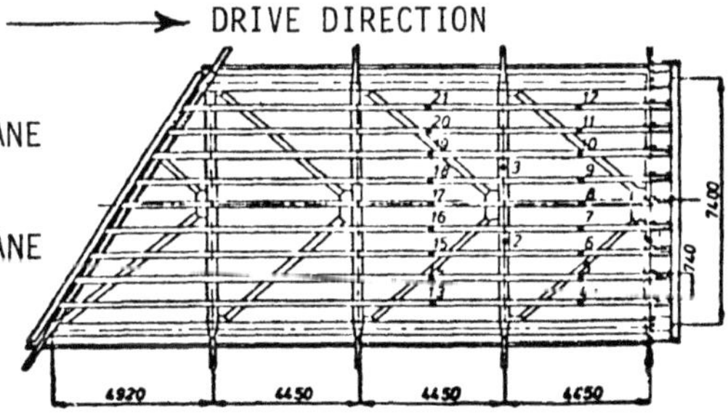

Fig. 12

Weightbridges and magnetic coils were installed a distance from the bridge in order to measure axle load, vehicle speed spacing, and to give all statistical traffic data.

All required data for the traffic load as well as the resulting loading effects are recorded and processed with a minicomputer. All the electronic equipment is situated in a measuring lorry in order to produce immediately 6 histograms of time function : Frequency distribution of amplitudes, level crossing counts.

Principal results :

We shall first see what happens when there is a single vehicle on the bridge, then in normal traffic conditions.

. A single vehicle :

- Determination of influence plans from static measurements :

The influence surface for each measuring point can be determined experimentally

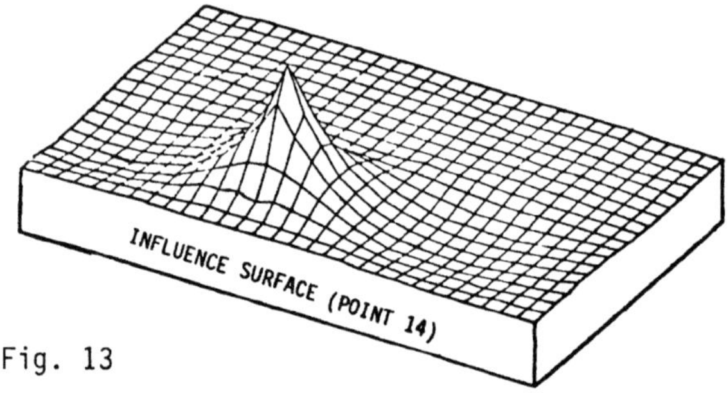

Fig. 13

A lorry with calibrated wheelloads was placed in different positions and the stress at the measuring point was recorded. The calculated stresses are about 30% higher than the measured values.

- Dynamic response of the bridge to moving vehicles
  - Moving vehicles transmit to the bridge-deck a local dynamic action which fluctuates around the static load. The direct impact action can be defined by an impact factor : maximum force between wheel and road divided by the corresponding static load ; the factor for trucks and trailers can vary from 1. 2 to 2.1
  - Moving vehicles generate vibrations in the bridge structure, and the resulting stress fluctuations at a measured point also depend on the general oscillation movement of the bridge even if the vehicle is far away.

This can be seen in the signal obtained by the passage of a truck and trailer over the bridge. This analogous recording enables us to clearly see the 4 maxima in spite of the superposition of higher frequency vibrations.

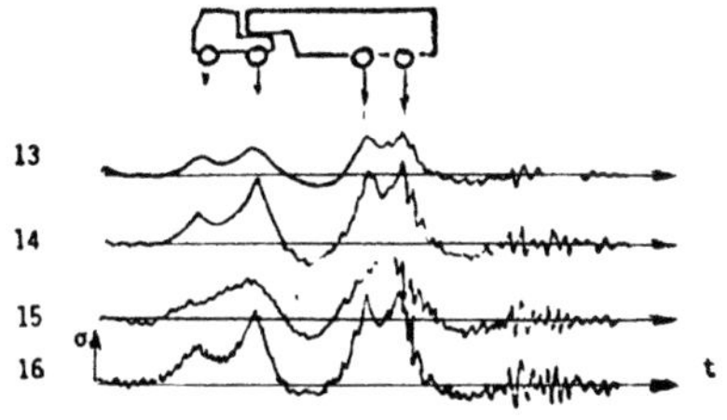

Fig. 14

Afterwards a more detailed analysis shows the influence of speed or lateral positioning of the vehicle. The ratio between the dynamic influence factor and the static influence factor reached 1.44.

.. Bridges subjected to all road traffic vehicles

The real problem in dynamic requirements of structures is the passage of vehicles at different speeds, lateral position, load, spacing.

Even though the axle-load of the traffic and stresses are measured simultaneously at any given moment, because of the random superposition principle, the mechanism of loading action-loading effects must be statistically studied. The vehicles composing the traffic sample have been grouped according to position and number of axles in 21 groups.

The main factor in the traffic statistics are vehicle group load, spacing, velocity and lateral position.

In regard to stress, the number of level crossings is of major importance. These measurements, carried out on the cross girder, clearly show that the slow lane bears the strongest stress.

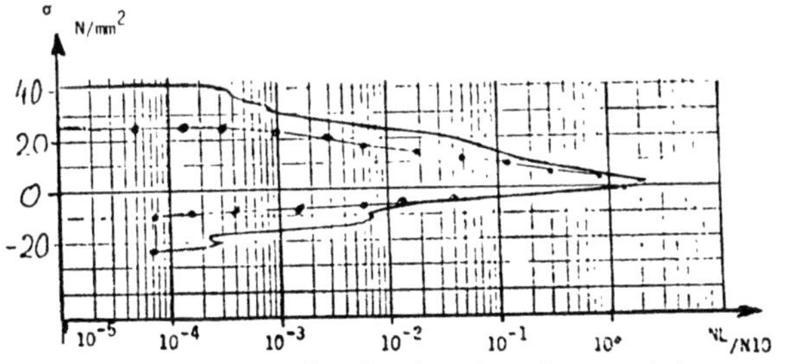

The number of levelcrosning of a given stress is measured at 2 points on the cross girder.

Fig. 15

The longitudinal stringers show very different stress values according to their lateral position.

Counting methods

For a given composition of traffic, the experimental stresses in the structure are analysed for a chosen time period. Two counting methods can be used in order to work out the stress-histogram. The classical levelcrossing method and the rainflow-counting method.

In the rainflow counting method, the collection of the stress variation with time (vertically represented) is similar to the different roof surfaces on which the rain flows. Each half fluctuation of the signal corresponds to a roof surface which leads the rainflow from a maximum to a minimum and inversely.

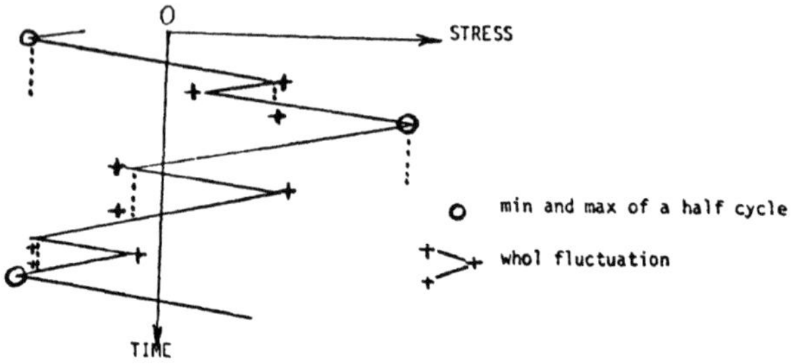

Fig. 16

In the rainflow counts, graphs are plotted in the range(s) in N/mm^2 against the number of oversteps of a range divided by the total sum of axle-loads greater than 10 KN.

Results and conclusions

Stress in material

Every passage of an axle-load produces a pulse with a maximum at a given measure point. The amplitude of this maximum is mainly proportional to the static axle-load. Different factors modify this equivalence.

Axle-load and stress : dynamic effect of the bridge suspension of the axle and profile imperfections of the road surface, lateral position of the vehicle, simultaneous action of other vehicles.

Computed level crossing and rainflow counts

For the design of steel bridges, in terms of fatigue, loading and durability, it is necessary to calculate the theoretical number of axle loads and from them to compute the stress frequency curve for a given period.

The influence lines associated with a measuring point and the axle-load count make it possible to establish mathematically model level-crossing and rainflow counts.

Comparison with the measured data

In order to test the above theoretical approach during a period of several hours, measurements of axle-loads and stresses have been made. The results of computed and measured level crossing and rain flow are in good agreement especially in the high stress cycles.

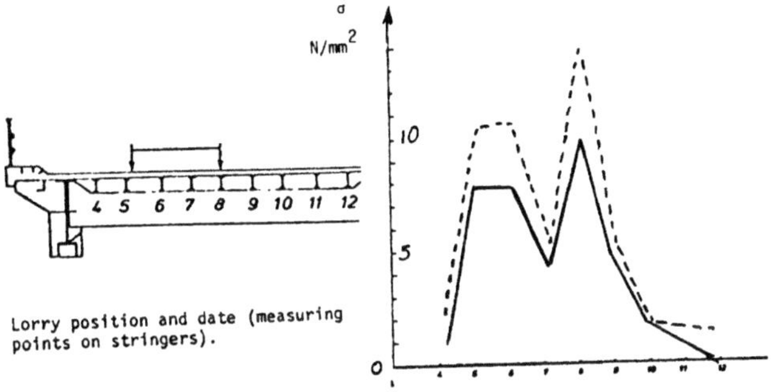

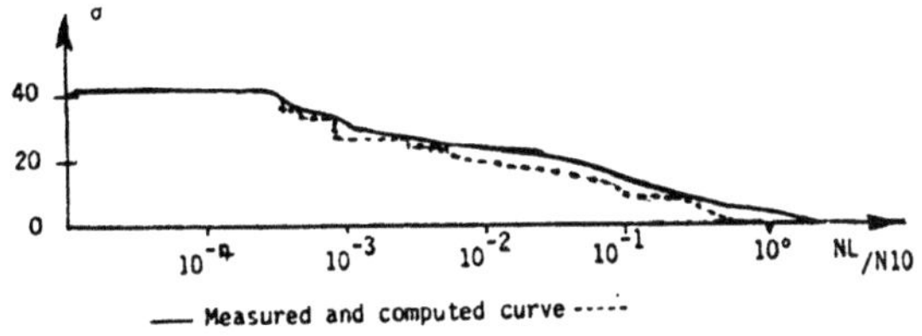

Fig. 17

Each vehicle type has a statistical frequency for crossing the bridge in the slow or fast lane in a given period of time, and it contributes to an increase in fatigue. In order to estimate with precision the damage potential, a new concept of prime importance has been used.

Characteristic axle-load

based on the equivalence of the fatigue damage potential between

[characteristic axle-load] N times = N axles of the axle load spectrum

For each vehicle, this value can be calculated. For a bridge having a defined structure and a given distribution of traffic the potential damage by each vehicle type can be estimated for a given duration.

$$\text{characteristic axle-load} \qquad \sqrt[\alpha]{\frac{\Sigma n l^{\alpha}}{\Sigma n}}$$

Obviously the sum of the potential damage induced by the different vehicle types increases with time. The time needed for this cumulative damage to reach the value 1 is nothing more than the theoretical durability of the bridge under the considered traffic data.

The application of the notion of characteristic axle-load is extremely viable. It shows that for this bridge, 91% of the accumulated damage in the slow lane is caused by only 5% of all traffic.

In particular, it points out the most damaging types of vehicles. Several other bridges have been studied by the same laboratory. Even though, the structure of the bridge and the traffic data are very different, the results are similar. The above power relationship made it possible to calculate the damage potential of a load range class or a stress range class with the Palgren Miner damage rule.

For a given vehicle type the characteristic axle-load can considerably differ for two different bridges, but also-for each bridge type, the fatigue life expectancy can be obtained. This datum is fundamental for future bridge codes.

b) Railway bridges

Some interesting research on bridges required for train-crossing has been realised by M. M.R. Pfeiffer at the Frauenhofer Institute of Darmstadt (W. Germany).

Axle-loads are measured on the rail itself due to an ingenious strain gage system which only delivers an electric tension when the wheel is situated between points 1 and 2. At the same time this system gives the count and 2 of these systems give the speed.

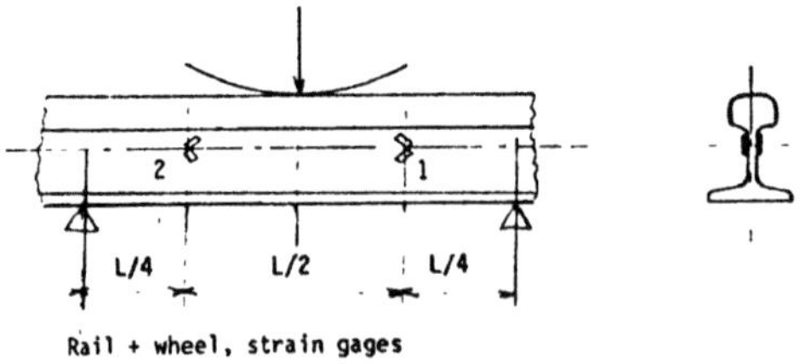

Fig. 18

The axle-loads are much more homogeneous and the axle-load spectrum, obtained by the passage of a passenger train is shown by Fig. 19.

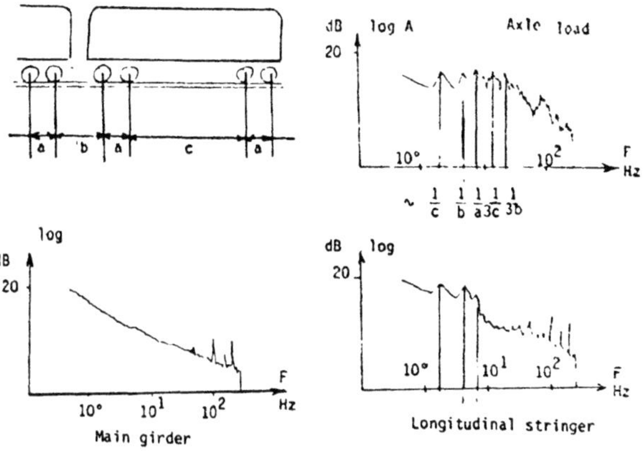

Fig. 19

Proportional to the inverse of the axle-distances, a,b,c, frequencies can be recognized on the curves, as the third superior harmonic.

The main girder chiefly provides a response for the grouped charges, while the longitudinal beam gives a punctual response for each passage of an axle-load.

As the dispersion of the axle-load is not as wide as for road-traffic, the distribution of the stress by the level crossing method is situated in a narrower range. For example, the level-crossing count of stress for one day, made on the 3 principal elements of a steel bridge (freight and passenger trains) is represented in fig. 20.

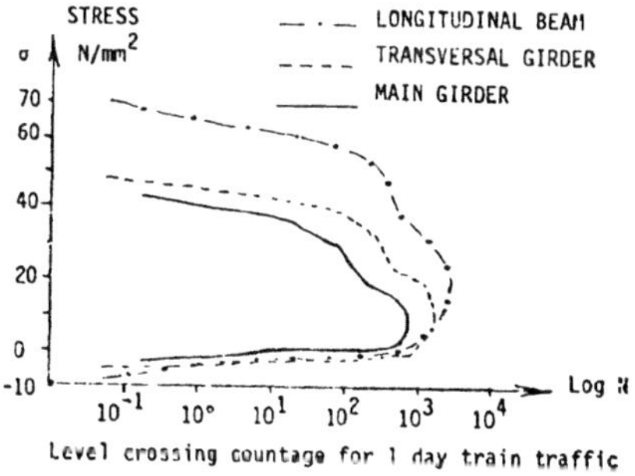

Fig. 20

Conclusion

For railway as for road traffic subjected bridges, the dynamic response can be written in terms of stress cycles. The bridge design method, in order to take into account the potential fatigue damage of structural elements is inseparable from the stress spectrum which must be converted into a design spectrum. Finally for a given traffic, it is the design spectrum which improves the fatigue life expectancy calculations.

This type of research on bridges subjected to traffic clearly shows the insufficiency of the classical notion of dynamic factor or homogeneous stress for ensuring safety

This research introduces a more efficient approach : it means attempting to level the cumulated damage for all points of the structure. With this concept all elements of a bridge structure have the same life time expectancy.

III. DYNAMIC TEST ON ACTUAL BRIDGES

1. Dynamic test with vertical impulse : Auscultation of full concrete slab bridge

It concerns a two-lane road-bridge, with short span. This full slab bridge (cast-in-place) lies on 4 elastomer bearings with a dead load of 120 metrical tons. The principal geometrical characteristics are given in fig. 21 and 22.

The dynamic test was made with a mechanical system. It consists in dropping a 50 Kg-steel plate from a height of 0.7 m on springs.

The dynamic characteristics have been obtained through the classical method : accelerometer, recorder, analysis.

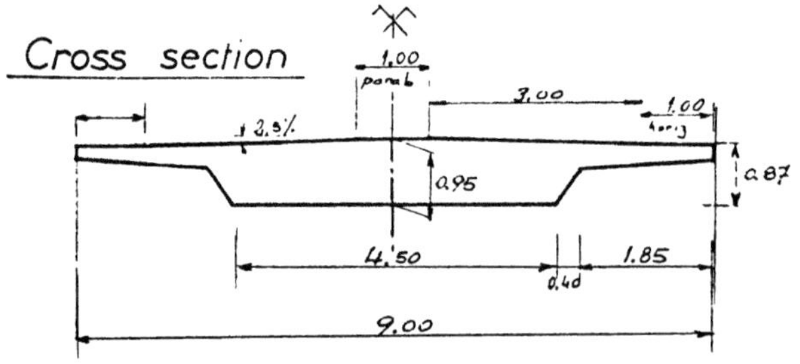

Fig. 21

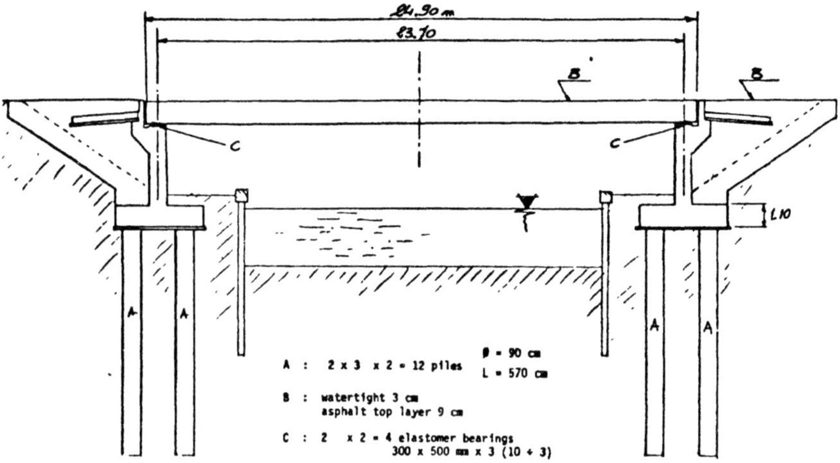

Fig. 22

The two main frequencies very clearly appeared, and whatever the conditions of measures might be, they have the following value.

Fundamental	mode	frequency	= 2,95 Hz	(calculated	3)
Second	"	"	= 11,0 Hz	("	10,3)
Third	"	"	= 23 Hz	("	23)

The damping obtained by logarithmic decrement has also given a constant value, similar for the 2 modes $\delta = 0{,}01$.

As in most concrete slab bridges, geometrical imperfections, errors of concrete quantity (mass), errors of steelbar positioning (inertia)... give experimental frequencies slightly different to the theoretical ones. The amplitude of the vibrations depends on the position of the measuring position M. Fig. 23 shows the response for 2 characteristic positions of M : in the middle and at 1/3 of the bridge span.

It would be extremely interesting to redo in identical conditions this test in order to follow the mechanical behavior through the years.

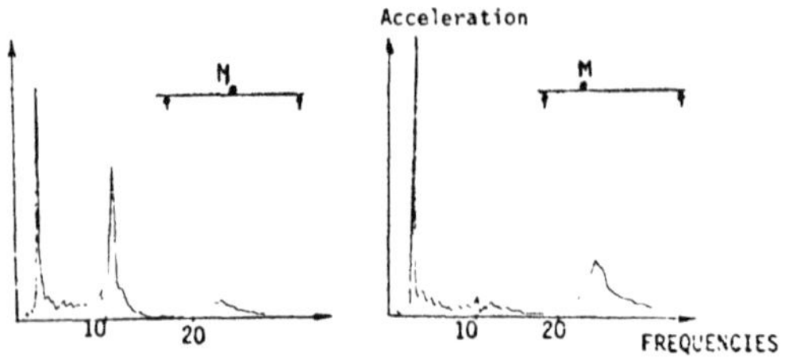

Spectral response for 2 positions of measuring point M

Fig. 23

The frequence spectrum evolution in the few seconds immediately after the impulse application is given in Fig. 24. This vibrational "survey picture" by itself sums up the main mechanical characteristics of the bridge. A further control carried out after n years can be compared with this initial survey picture taken as reference so that any modification in the response data can be detected immediately.

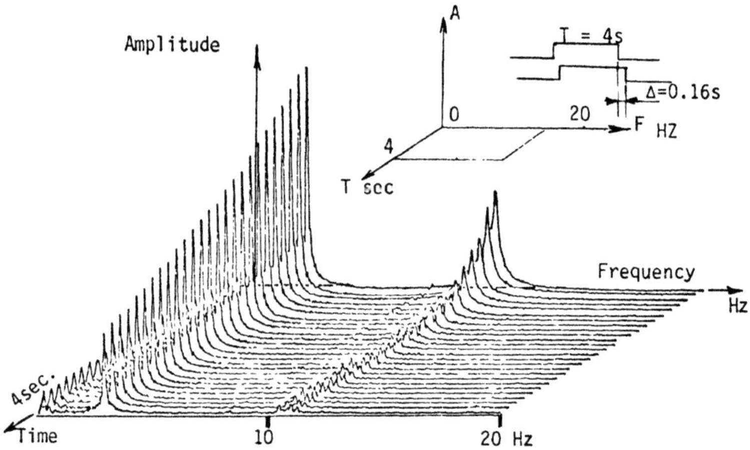

A three dimensional figure showing the evolution of the frequency spectrum with the time

Fig. 24

2. Sinusoidal excitation test

Comparison between normal and light concrete deck on mechanical properties of 2 identical bridge-structures.

In this test, dynamic excitation is obtained by a hydraulic vibrator which produces a sinusoidal force applied at mid span. Two types of methods can then be used.

logarithmic decremental method :

When the vibrator works with a constant frequency close to the fundamental frequency of the bridge (forcing vibration), the amplitude is much greater than in the impulse method.

The excitation is stopped and the natural response is obtained. If A_i and A_{i+1} are two successive maxima then

δ : Log $\frac{A_i}{A_i+1}$

* The structural damping is written $\nu = \frac{\delta}{\pi}$

* the viscous damping $\xi = \frac{\delta}{2\pi}$

. resonance method

A constant sinusoidal force with a progressive variable frequency is applied. For each frequency N_i the corresponding amplitude A_1 is measured. For the resonance, frequency N_2 , the maximum of amplitude reaches A_r .
If 2 different frequencies N_1 and N_2 (so that $N_1 < N_2 < N_2$) produce the same amplitude $A_1 = A_2 = \frac{A_r}{\sqrt{2}}$ the damping can be obtained :

$$\frac{N_r}{N_2 - N_1} = Q \text{ surtension factor}$$

the structural damping $\nu = \frac{1}{Q}$

the viscous damping $\xi = \frac{1}{2Q}$

With these two methods, a very interesting experiment allowed the study of the influence of the concrete-slab nature on the dynamical behaviour of a 77 m span steel bridge structure.

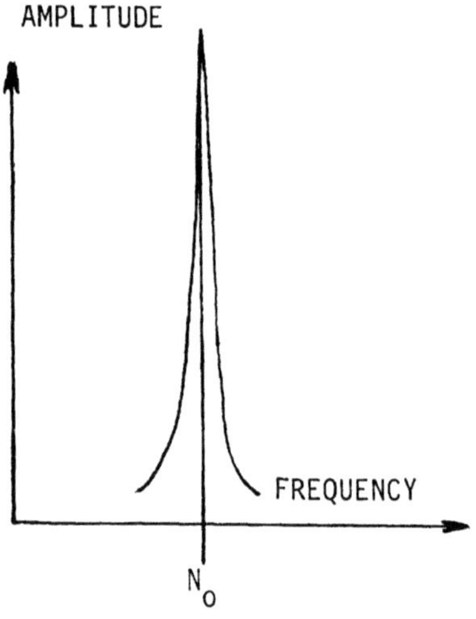

Impulse spectrum

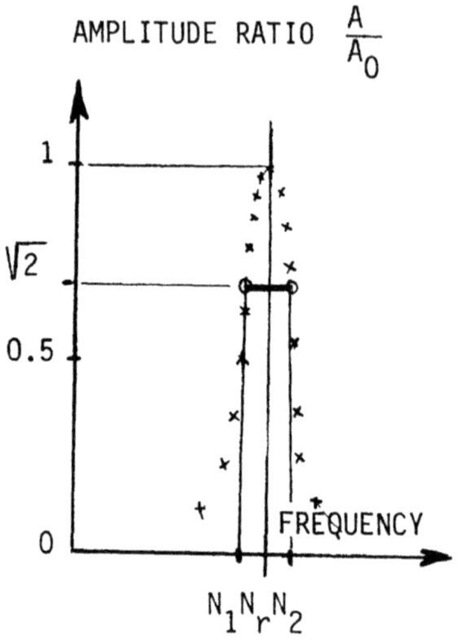

resonance

. the main structure of each bridge is constituted of 2 boxgirders with variable inertia and 2310 to 2330 mm in height.

. The bridge deck is constructed in a concrete slab 9.360 m wide and 12 cm thick in 2 versions (normal and light concrete).

Cross section :

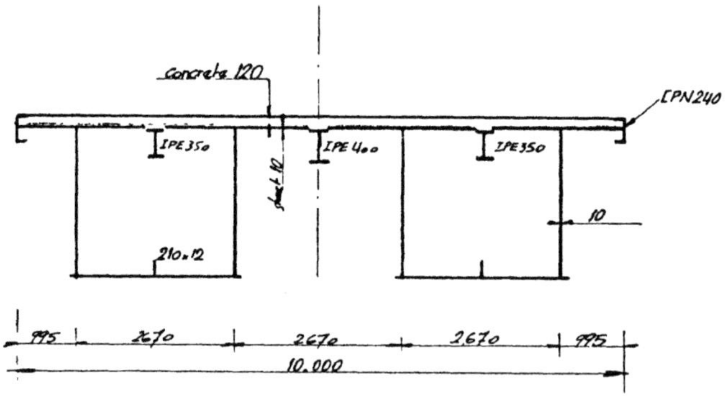

Fig. 25

The bridge was sinusoidally excited. Measures have been made for the main girder without deck and with the 2 versions of the bridge deck.

	FUNDAMENTAL FREQUENCY (N_0)		DAMPING %	
	measured	computed	log. decrement	resonance
steel girder	1.19	1.23	2.35	2.24
- with normal concrete deck	1.19	1.15	2.60	2.84
- with leight concrete deck	1.18	1.15	2.50	2.50

Fig. 26

The bridge was excited by a hydraulic vibrator and the natural frequencies have been obtained with an accelerometer. The damping was measured with the logarithmic decrement method and also with the forming resonance.

The 2 methods have approximately given the same results and if δ decrement log differs, it means that the decreasing amplitudes are not strictly linear and in fact, depend on the considered number of oscillations.

The damping value obtained with 20 oscillations is given in Fig. 27.

It is interesting to note that the calculated frequencies, with the method of lumped masses and with the equivalent beam method, give values near to those measured.

The bridge has in fact a slenderness of $\frac{77}{9.75} = 7.5$ which is chiefly supported by the steel girder, the participation at the deck is sufficient to keep frequency at the same value in spite of its mass. In fact in the case of light concrete, the structure is overdimensioned.

The resonance method needs more implements than the impulse method, also much more time and the performances are not better in precision ; but the injected energy can be higher with a changeable moving load of the vibrator from 0.365 to 3.4 tons. With an increase of the vibrational energy, the amplitude grows so that the observed damping mechanism of the bridge corresponds effectively more to that which works in traffic conditions. In spite of the relatively high vibrational energy level, the evolution of the damping value with the decreasing amplitudes of the bridge shows dispersion.

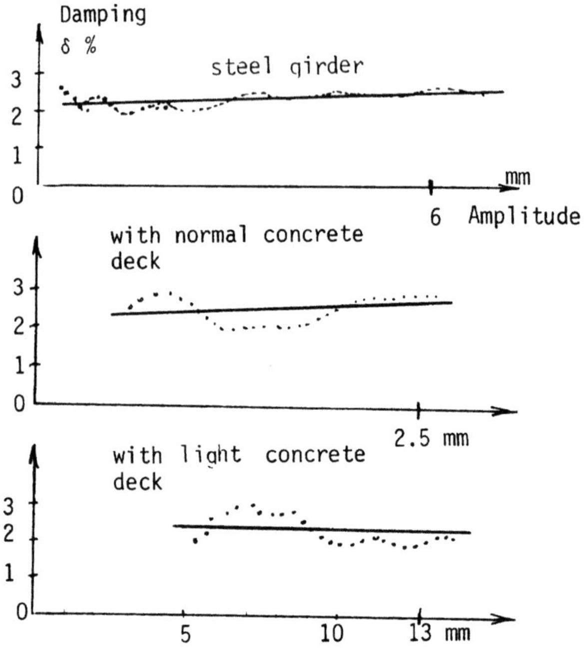

Fig. 27

IV. DYNAMIC TEST IN LABORATORY

The dynamic behaviour of a large number of bridge elements can be usefully studied inthe laboratory with dynamic tests to assess durability.

1. Accumulative damage

If n_1 n_2 ... are the number of stress cycles applied to the material at amplitudes of stress σ_1 σ_2 ... The σ-N Miner relationship give us the linear accumulative damage criterion D in the discrete form

$$\frac{n_1}{N_1} + \frac{n_2}{N_2} + \dots \frac{n_i}{N_i} = D$$

N_i is the number of cycles (with a stress level σ_i) to failure. Often D can be written in a continuous form (narrowband system)

$$D = \int_0^{\infty} \frac{n(\sigma)}{N(\sigma)} \, d\sigma$$

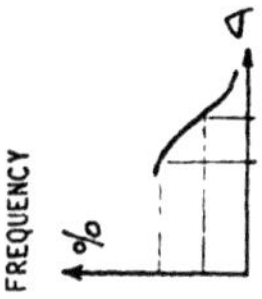

stress spectrum

Fig. 28

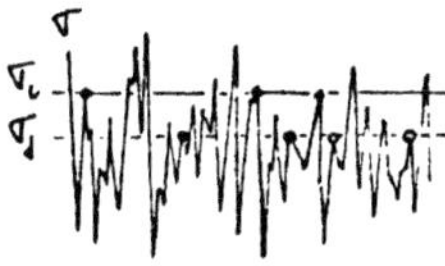

stress histogram

Fig. 29

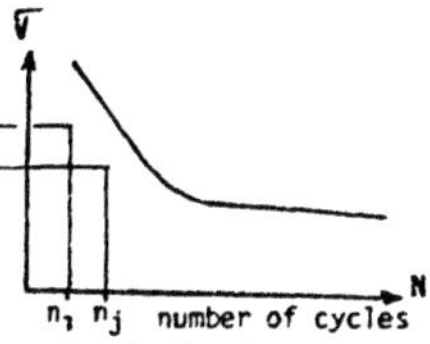

cumulative damage

Fig. 30

In some cases the modified Miner law gives a better approximation of the damage D_i obtained by the application of n_i cycles with a stress σ_i : the Miner element D_i of the Miner sum can be written $D_i = \left(\frac{n_i}{N_i}\right)^x$ with $x > 1$.

Fatigue testing methods for traffic simulation.

For a structural element of a bridge, the engineer can only make the design with a given security and life expectance if he can solve the problem of damage under vibrational random forces. For this, different experimental tests allow the determination of the fatigue line curves.

The statistical estimation of the traffic data can be translated in a load spectrum applied by the vehicles running the future bridge. The service inputs are then applied by dynamical tests using different types of cycles in order to obtain the actual service behaviour of the bridge element and especially the accumulation damage.

The single and constant amplitude cycles method, the most popular, is nowadays replaced by a better approach allowed with the new test machines : constant amplitude cycle blocks which can be programmed in advance fixed sequences or in a random process.

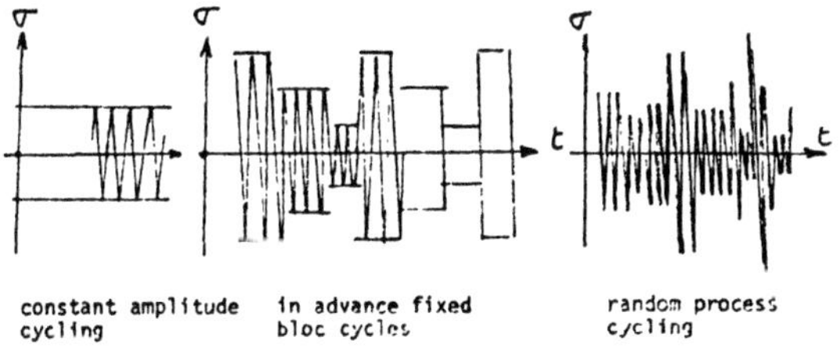

Fig. 31

Markov matrix method

The forecasting of durability and damage provision on the basis of the loadhistogram can be more explicit if the process respects the distribution of the different stress levels and also the mean stress random. For this the total stress range is divided in n levels. The number of passage from a given minimum stress level σ_i for example to a given maximum level σ_j is called the transition frequency T_{ij} from the initial to the final considered levels.

For each transition possibility from a low criterion to a higher criterion, the frequence is written in the corresponding case in the transition matrix (with n rows and n columns) on the right of the diagonal. The reverse transition possibilities figure under the diagonal. The use of this Markov matrix allows the simulation for whole statistical data of any dynamical loading process.

In figure 32 for an initial stress level σ_i , with $\sigma_1 < \sigma_i < \sigma_n$ in the row i , all corresponding transition frequencies possibly one to the higher stress level (from σ_j to σ_n) figure in the rows i in the right part of the diagonal. A statistical process allows us to explore the totality of the matrix elements as shown in the figure. The stress levels in regard to the corresponding transition frequencies can then be applied in a structural element with a hydraulic load system.

A more schematic and concise representation of the transition frequencies can be made and for this 12 stress level numerical matrix in Fig. 33 a hachure map can be obtained. For this we can choose for example six classes of transition frequencies such as for the class $10^{\alpha} < T_{\alpha\beta} < 10^{\alpha+1}$ the corresponding zone can be drawn with a unique hachure surface. With a little experience, this map gives immediately some indication of the severity of a cyclic loading process; and for different tested elements, a classification of cumulative damage in fatigue can be made. With the transition frequency method, the whole characteristic of a dynamic loading process can be simulated in a laboratory in order to provide qualitative observation of actual behaviour and quantitative data to the design engineer.

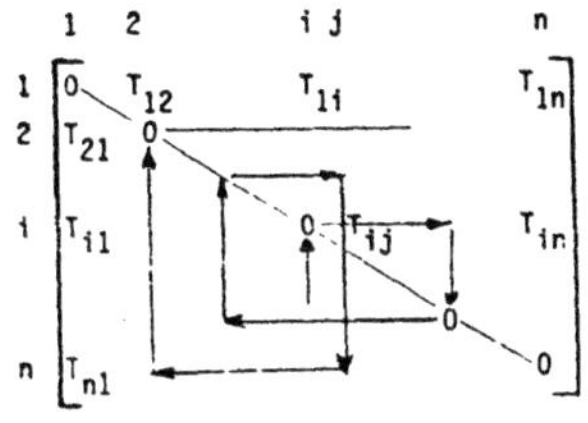

Markov matrix with frequencies of transition between stress levels

Fig. 32

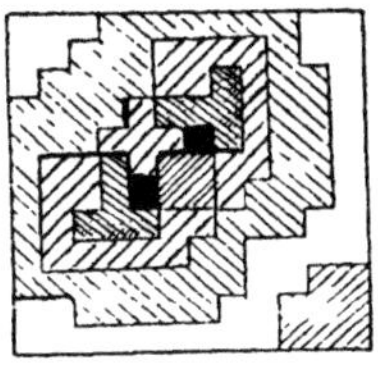

Fig. 33

The numerous research studied on damage of concrete under cyclic or random loading made with acoustic emission control, give similar results in structural behaviour.

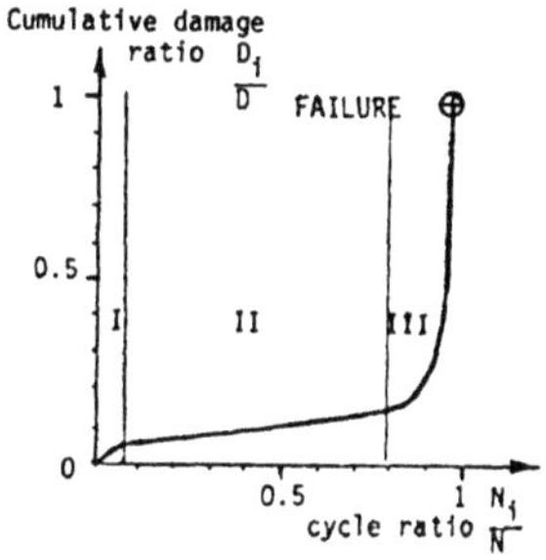

Fig. 34

Phase I : microcracks appear immediately corresponding to structural modification

Phase II : a linear but moderated variation of the material damage with the number of cycles can be observed. The corresponding acoustic emission showed a low level of acoustic emission

Phase III : after the limited damage evolution of the extended phase II the damage in phase III is growing quickly to the failure. The corresponding high emission show a instability of the crack evolution which shorten the life time.

2. Damping

The modal analyses of a bridge furnish the frequencies and mode from the mass and stiffness matrixes without necessitating other hypotheses. On the contrary the damping value has to be introduced arbitrarily. This value is based on an experimental damping behaviour in similar bridges, generally small. The errors of the damping estimation have little effect on the value of the different modal frequencies, the pseudo-period remaining always close to undamped period.

During vibrations the alternative transfer from kinetic to potential energy is carried out with internal loss. This damping effect can be obtained in two different ways:

In the case of forced vibration the energy dissipation per cycle is delimited by the hysteresis curve. This latter shows the deviation from Hooke's law according to the type of damping.

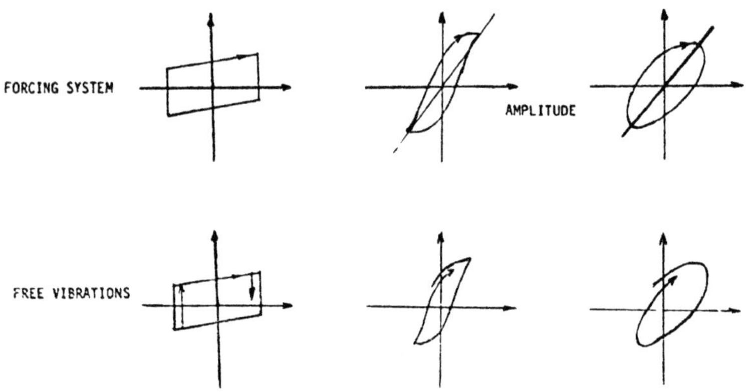

Fig. 35

The system of forces can be written

$$m\ddot{x} + c\dot{x} + kx = F_0 \sin \omega t$$

and the corresponding vectorial representation directly gives the phase angle.

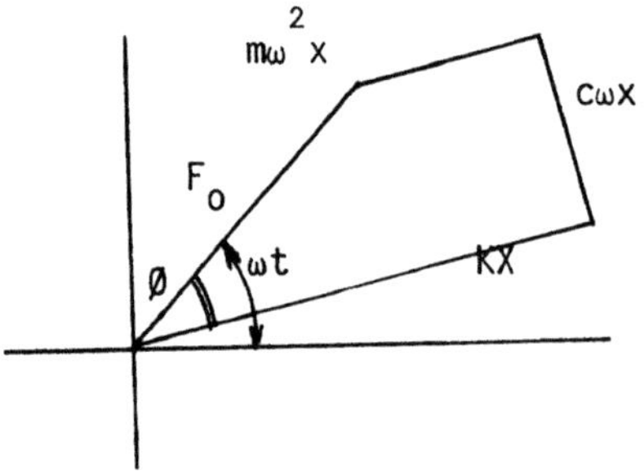

Fig. 36

When the exciting forces cease, the amplitude evolution with the time can be followed by the logarithmic decrement δ . For a complex structure, δ results in general from the superposing of the different damping types : Coulomb friction viscous and structural damping. Considering the envelope of the successive maximum of amplitude according to the order n, different curves are obtained. Fig. 37.

Moreover, δ is also a function of the emplitude, Fig. 38.

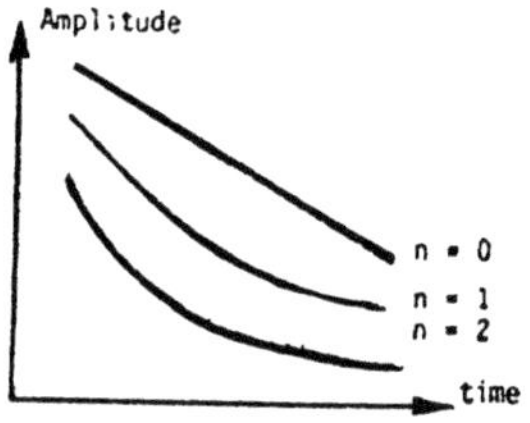

Fig. 37

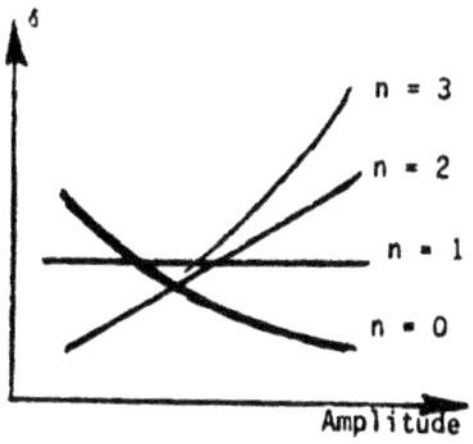

Fig. 38

Considering now the influence of the prestressing stress on the value of δ, tests on the prestressed beams show clearly that from a certain stress level σ_p stabilisation of δ is obtained, Fig. 39.

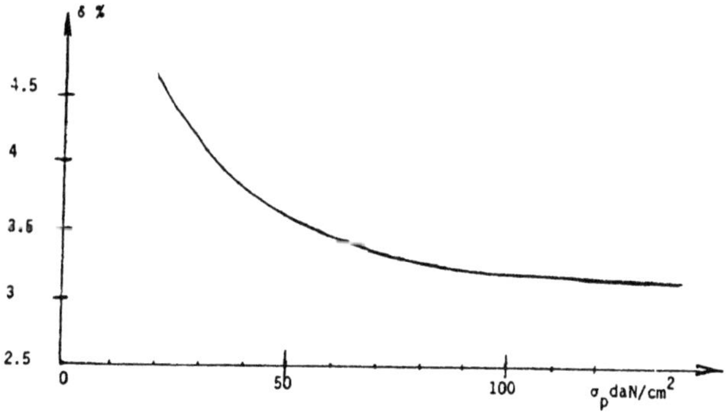

Variation of δ with the prestressing σ_p

Fig. 39

δ is also dependent on this amplitude but less, as long as the elastic domain is explored.

Concerning the damping, the prestressed and poststressed beams have identical behaviour. δ decreases when the stress increases (pre or post). When $\frac{R+}{2}$ or $\frac{R-}{2}$ are exceeded δ increases suddenly ; this phenomenon is related to the microcrack extension, especially in the areas under tension.

The correspondence between δ and the state of the microcracks in the material can be presented schematically in Fig. 40.

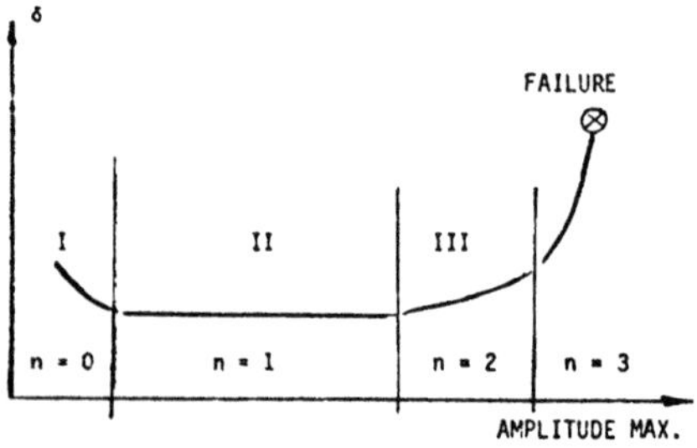

Fig. 40

In the domain I, the level of stress is low, the damping results from Coulomb friction. Progressively with the increase of the amplitude, the viscous damping, as the structural damping, appear in the domain II of visco-elasticity the deformation law is quasi-elastic. With higher stress application the internal microcrack propagation grows as the damping which reaches the order n = 2 (hydrodynamic). With only a small number of higher stress cycles the damping grows to order 3. Simultaneonsly the microcrack development generates visible cracks and the cumulative damage reaches the value 1 (failure).

For concrete elements tested in laboratories, the damping evolution is a useful indicator of the internal microcracking mechanism. Two complimentary procedures allow currently to follow directly the mechanical and structural state of the material : In the acoustic emission, damage is assumed, if a certain level of emitted acoustic pulses is exceeded. In the volume alteration measuring system, a constant amplitude repeated load first produces a decrease in volume domain I in the increase in damping. Then, the volume decrease diminishes rapidily, as soon as a low level of cumulative damage is attained. This corresponds to the passably large domain II. After the transition from the domain II to III the crack propagation is so intense that it produces an increase in volume and the damage progress is then rapid.

Acoustic emission, volume alteration and damping variation are indicators of the microstructure concerning cracks, that means of the cumulative damage. In the laboratory microelastic as well as thermodynamic approaches tend to define the energy dissipation in the material.

From the laboratory to the bridge

The damping is the macro indicator for the evaluation of the level of internal microcracking state of the concrete or its cumulated damage. For a bridge, in so far as domain II is explored, small alterations of δ risk being hidden by measurement imprecision, affecting the damping. However, if in a given area, a diminution of the constraining force or an increasing of an external load occurs; so that, at minimum, one of the stress value in the extreme fiber leaves the elastic domain, the measure of the corresponding damping shows an important increase.

Additionally the position of this corresponding area can be obtained : the damping evolution affects firstly modes for which the altered area is located at the antimode. A high prestress value delays the appearence of cracks. It is important to note the similarity of curve deformation-prestressing load with the curve prestress value-crack width.

The damping survey equally for a bridge span as for a local sections allows the estimation of the structural damage consequence of high stress cycles, or a loss of cable tension.

Current methods of control of the vibrational properties should be completed by a damping diagnostic in order to estimate the damage before external cracks can be visually observed.

Thus a prestressed box girder bridge 180 m long with 3 spans has been tested. The vertical cracks (1 to 3 mm) in the bottom part between the adjacent elements and the horizontal cracks between bridge deck and girder are the result of the combined action of the daily temperature gradient with strong stress cycles.

The obtained damping value of the damaged bridge was $\delta = 10\%$.

3. Mechanical interaction railway-longspan suspension bridge

Much research has been accomplished on the problem of bridges required by railway traffic. However, for suspension bridges, the low rigidity causes great deformations at the time of the passage of trains. Otherwise the mechanical interaction - bridge - carriage can cause vertical and horizontal vibrations :

Tracks supported by suspension bridges are so flexible, that they may be influenced by the running vehicle itself (as by wind or seismic actions).

Some very important research has been conducted in this direction by the Professors Yasoshima Y. and Matsumoto Y. of the Civil Engineering Department-University of Tokyo. The question was (for Japan Railway Construction Corporation) to improve carriage suspension, and at the same time, to study the mechanical behaviour of a long span suspension bridge, exposed to the passage of an express train.

Vibrational behavior of trains

Railway vehicles, running on tracks, are always affected by displacements of the positions of the wheel axles. Track irregularities, especially of a constant wave length, affect all axles with a different phase. However, track alignment which has a longer wavelength , affects a great portion or the total number of axles simultaneously.

Railway vehicles, running at a constant velocity on a track with irregularities of a constant wavelength produce a periodical force on the track and the bridge deck.

Lateral instabilities also have to be taken into consideration, especially wheelsets and flange action behaviour which can induce transversal periodic acceleration.

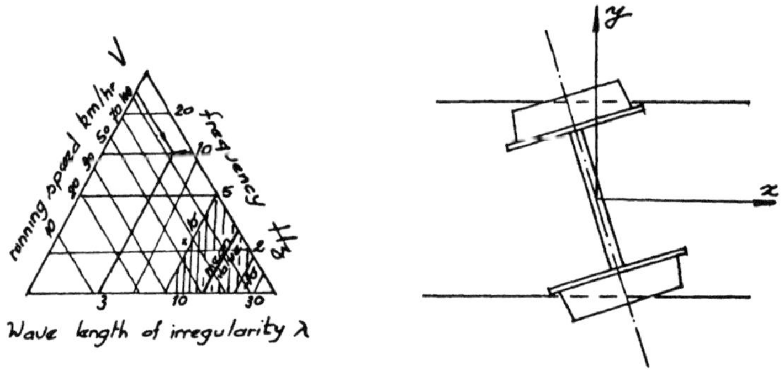

Fig. 41

Characteristics such as rail-wheel creep coefficient, stiffness of the vertical and horizontal wagon suspension, hysteresis damping must be taken into account.

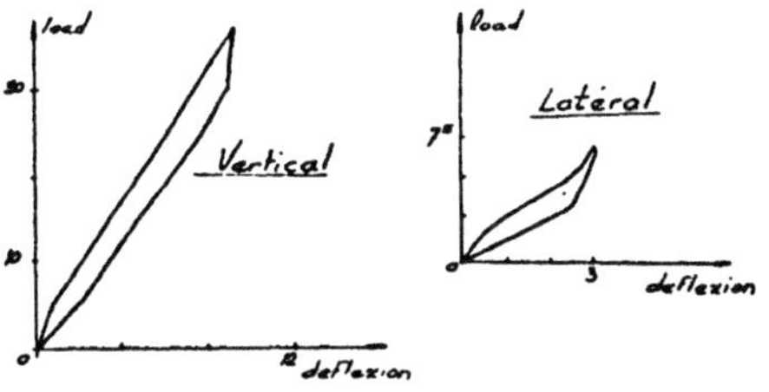

Fig. 42

Stiffness and hysteresis of the bogie suspension

The whole dynamic behaviour of the wagon on rails has been tested on a two dimensional, 36 m long shaking-table. The model of test vehicle could be accelerated laterally and vertically in a frequency range up to 10 Hz.

The tests have made it possible to determine all the mechanical characteristics of the movement (amplitude and frequency of stationary lateral snake motion, of rolling motion, response of wheelset with velocity increase ...).

The experimental system was carried out by taking the vibration modes of the suspension bridge (1st to the 6th mode).

Feedback to the shaking-table, made it possibleto take into account the calculated response of the suspension bridge. The vertical acceleration of the bridge does not have a so great importance, as the lateral one. The lateral acceleration possibilities have been simulated, in orderto produce vehicle instability for a better knowledge of the derailment conditions.

- lateral action : for a 1000 m long span bridge, the lateral vibration is small because of the small mass ratio and the great difference in frequencies. A small amplitude produced by wheelset appeared only for the 6th vibrational mode of the bridge.

- vertical action : the running vehicles produce

  . impulsive vibration due to the irregularities of tracks, railjoints, movements of vehicles (rolling, snake)

  . dynamic motion of the trusses which affects the complete bridge system

The bridge

Several types of suspension bridges have been studied. The 1700 m suspension bridge project, joining Honshu to Shikoku, has been investigated in order to improve the serviceability of running trains and roadcars . With a main span of 1100 m and two side-spans of 300 m, the stiffening truss has a constant depth of 10,37 m ; a width of 33 m with a sectional area of 0.68 m^2 and an inertia of 2.687 m^4.

When a train crosses this suspension bridge, the stiffening trusses undergo significant deformation:

- vertical deflection
- elongation and contraction of truss extremities due to deflection and temperature change (117 cm)
- vertical curvature at main towers with deflection change.

The deflection of stiffening trusses depends greatly on the type of the bridge suspension. Thus, besides the classical cable suspension bridges (type 1), two other solutions have been proposed : auxiliary cable, continuous suspension bridge having hangers at branching points (type 2) or having no hangers at branching points (type 3).

In the auxiliary cable continuous suspension bridge, the stiffenening trusses are not supported on the main tower bearings but only on the auxiliary cables and on the main cables.

Vertical deflection of a concentrated moving load

A theoretical analysis shows that the velocity effect on dynamic deflection under distributed loads (road traffic) is far less than under a concentrated load.

By considering the local deformation of the bridge, there are two different gradients to consider : the gradient in the direction of the vehicle movement, the gradient of the track under the vehicle. For a concentrated load of 100 t, the running and track gradient for a bridge type 1 with similar length are given in figure 43.

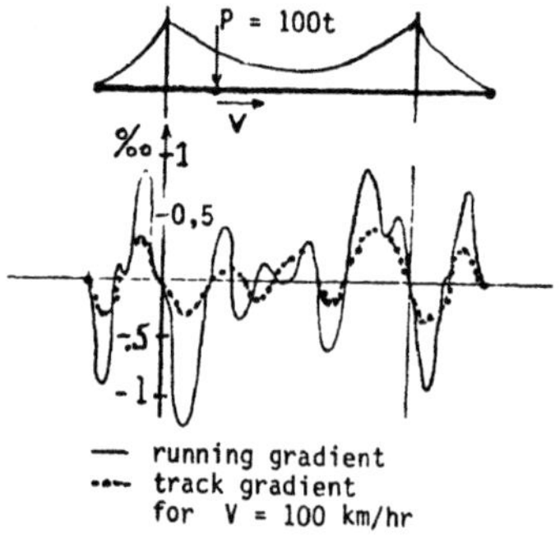

Fig. 43

So it appears that, if the track gradient is an angle α , the trajectory of the moving load is in the direction β . For a simple supported beam, the running gradient is double the track gradient $\beta = 2\alpha$. In addition these gradients are greatly affected by the velocity of the moving load : Fig. 44. The dynamic amplification factor can reach 2. The vertical acceleration of the stiffening truss caused by the moving load with only smooth running is given in Fig. 45.

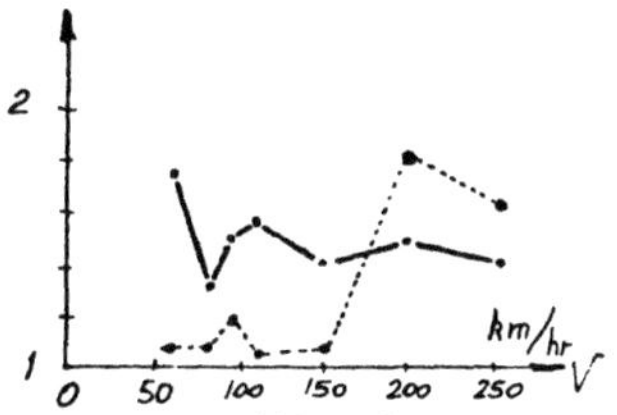

Dynamic amplification
— running gradient
....track gradient

Fig. 44

Speed	Max vert. acc.
65 km/hr	0.022 g
110	0.030 g
250	0.162 g

Fig. 45

Deflection change due to train passage

In the case of a train with a length of 400 m and a linear load of 1.2t/m a numerical application was done to investigate the essential behaviour of auxiliary cable suspension bridge. The experimental results are in good agreement with the theoretical values for all bridges types and all load conditions. The deflection change due to train passage is given in Fig. 46.

Vibration tests

The vibration modes and natural frequencies have been studied experimentaly (bridge model of 48 m) and analytically (multimass system). The general vibration properties of an auxiliary cable suspended bridge are similar to those of a classical type 1 except the damping effect which is better for the type 2 and 3.

Vertical resonance : with increasing the train velocity in the experiment it was found that the first symmetric mode is obtained at 328 km/hr,far from real traffic conditions.

The very difficult problem of mechanical interactions bridge-traffic has been followed up in this research without needing to refer to hypotheses. All basic effects have been simulated including earthquake shaking and wind excitation (wind tunnel). An optimum auxiliary cable geometry improves greatly the serviceability of train and highway traffic :

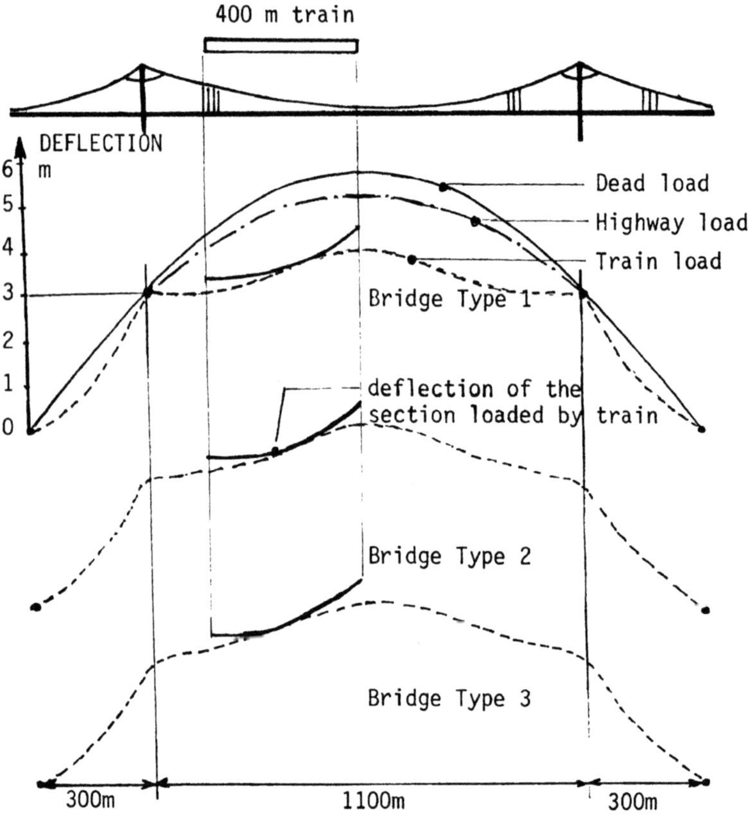
400 m train
DEFLECTION
m
6
5
4
3
2
1
0
Dead load
Highway load
Train load
Bridge Type 1
deflection of the
section loaded by train
Bridge Type 2
Bridge Type 3
300m
1100m
300m

Fig. 46

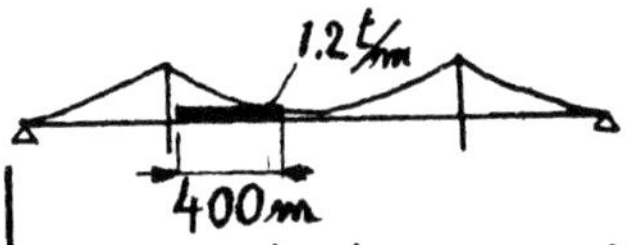

Bridge type	modes order	frequencies computed	frequencies measured	Damping	frequencies computed	frequencies measured	Damping
1	1	0.686	0.683	0.0176	0.654	0.700	0.0265
	2	0.827	0.867		0.776	0.867	
	3	1.310	1.367		1.144	1.333	
2	1	0.733	0.700	0.0225	0.697	0.720	0.0288
	2	0.839	0.867		0.799	0.840	
	3	1.363	1.383		1.297	1.300	
3	1	0.663	0.650	0.0265	0.638	0.700	0.0427
	2	0.830	0.867		0.761	0.850	
	3	1.151	1.217		1.088	1.233	

Fig. 47

. a milder change in deflection of stiffening trusses and compared with a classical suspension bridge only a third of the maximum negative bending moment.

. In the central part of the bridge, the mechanical behaviour is similar and the damping value is higher.

V. SIMULATION AND MODELING

1. Choice of mechanical properties of bearings in order to improve the safety of a boxgirder bridge subjected to seismic vibrations.

a) Mechanical characteristics of bearings

For the vertical transmission of bridge-deck load, restrained elastomer bearings are used more and more. They are composed of a succession of elastomer layers vulcanised with steel plates (rectangular or circular type).

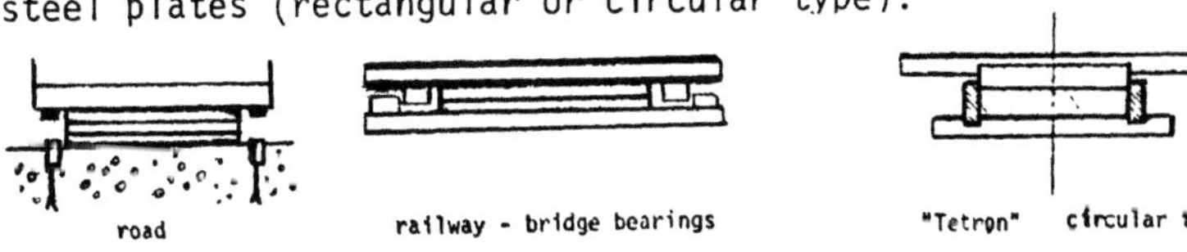

Fig. 48

In France, these bearings have been sized according to a static design calculation. The experimentation and the normalisation of the different types of bearings have been led by ORE* and SETRA *.

The mechanical characteristics of the elastomer material vary with the type of chemical formula. For the bearings, the recommended shore hardness reaches 60 and the advised modulus is G = 0.8 MPa for road bridges and G = 1.1 MPa for railwaybridges.

Vertical compression :

A vertical load N on a bearing produces an average stress for a rectangular plate (a x b) $\sigma_M = \frac{N}{a \times b}$ and a shearing τ_N at the interface steelplate-elastomer. The maximum value is obtained in the middle of the length a.

$$\tau_N = \frac{1{,}5\ \sigma_m}{\beta}$$

β is a shapefactor obtained experimentally. The shearing stress has increasing values from the middle of the bearing plate to the sides where it reaches the maximum values. If there are local damages for a rectangular bearing, they always appear towards the middle of the length. (τ_N = max)

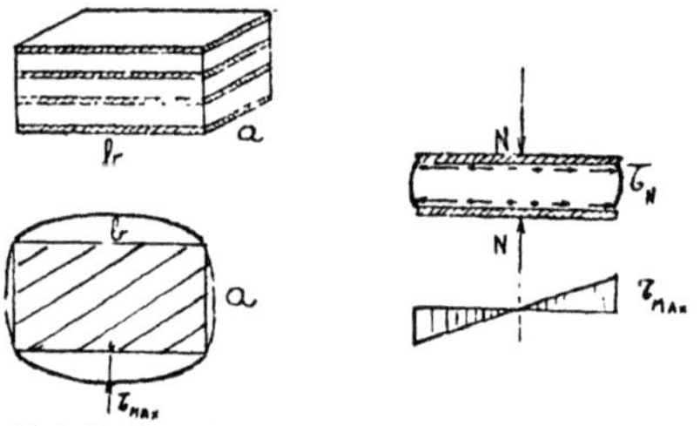

Distribution of shearing stress τ of a bearing subjected to a vertical compression

Fig. 49

---

* ORE Office de Recherches et d'Essais de l'Union Internationale des Chemins de Fer

* SETRA Service d'Etudes Techniques des Routes et des Autoroutes

Horizontal forces

All the dimensional variation of the bridge (temperature, creep ...) with horizontal forces applied produce distortion in the bearing and a shearing stress σ_H is developed.

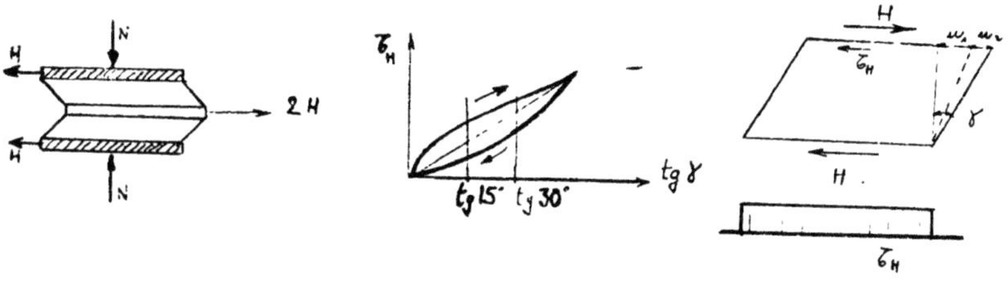

Fig. 50

Rotation

In addition an elastomer plate can be a subjected to a rotation α and the corresponding stress is τ_α. When the linear or angular displacements are too large, the bearing has to slide so that overly larger efforts can be avoided on the bridge deck as on the foundation ; the friction coefficient on a PTFE plate is experimentally given (approximately a few percent).

Dimensioning

Whatever the types of actions applied to the elastomer bearings the shear-stress is limited $\tau = \tau_N + \tau_H + \tau_\alpha < 5G$

as well as normal (σ)stress $\sigma_M < 15$ MPa .

For a railway bridge the values $\tau = 1{,}5\ \tau_N + 1{,}5\ \tau_\alpha$ are recommended because of the fatigue phenomenon resulting from the great variation of the vertical loads.

By their elasto-plastic adaptation, the bearing allows a good distribution of loads even with small differential movements of the foundation ; after an initial static deformation, a supplementary adaptation must be taken into consideration.

Dynamic Behaviour

The visco elastic properties interfere at the time of the transmission of dynamic charges. However, in case of road traffic the dynamic components remain limited, even if there is a small vertical oscillation around the average position, the amplitudes remain low compared to the dynamic bridge deflection. The influence on proper frequencies of the bridge consequently remains insignificant. In any case the damping effect on the bridge structure will be limited and can not easily be separated from the foundation damping.

The elastomer bearings are designed for long service in normal conditions. However, in case of seismic vibrations, the stress and deformation might be much stronger and can reach the point of damage.

In the limit conditions, the choice of the mechanical characteristics of the bearings has an influence on its own serviceability , as we shall show in a real example on mechanical coupling between foundations abutment and bridge-deck movements.

b) Choice of elastomer bearing of a boxgirder bridge (in prestressed concrete) in order to improve the mechanical behaviour under seismic action.

The bridge

It is a big highway bridge with several spans, achieved by prefabricated concrete box-elements. The elements situated in the central zone of the spans are made of light concrete (fig. 52).

Seismic risk, Shake-program

The bridge is situated in a low risk seismic area. The spectrum has been obtained from surface recorded seismic movement in a nearby area with similar seismicity. The characteristics of the seismic wave have been computed for a height, midway up the foundation piles (fig. 51).

The ground has the following characteristics :

- powdery soil
- C Menard $\alpha = 3,5$
- Pressiometric modulus 285 kg/cm^2
- G = 357 kg/cm^2 horizontal
- E = 1000 kg/cm^2 $\nu = 0,4$

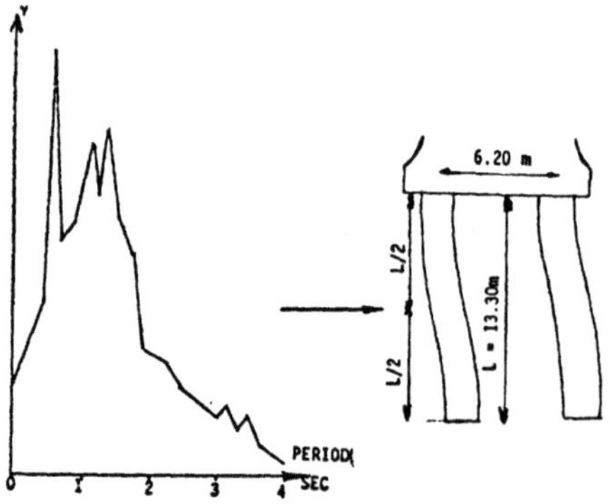

Fig. 51

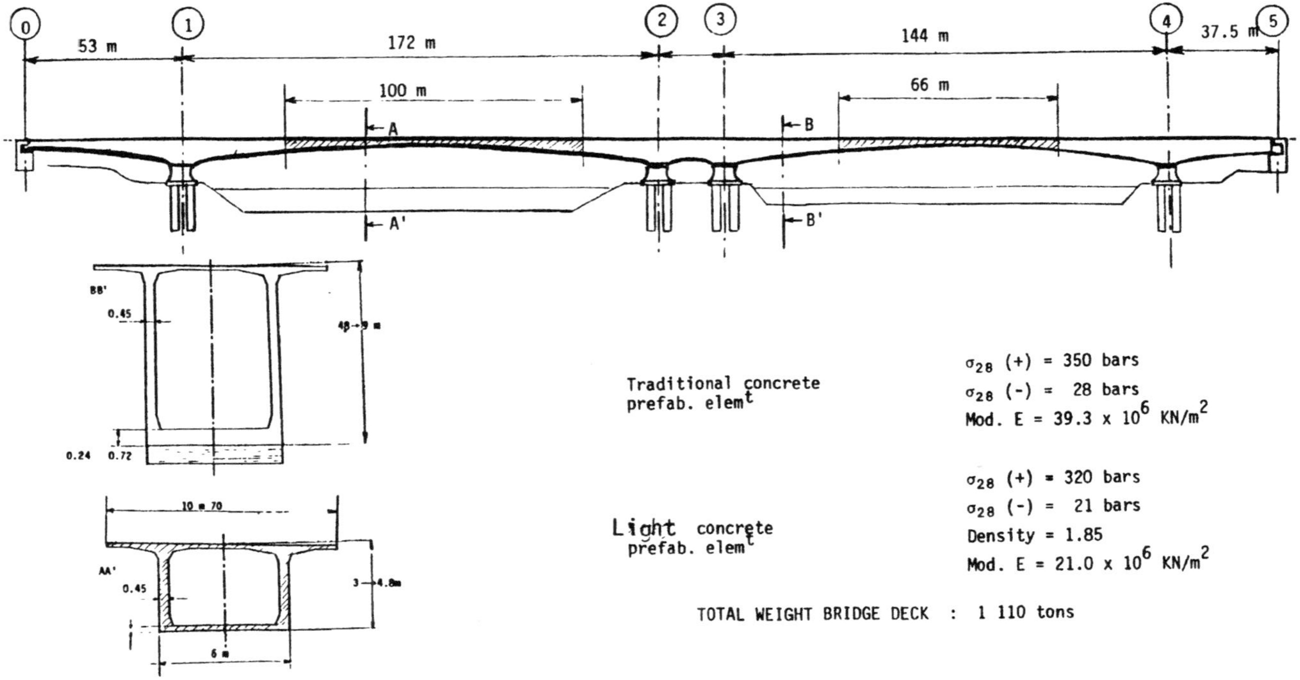
0
1
2
3
4
5
53 m
172 m
144 m
37.5 m
100 m
66 m
A
A'
B
B'
BB'
0.45
0.24
0.72
AA'
0.45
6 m
Traditional concrete prefab. elemt
σ28 (+) = 350 bars
σ28 (-) = 28 bars
Mod. E = 39.3 x 10^6 KN/m^2
Light concrete prefab. elemt
σ28 (+) = 320 bars
σ28 (-) = 21 bars
Density = 1.85
Mod. E = 21.0 x 10^6 KN/m^2
TOTAL WEIGHT BRIDGE DECK : 1 110 tons

Fig. 52

. For horizontal action :

It appears that the acceleration $\gamma_0 = 0{,}4$ g is not easily transmitted to topsoil Layer. A more realistic acceleration has been applied $\gamma_0 = 0{,}15$ g . It corresponds to a 6.5 seismic magnitude with its epicenter 40 Kilometers away.

. For vertical action :

It is supposed that the vertical charges are diffused from the bottom of the abutment (6 x 10 m) into a pyramidal slope model of 45°.

The Shake-program yielded the following G moduli from the top to the bottom of the pier : G varying from 550 to 770.

Dynamic behaviour of the bridge

All characteristics of the real bridge are well-known (dimensions, masses...). A dynamic modulus for the concrete of the pier $E = 30 \times 10^6$ KN/m^2 = 300000 kg/cm^2 has been admitted.

The ANSYS program was used with the specific purpose of determining the influence of the mechanical characteristics of bearings on the stresses in the foundations as well as on the bridge deck.

The first oscillatory modes are represented in Fig. 53.

Comparative test of two solutions of bearings

The ANSYS program has made it possible to determine all the actions on the different elements of a bridge with the results of the ground analysis (Shake). The problem was to determine which of the 2 solutions I and II for the bearing makes the foundations and bridge a safer unit.

The 2 analyses differ only in the stiffness matrix of the bearings. The "tetron" sliding bearing types (bi or multidirectional) are supposed to have a grip coefficient of 2 % (constant with temperature and vertical load).

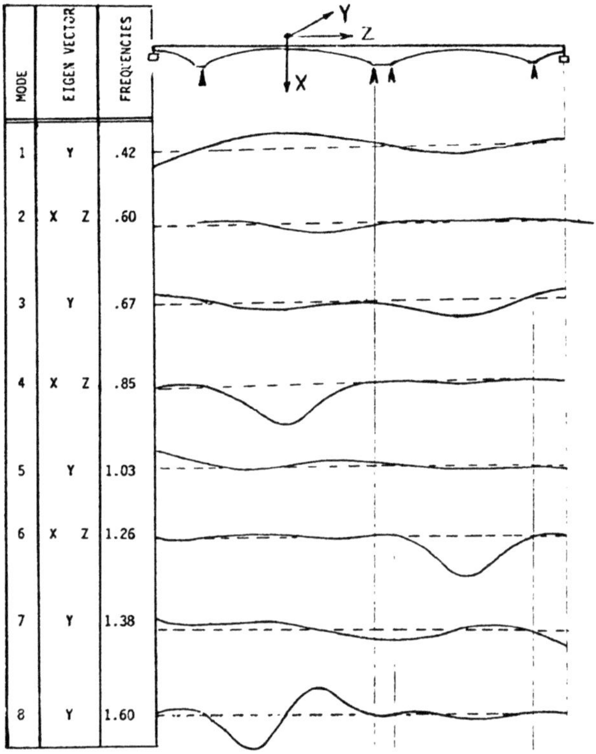

Fig. 53

The types of bearings used are shown in Fig. 54.

Results

The forces and moments produced by the precedent seismic action are plotted for solutions I and II for the different bearings (from 0 to 5) and given in Fig. 55.

Analysis and comment

Solution I :

The bearings of pile 2 and 3 are over-loaded compared to the maximum tolerated. As the neoprene bearings are not strong, the maximum seismic acceleration values acceptable are : γ_0 = 0.08 g.

N° abutment	TYPE OF BEARINGS		PERMAN^t LOAD tons
	SOLUTION I	SOLUTION II	
0	2 TETRONS D2E 400	identical I	520
1	1 TETRON D2E 2500 1 TETRON D2E 2500	4 Neop. blocks 900 x 900 9 [20 + 3]	4090
2	4 Neopren blocks 800 x 800 4 [12 + 3]	identical I	2450
3	identical 2	identical I	2500
4	1 TETRON D2E 2000 1 TETRON D2E 2000	4 Neop blocks 500 x 500	2910
5	2 TETRONS D2E 325	identical I	300

Fig. 54

N°	SOLUTION	F_x	F_y	F_z	M_x	M_y	M_z
0	I	1809	0	0	0	2	15 403
	II	1778	0	0	0	2	826
1	I	2305	6472	0	0	40	63 283
	II	2324	2257	1892	175	181	12 116
2	I	5734	2667	4804	289	265	15 014
	II	5697	2441	4429	190	261	12 318
3	I	6458	2276	4810	337	356	15 017
	II	6460	2425	4432	239	357	13 261
4	I	3701	7609	0	0	27	26 173
	II	3825	2475	1691	226	118	12 757
5	I	2430	0	0	0	0	19 795
	II	2418	0	0	0	0	1 338

Dynamic action on bearings F_x : kN ; M_x : kN.m

Fig. 55

The foundations at 1 and 4 are strongly crosswise loaded because of the great stiffness of the "tetron" bearings, which do not allow transversal sliding.

Lengthwise, by contrast, bearings 2 and 3 transmit the quasi-totality of the charge. The bridge-deck will be submitted to relatively weak tractions in a zone normally under compression. The fissured zone will remain very limited without being able to provoke rupture.

Solution II

A sensible modification of the dynamic behaviour is afforded by elastomer block-bearings. The system having a more homogenous transversal flexibility, the dynamical behaviour is outrightly improved transversally(G dyn = 160 t/m^2). Lengthwise, supports 1 and 4 must underpin the dimensional variations of the bridge-deck. To avoid a too large distorsion in elastomer a larger height of the bearing is necessary.

A weaker longitudinal stiffness results from this and it is again piles 2 and 3 which take the majority of the longitudinal seismic force.

Conclusion

The most important benefit of solution II is obtained on the foundations. The forces on the different abutment piers of the bridge are considerably reduced compared to solution I. We can outline the most damaging vertical and horizontal actions due to dead-load and seismic-load for each solution, in the table of Fig. 56.

BEARING SOLUTION	STATICAL + SEISMIC ACTION/PIER	1	2	3	4
I	↓ N_+ ton	1807	1433	1471	1604
	↑ N_-	203	409	463	386
	T ←→	163	142	144	191
II	↓ N_+	1632			1376
	↑ N_-				
	T ←→	114			112

Fig. 56

Solution II shows a better distribution of the horizontal charges, which are brought down to an acceptable level.

In both solutions, the bearings situated in 0 and 5 are subjected to the same actions (pendular counterweight) and the vertical loads applied on the different abutments have similar values.

The dynamic behaviour of the girder-foundations unit is inseparable from the mechanical characteristics of the bearings, especially in case of seismic action.

The dynamic analysis of the mechanical system with solution I and II makes an advised choice of the bearings possible. Safety is increased by limitation of the forces in the whole structure.

2. Effect of profile imperfection of the bridge subjected to moving loads

The dynamic effects due to traffic are generally introduced in calculations with a dynamic amplification coefficient.

This coefficient is not always the consequence of a dynamic approach of mechanical response or of experimental measures.

These coefficients depend on different parameters, the different national code values show significant differences because they are not based on the same parameters.

Most of the codes give a global dynamic coefficient which allows the obtention from a static response of the so-called dynamic moment and deflection of the bridge.

A more subtle approach has been conducted by Mr.Palamas (L.C.P.C. Paris). Mr. Palamas shows in fact, that the coefficient has to be defined for each section of the bridge. The dynamic majoration coefficient of the vertical deflection (Cd) and of the moments (Cm) for a given section then depends not only on the mechanical characteristics of a bridge and of the moving load, but is also greatly influenced by the profile imperfections and the vehicle speed.

Calculation of the bridge-moving load system was realised by a computer program "Palmares", the numerical application was done for a classical concrete bridge and one vehicle.

The essential characteristics are given in Fig. 57. When the profile is without irregularities and when the vehicle arrives at the entrance of the bridge without vertical oscillation, the force exerced is constant and C_d at midspan, scarcely exceeds the unit value (Fig. 58).

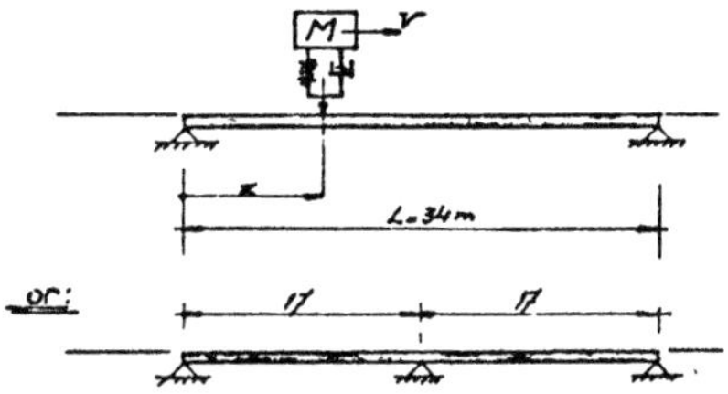

mechanical charateristics :

Bridge span 34_m or 17 x 2 m

fondamental Freq. N_{34} = 4HZ N_{17} = 16.7 HZ

Rigidity EI = 9.92×10^{10} N/m^2

Dead Load 11.4 t/m

Vehicule mass 31,7 t

suspension stiffness k = 9.123 200 N/m

suspension damping n = 85 600 N sec/m

vertical period T_o = 0,37 sec

velocity V $0 < V < 100$ Km/hr

Fig. 57

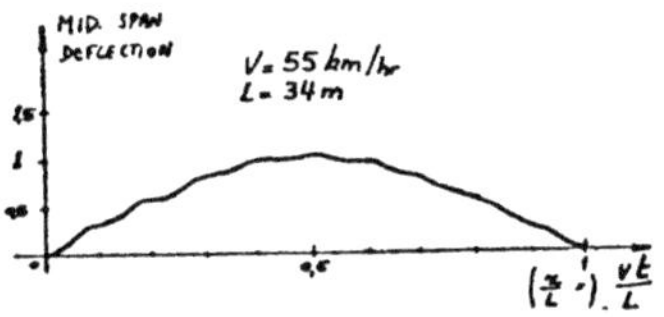

Fig. 58

If before arriving on the bridge-deck the vehicle runs over an obstacle which induces oscillations, the bridge is also subjected to this effect and Cd reaches about 2 for 100 km/hr (Fig. 59).

If the bridge-deck presents a global imperfection, the bridge-motion is attenuated compared to the precedent case (Fig. 60).

If a two-span bridge is affected by global defects, the influence of these two imperfections on the deflection of the span are given in Fig. 61.

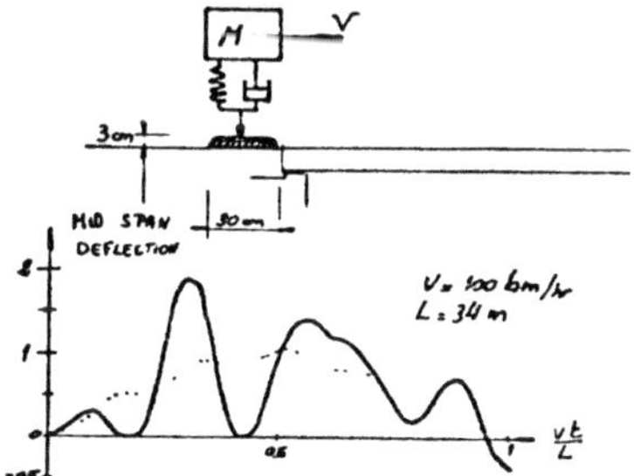

Fig. 59

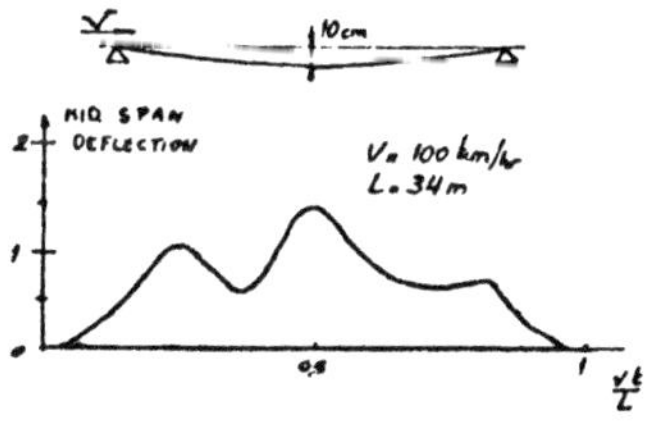

Fig. 60

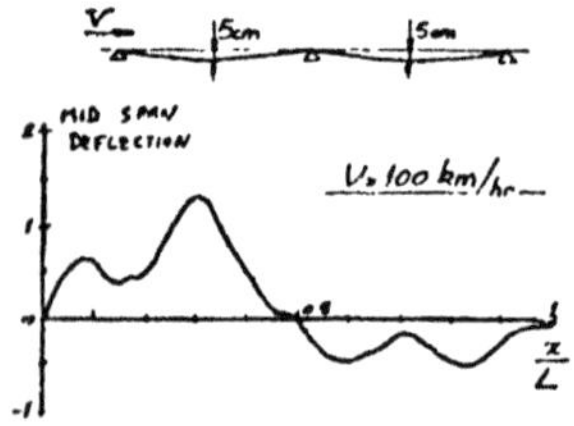

Fig. 61

A local imperfection at mid-span

The vehicle, this time, gives a great impulse and the bridge shows an important couplage. The graph-representation of the dynamic coefficient variations in the different transversal sections of the bridge shows clearly how a unique value (generally accepted in codes) is of poor signification.

For a local obstacle, situated at mid-span, the velocity at the vehicle has a great influence on the dynamic moment C_M (Fig. 62 and 63).

As it concerns an energy transfer from the vehicle to the bridge, the vehicle damping as well as the bridge's has great effect on the value of C_M . The velocity of the vehicle is also important as shownin Fig. 64 and Fig. 65.

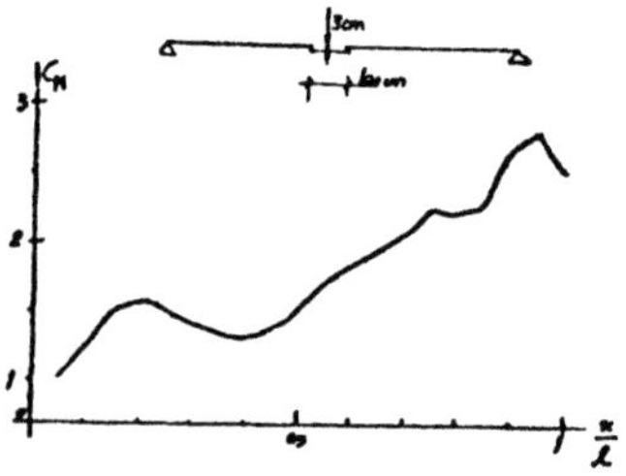

Fig. 62

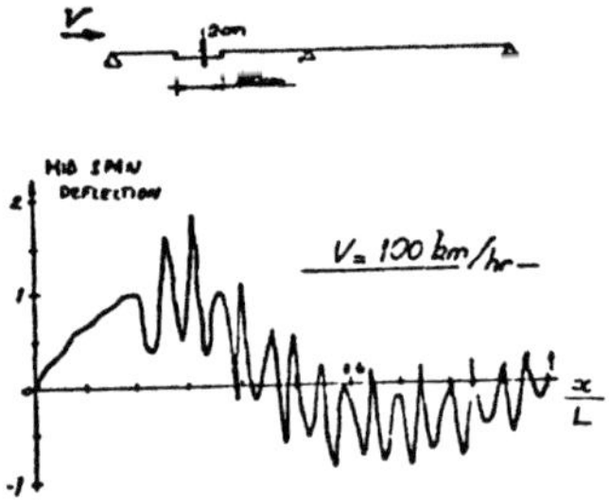

Fig. 63

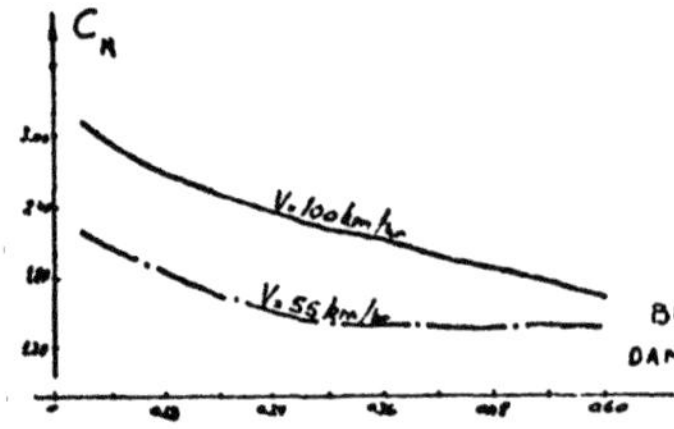

Fig. 64

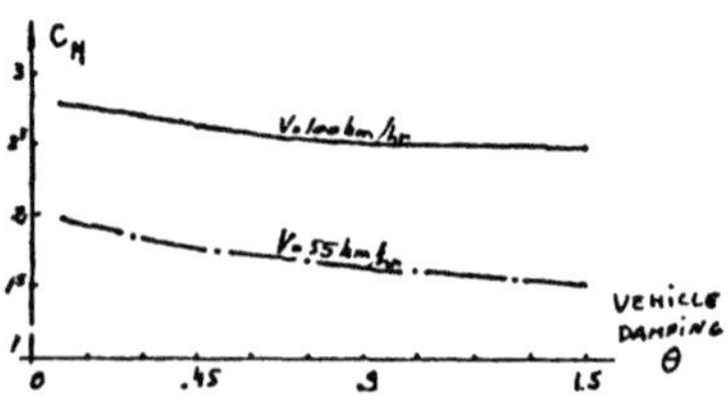

Fig. 65

This study shows how the profile imperfections can affect the dynamic bridge behaviour. A local imperfection can increase deflection and moments much more than the dynamic code coefficients let us suppose. The immediate mechanical consequence of profile imperfection consists in a more severe stress cycle (level and number of cycles) so that the damage process can affect considerably the life time of the bridge.

During the final stage of a bridge construction heavy vehicles crossing the uneven deck can develop higher stress than in the future life time.

Moreover, in certain conditions road surface profiles can be modified, for example when differential melting snow occurs. This is why in different circumstances, in certain countries, in addition to the classical dynamic coefficient curve a second curve with higher values has been introduced in the codes.

CONCLUSION

The problem of the dynamic behaviour of a bridge is complex. Different methods of measurements and analysis process have been presented. The control of the bridge response under traffic demands important infrastructures. Whereas the dynamic tests, such as vertical impulse or sinusoidal excitation which can be repeated in the same conditions, are the indicators for survey and maintenance.

This paper shows that on the one hand different bridge types must be tested in situ in order to establish the diagnostic, and on the other hand, the dynamic behaviour of new elements and material must be evaluated in advance by a more scientific approach through laboratory tests and mathematical simulation. These 2 aspects are complimentary. In the stage of bridge design the life time prediction must be obtained with an acceptable level of threshold. That means that the whole structure must resist simultaneously two types of phenomena :

- mechanical loading, which introduces stress cycles with the consequence of fatigue and cumulative damage
- physical and chemical environmental effects and their interaction with the first one.

The mechanical characteristics can be considered as indicators of the vibrational behaviour so that the durability of a bridge or bridge element is nothing more than the maintenance of performance in time. The value of frequencies and damping of the different modes are with the stress the main indicator. Of course their evolution depends on the environmental characteristics such as temperature effect, chemical action, corrosion, time.

In laboratories,these characteristics can be more or less simulated but the correlation with the in situ behaviour is not easy when the time factor is of great importance.

In fact, concerning the environmental agents which can attain the mechanical performance, the life time cannot be correctly estimated. Exterior conditions, bridge type, load cycles... vary to such a great degree that present experience is almost always founded on numerous spot observations made in situ.

More precise methods of damping control must be established to facilitate the appreciation of the state of cracks and the cumulative damage. These methods work well under laboratory conditions at the present time as in situ for damaged bridges. A methodology situated between the 2 extremes has yet to be elaborated for a better survey of a given section or of the whole bridge.

So the measure of the evolution of all indicators - amplitude, frequency, damping - must be accomplished to improve the in situ diagnostics and permit recognition of the cause in order to choose the appropriate maintenance intervention : girder bearing foundation... This also contributes to improvement of codes and standards concerning safety and life-time.

REFERENCES

CHAPTER I

- Bendat J.S., Piersol A.G. - Random data : analysis and measurements procedures. Wiley Interscience, New York, 1971.

- Guillard Y. - Vibration dues au trafic routier, dégradation de l'environnement. Thèse 3e cycle. Fac. Science, Université Le Mans, 1978.

- Zeller E. - Comportement vibratoire d'une structure métallique test impulsion et excitation sinusoïdale. Etude Juin 1982.

CHAPTER II

- De Back J., Kolstein M.H., Van Leeuwen J.L. - Measurements and interpretation of traffic loading in bridges. Lewin Inst. Univ. Technology Civil Eng., DELFT, 1981.

- R. Pfeiffer, J.M. Zaschel - Ermittlung der Verkehrslasten und der Verkersbeanspruchungen von Stahlbrücken als Grundlage für künftige Bemessungsregeln, LBF Fraunhofer Institut, Darmstadt FhG, Berichte n° 4, 1979.

CHAPTER III

- J. Caracilli - Constatation sur ouvrages métalliques en construction et en service. Etude Paris 1980.

- L. Ferron, E. Zeller - Comportement vibratoire d'un plancher industriel, du calcul aux tests dynamiques. Construction Métallique n° 4 1982.

CHAPTER IV

- H.P. Lieurade - Essais de fatigue sous sollicitation d'amplitude variable, RFP 255, IRSID 1978.

- C.P. Heins - Bridge response and damage, J. Washington Ac. Science 1979, 69 n° 1.

- M. Gregoire - Interprétation des résultats d'analyse statistique de sollicitations relevées en service, CETIM. 1961

- B.T. Yen - Minimizing fatigue and fracture in steel bridges, ASME 1979 Conference, Washington.

- J. Caracilli, A. Aloustaz - Influence de la précontrainte sur l'amortissement des vibrations transversales des poutres en Béton Précontraint, Bulletin Liaison Labo P.C., Mai-Juin 1978.

- J.L. Van Leeuwen, AJM Siemens - Miner's rule with respect to plain concrete, Heron vol. 24, 1979.

- H. Van Koten - Structural damping, TNO CUR report n° 75, 1977.

- Y. Yasoshima, Y. Matsumoto, T. Nishioka - Studies on the running stability of railway vehicles on suspension bridges. Journal of Fac. of Eng. University of Tokyo, n° 1, 1981.

CHAPTER V

- Appareils d'appui en elastomère fretté, Bulletin technique n° 4, 1974.

- Palamas J. - Imperfections du profil d'un pont et vibrations sous trafic, recherche effectuée au LCPC Paris, Thèse 1982.

- Y. Guillard, E. Zeller - Initiation Dynamique des Structures, constructions parasismiques; mesures sur ossatures industrielles et ponts sous trafic routier. Cours Ecole d'Ingênieurs Géophysiciens - Institut de Physique du Globe, Strasbourg.

V. EARTHQUAKE AND WIND EFFECTS

SEISMIC DESIGN OF HIGHWAY BRIDGES

C. P. Heins, Professor
Institute for Physical Science and Technology and
Civil Engineering
University of Maryland, College Park, MD. 20742

INTRODUCTION

With the advent of the 1964 Alaskan Earthquake, the 1971 San Fernando Earthquake and more recently the 1978 Santa Barbara Earthquake (1), bridge structures in the USA have undergone considerable destructive forces.

The earthquakes have caused the bridge professionals to reassess the design techniques that have been applied, up until that time, for seismic design.

The present 1977 AASHTO bridge code (2), as related to seismic design, suggests an equivalent static force method for simple structures and when the structure is complex, for example curved bridges, a computer based response spectrum or dynamic analysis should be considered.

In this present 1977 AASHTO code, most engineers would utilize the seismic coefficient method (SCM), however, the utilization of the SCM, may give erroneous results when designing under seismic conditions (3).

Because of such errors and from experience gained from recent earthquake the 1975 AASHTO code was reassessed in 1977 by the ATC (4) research group. The work of this council, developed a new specification (5), which will be given in part herein, relative to straight girder bridges.

Although the ATC-6 design guide (5) has advanced the state of seismic design of highway bridges, it was recognized (6) that additional research was necessary to determine the seismic response of bridges which are influenced by geometry and curvature. Such a study has been conducted (7), resulting in a proposed design guide for curved steel box girder bridges. These results will be given in detail in the following section.

STRAIGHT HIGHWAY BRIDGES

The analysis techniques, as recommended in the proposed design guide (6) are:

1. Single-Mode Spectral Method (SMSM)
2. Multi-Mode Spectral Method (MMSM)

The (SMSM) procedure can be applied by considering;

i) Distributed mass and equivalent static loading (5)

ii) Equivalent structural stiffness (k) and equivalent static loading (8, 9)

in which both methods can employ simple analytical techniques not requiring dynamic computer programs (10).

The (MMSM) requires modeling of the structure as a lumped mass or distributed mass continuum and application of a given response spectrum or time history. The difficulty involved in evaluating the various mode shapes and eigen vectors and actions requires use of a computer program (1). Such application is generally warrented when the structure has an unusual configuration.

The details to be presented herein will only be relative to the (SMSM) technique and direct design criteria.

General - The development of the (SMSM) technique requires utilization of the generalized displacement function concept. In order to explain this idea, consider the static displacement of a general girder given as;

$$\Delta(x) = \sum (b_n) \sin \frac{n\pi x}{L} \quad (1)$$

Eq. (1) can also be written as;

$$\Delta(x) = \sum (Z_n) \quad \psi_n (x) \quad (2)$$

where; Z_n = Amplitude terms

ψ_n = displacement functions

In the case of dynamic response the displacement function can be described as;

$$\Delta(x,t) = \psi(x)\, Z(t) \quad (3)$$

where: $Z(t)$ = Amplitude of motion

$\psi(x)$ = Shape function

The generalized displacement function as given by Eq. (3), will now be used to develope design oriented equations.

Formulation of SDOF Equations - Consider a SDOF system, as shown in Fig. 1, subjected to a ground motion. Summation of the forces gives;

$$f_I + f_D + f_S = P(t) = m\ddot{\Delta}_g \tag{4}$$

where:
$$\begin{cases} f_I = m\ddot{\Delta} \\ f_D = c\dot{\Delta} \\ f_S = k\Delta \end{cases} \tag{5}$$

The virtual work method will now be used to obtain the general equations of motion; and is given by;

$$\delta W = \int F \cdot \delta\Delta = 0 \tag{6}$$

where:
$$\begin{cases} F = \text{Applied Forces} \\ \delta\Delta = \text{Virtual Displacement} \end{cases}$$

The generalized displacement functions, including ground displacement, can be written as;

$$\Delta^t(w,t) = \psi(x) \quad Z(t) + \Delta_g(t) \tag{7}$$

Substituting Eq. (7) into Eq. (5) gives;

$$\left.\begin{aligned} f_I &= m[\psi(x)\ddot{Z}(t) + \ddot{\Delta}_g(t)] \\ f_D &= c[\psi(x)\ddot{Z}(t)] \\ f_S &= k[\psi(x)Z(t)] \\ \delta\Delta &= \psi(x)\delta Z(t) \end{aligned}\right\} \tag{8}$$

Substituting Eq. (8) into Eq. (6) gives;

$$\delta w = \int_0^L m(x)[\psi(x)\ddot{Z}(t) + \ddot{\Delta}_g(t)]\psi(x)dx\delta Z(t)$$
$$+ \int_0^L c(x)[\psi(x)\ \dot{Z}(t)]\psi(x)dx\delta Z(t)$$
$$+ \int_0^L k(x)[\psi(x)Z(t)]\psi(x)dx\delta Z(t) = 0$$

Since $\delta\Delta = \psi(x) \cdot \delta z(t)$ is arbitrary, then set equal to any value and also assume;

$$\left\{\begin{aligned} \gamma &= \int_0^L (m(x))\ [\psi(x)^2]dx \\ \beta &= \int_0^L m(x)\ [\psi(x)]dx \\ c^* &= \int_0^L c(x)\ [\psi(x)^2]dx \\ k^* &= \int_0^L k(x)\ [\psi(x)^2]dx \end{aligned}\right\} \tag{9}$$

therefore;

$$\gamma \cdot \ddot{Z}(t) + c^*\ \dot{Z}(t) + k^*\ Z(t) = -\ddot{\Delta}_g(t)\ \beta$$

or

$$\ddot{Z}(t) + (\frac{c^*}{\gamma})\ \dot{Z}(t) + (\frac{k^*}{\gamma})\ Z(t) = -\ddot{\Delta}_g(t)\frac{\beta}{\gamma} \tag{10}$$

Let:

$$\omega^2 = \frac{k^*}{m(x)} \quad ; \quad \xi = \frac{c^*}{2m(x)\omega}$$

Then the general equation of motion is;

$$\ddot{Z}(t) + 2\xi\omega\dot{Z}(t) + \omega^2\ Z(t) = \ddot{\Delta}_g(t)\frac{\beta}{\gamma} \tag{11}$$

The solution of Eq. (11) is obtained using Duhamel integral;

$$Z(t) = \frac{\beta}{\gamma} \cdot \frac{\bar{\Delta}(t)}{\omega}$$

where: $\Delta(t) = \frac{\bar{\Delta}(t)}{\omega}$

and $\bar{\Delta}(t) = \int_0^t \ddot{\Delta}_g(\tau)\exp[-\xi\omega(t-\tau)\sin\omega(t-\tau)dx$

which is the earthquake response integral.

The local displacements are therefore;

$$\Delta(xt) = \psi(x) \quad Z(t) = \psi(x)\frac{\beta}{\gamma} \cdot \frac{\bar{\Delta}(t)}{\omega} \tag{12}$$

Local Forces:

Neglecting damping:

$$f_I(x,t) + f_s(x,t) = 0$$

$$f_I(x,t) = m(x) \quad \ddot{\Delta}(x,t)$$

for harmonic motion;

$$\begin{cases} \Delta = A \sin\omega t \\ \dot{\Delta} = \omega A \cos\omega t \\ \ddot{\Delta} = -\omega^2 A \sin\omega t \end{cases}$$

Substituting $\Delta = A \sin\omega t$ gives

$$\ddot{\Delta} = -\omega^2\Delta$$

therefore $f_I(x,t) = -m(x)\omega^2\Delta(x,t) = -f_S(x,t)$

or $f_S(x,t) = [m(x) \quad \omega^2]\psi(x)\frac{\beta}{\gamma} \cdot \frac{\bar{\Delta}(t)}{\omega}$

$$f_S(x,t) = m(x) \quad \psi(x)\frac{\beta}{\gamma}\omega\ \bar{\Delta}(t) \tag{13}$$

Maximum effects:

Displacements

Using the response spectrum, the spectral velocity, which by definition is a maximum of the response integral or

$$S_v(\xi,T) \equiv \overline{\Delta}max(\xi,T)$$

also from the response spectrum

$$\Delta(x,t)_{max} = \frac{1}{\omega} S_v = S_d$$

$$f_S(x,t)_{max} = m\omega S_v = mS_a = S_d$$

therefore:

$$\overline{\Delta}_{max} = S_v = S_{a/\omega}$$

and

$$S_a = g \cdot C_s$$

$$\therefore \overline{\Delta}_{max} = C_s g/\omega$$

$$\Delta(t,w)_{mx} = \psi(x)\frac{\beta}{\gamma}\frac{C_s g}{\omega^2} \quad (14)$$

Forces

$$f_S(x,t) = m(x)\psi(x)\frac{\beta}{\gamma}\omega\overline{\Delta}(t)$$

$$f_S(x,t) = m(x) \cdot \psi(x)\frac{\beta}{\gamma} \cdot \omega\frac{Sa}{\omega}$$

$$f_S(x,t) = m(x) \cdot \psi(x)\frac{\beta}{\gamma} \quad S_a$$

However

$$\begin{cases} Sa = g \cdot C_s \\ m(x) = \dfrac{w(x)}{g} \end{cases} \quad \text{and}$$

$$f_s(x) = \frac{\beta}{\gamma} C_s w(x)\ \psi(x) \tag{15}$$

where:

C_s = Elastic Seismic Response Coefficient

Natural Frequency:

The natural frequency can be determined by equating external and internal work. Consider a simply supported beam subjected to a load P_o, as shown in Fig. 2, the external work is;

$$W_E = \int \frac{F \cdot \Delta}{2} = \frac{P_o}{2} \int_0^L \Delta(x)\,dx = \frac{P_o}{2} \cdot \alpha$$

where: $\alpha = \int_0^L \Delta(x)\,dx$

the kinetic energy is given by;

$$T_{max} = \int \tfrac{1}{2}\, m(x) v^2 \omega^2 dx$$

$$T_{max} = \tfrac{1}{2} \frac{\omega^2}{g} \int_0^L w(x) (\dot{\Delta}_s)^2 dx$$

$$T_{max} = \frac{\omega^2 \gamma}{2g} \quad ; \quad \gamma = \int_0^L w(x) (\dot{\Delta}_s)^2 dx$$

Equating $W_E = T_{max}$ gives;

$$W_E = U_{max} = T_{max}$$

$$\frac{P_o}{2} \alpha = \frac{\omega^2 \gamma}{2g} \qquad \text{and} \quad \omega = \frac{2\pi}{T}$$

Therefore the natural frequency is

$$T = 2\pi\left[\frac{\gamma}{P_o\, g\alpha}\right]^{1/2} \tag{16}$$

Elastic Seismic Response Coefficient C_s

Computation of the SRC is utilized in conjunction with the displacements given by Eq. (14) and the local forces given by Eq. (15). The coefficient C_s, given as follows, is a function of the acceleration coefficients, site conditions and natural frequency (given by Eq. (16)), and is;

$$C_s = \frac{1.2\,AS}{T^{2/3}} \quad (17)$$

where:

A = Acceleration Coefficient (Fig. 3)

S = Site Coefficient Table

T = Natural Frequency (Eq. 16)

General Coefficients A, S:

The coefficient A and S given in Eq. (17) are prescribed as follows:

Acceleration Coefficient A. - The acceleration coefficient A, is given in Fig. 3, as applied to the United States.

Site Coefficient S. - The site coefficient approximates the effects of the site conditions on the elastic response coefficient and is;

(S) Coefficient

	Soil Profile Type		
	I	II	III
S	1.0	1.2	1.5

The soil profile type identification I, II, III is prescribed as follows:

SOIL PROFILE TYPE I is a profile with either

1. Rock of any characteristic, either shale-like or crystaline in nature (such material may be characterized by a shear wave velocity greater then 2,500 ft/sec (762 m/sec), or by

other appropriate means of classification); or

2. Stiff soil conditions where the soil depth is less than 20 (61 m) and the soil types overlying rock are stable depos sands, gravels, or stiff clays.

SOIL PROFILE TYPE II is a profile with stiff clay or deep cohesion conditions where the soil septh exceeds 200 ft (61 m) and the soil types overlying rock are stable deposits of sands, gravels, or stiff clays.

SOIL PROFILE TYPE III is a profile with soft to medium-stiff clays sands, characterized by 30 ft (9 m) or more of soft to medium-stiff clay with or without intervening layers of sand or other cohesionless soils.

Design Procedure - The design technique utilizing the previously gi equations, is performed as follows (5):

STEP 1

Apply unit uniform loading P_o in transverse direction Fig. 4 and evaluate displacement functions $\psi(x)$, similarly apply unit longitudinal loading, Fig. 4(b), and evaluate $\psi(x)$.

STEP 2

Calculate parameters α, β, and γ ;

$$\alpha = \int \psi(x)dx$$

$$\beta = \int m(x)\ \psi(x)dx \quad \text{or} \quad \int w(x)\psi(x)dx$$

$$\gamma = \int m(x)\psi(x)^2dx \quad \text{or} \quad \int w(x)\psi(x)^2dx$$

where $m(x)$ is the mass and $w(x)$ weight of dead load of bridge su structure.

STEP 3

Calculate natural period;

$$T = 2\pi[\gamma/P_o g\alpha]^{1/2} \quad , \quad P_o = \text{unit load}$$

STEP 4

Calculate equivalent static earthquake loading:

$$f_s(x) = \frac{\beta}{\gamma}C_s w(x)\psi(x) \qquad \text{(force/unit length)}$$

STEP 5

Apply loading $f_s(x)$ to structure and determine resulting member faces and displacements.

CURVED BOX GIRDER HIGHWAY BRIDGES

General - As indicated previously, the proposed highway design code (5) does not include curved girder bridges. Therefore a study (7) has been undertaken to develope such a design technique, which will now be described.

THEORY

The general static response of curved bridge structures requires incorporation of the interaction between the bending and torsional forces. Such interaction can be considered by solving Vlasov equations (11), or by development of ths stiffness matrix (10) and appropriate restraint conditions.

The matrix oriented technique, however, is more versatile, in that a three dimensional model (space frame) can be considered. This then permits modeling of the structure such that the support conditions can represent the physical restraints. Also, the induced actions can be applied in three directions, and thus simulate the various earthquake induced actions.

Therefore, the study of the induced actions on a structure, when subjected to earthquakes, was confined to utilization of a space frame matrix simulation. The basic modeling consists of series of typical elements attached rigidly together to form a continuous curved box girder bridge.

The basic properties of each beam element will consist of I_z, I_y and K_T . Although warping and distortional properties (I_w, W_n, W_a) can be computed, they were not considered in this study, as it has been shown (12), that with proper bracing of the box girder, warping and distortional effects are negligible.

Therefore, by proper evaluation of the stiffness of each beam element, and identifying each joint load, the static response of the continuous curved girder can be determined. The static response can then be utilized to determine the effective earthquake effects, using the response spectrum curves. The general procedure using this method will now be described.

The support restraints, to be imposed on the bridge model, can be identified as releases in the computer model. Because a space frame model is being utilized, six releases or restraints were considered.

Equivalent Dynamic Analysis.-

The natural frequency response of a single degree of freedom system can be predicted by the following;

$$\omega = \sqrt{\frac{k}{m}} \tag{18}$$

where

k = spring constant

m = mass (w/g) or mass moment of inertia .

If the natural frequency ω of the system, as computed from Eq. (18) is reliable, the the corresponding induced acceleration of the the mass m created by an earthquake can be predicted by using the response spectrum curve given in Figures 5, 6, and 7. The resultant accelerations obtained therein are then used to determine the induced dynamic force

$$F = M \cdot a_{R.S.} \tag{19}$$

where

$a_{R.S.} = \ddot{y}$ = linear acceleration obtained from response spectrum curves.

If the system is subjected to angular accelerations $\ddot{\theta}$, then the induced dynamic torque is

$$M = \overline{I}\ddot{\theta}_{RS} \qquad (20)$$

where:

$\overline{I}$ = mass moment of interia = $\int \rho r^2 dA = \rho_{steel}\int (x^2+y^2)dA + \rho_{concrete}\int (x^2+y^2$

$\ddot{\theta}_{RS}$ = rotational acceleration obtained from the response curve

ρ = mass per unit area.

Natural Frequency.-

As indicated by Equation (18), the natural frequency of a SDOF system is given as a function of the spring stiffness (k) and the spring mass (m). If the system is a bridge structure the spring constant can be represented by:

$$k = \left(\frac{1}{\Delta}\right) \qquad (21)$$

where:

Δ = induced maximum displacement caused by a unit load.

Therefore by determining the response of a given box girder bridge, when subject to a unit load, an equivalent spring constant can be obtained. This constant, k, and the total mass of the bridge, will then permit evaluation of the natural frequency as given by Eq. (18).

In the instance of curved structures, the dynamic action can occur in three principal directions, and one primary rotation as shown in Figure 8. The esulting maximum displacement, induced by these unit loads, will then give the orresponding equivalent stiffnesses k_x, k_y, k_z and k_T.

he corresponding natural frequencies can then be determined by applying Eq. (18).

BRIDGE STUDIES

Typical Sections.-

In order to develop a simplified design technique, the response of arious curved box girder bridges must be examined. Such box girders, hich have been used in previous studies (13), were utilized in this arametric study. In this study only the three-lane, three girder ystem were considered because this is most typical of the curved box irder structures (12).

The basic span length configurations that were examined, are shown n Fig. 9, where L=50' to 150' and N=1,2. The radius for these various tructures that were used, varied from 200 feet to infinity.

eneral.-

Utilizing the basic box geometry and the support spring constants of

$$k_x = 0 \sim 2 \times 10^3 \text{ k/in; } K_y = \infty \text{ ; } k_z = 0.5 \times 10^3 \text{ k/in.} \sim \infty$$

he equivalent seismic resonse of the single, two, three, and four span tructures were examined. The resulting natural frequencies (ω_x, ω_y, ω_z, ω_t) or all bridge spans and their corresponding induced accelerations were en obtained (7). For the continuous spans, the pier flexibilities, as ven by k_x and k_z were also included as a variable.

The induced accelerations , as determined from the response spectra have been plotted as a function of $(\frac{L}{R})$ and the span length (7), for the single span structure. The relationship between the corresponding accelerations for the continuous spans and the single span however have plotted as a function F , where

$$F_x = \frac{A_x \text{ (continuous span value)}}{A_x \text{ (single span value)}} \quad (2$$

versus the number of spans, radius, stiffness and span lengths.

For all the span lengths of 50', 100', and 150', similar relationsh between F_y, F_z, and F_t have been plotted (7). These data were then used develop appropriate design criteria.

DESIGN CRITERIA

Trends.-

The seismic design of continuous curved box girders were related t the response of single span curved girders. Therefore, the single span accelerations (A_z, A_y, A_z and A_t) were determined with respect to the b bridge geometry, giving the following:

A_x

for 100' < L ≤ 150' $\quad A_x = 2.2\ (\frac{L}{R})^2+0.011L+0.45$ (2

A_y

for L ≥ 100' $\quad A_y = -0.016(L)+4.7$ (2

A_z

for L ≥ 100' $\quad A_z = 3.8$ (2

$\underline{A_t}$

for $L \geq 100'$ at = 1.5 (26)

The continuity factors F have similarly been determined in analytical form, resulting in the following four continuity factors:

$\underline{F_x}$ (longitudinal)

$$F_x = -0.02\ (L) + 3.75 + K \qquad (27)$$

where: $K = 0.00125\ R$ for $R \leq 600'$

$K = 1.0$ for $R > 600'$

$\underline{F_y}$ (vertical) (28)

$$F_y = -0.125\ (NS) - 0.002L + 1.35$$

where: NS = number of spans (2, 3, or 4)

$\underline{F_z}$ (transverse) (29)

$$F_z = -0.005L + 1.5$$

$\underline{F_t}$ (torsion) (30)

$$Ft = -0.075\ (NS) + 1.15$$

where: NS = number of spans (2, 3, or 4)

Design Approach.-

The equivalent seismic design of curved box girder bridges will incorporate the primarily developed equations and the effective peak acceleration map (K_p), given in AASHTO. The general design equation is of the form:

$$EQ_n = F_n \cdot A_n \cdot m \cdot K_p \qquad \text{Translation} \qquad (31)$$

$$EQ_n = F_n \cdot A_n \cdot I \cdot K_p \qquad \text{Rotation} \qquad (32)$$

where:

F_n = continuity factor in x, y, z or t directions

A_n = single span acceleration

K_p = effective peak acceleration modifying factor

$m = w/g = \frac{\text{total weight of structure}}{\text{gravity}}$

$\bar{I}$ = rotational mass moment of section = $\rho \int (x^2+y^2)\,dA$

ρ = mass per unit area

EQ_n = total applied seismic force in x, y, z, or t directions

for the specific direction n, the continuity factor F and single span acceleration A_n are given by:

Longitudinal Direction (x)

$F_x = -0.02\ (L) + 3.75 + K$

$K = 0.00125\ R$ for $R \leq 600'$

$K = 1.0$ $R > 600'$

$A_x = 2.2\ (\frac{L}{R})^2 + 0.011\ L + 0.45$

Vertical Direction (y)

$F_y = -0.125\ (NS) - 0.002L + 1.35$

$A_y = -0.016\ (L) + 4.7$

Transverse Direction (z)

$F_z = -0.005\ (L) + 1.5$

$A_z = 3.8$

Torsional Direction (θ)

$F_t = -0.075\ (NS) + 1.15$

$A_t = 1.5$

EXAMPLE

In order to illustrate the application of these proposed equations, consider a two span curved box structure of spans 100'-100' and

radius R = 600'

the single span accelerations are:

$\underline{A_x}$

$$A_x = 2.2\left(\frac{L}{R}\right)^2 + 0.011\,L + 0.45$$
$$= 2.2\left(\frac{100}{600}\right)^2 + 0.011\,(100) + 0.45$$
$$= 1.61$$

$\underline{A_y}$

$$A_y = -0.016\,(L) + 4.7$$
$$= -0.016\,(100) + 4.7$$
$$= 3.1$$

$\underline{A_z}$

$$A_z = 3.8$$

$\underline{A_t}$

$$A_t = 1.5$$

the continuity factors can now be computed as:

$\underline{F_x}$

$$F_x = -0.02\,(L) + 3.75 + K \qquad (K = 0.00125\,R,\quad R \leq 600')$$
$$= -0.02\,(100) + 3.75 + 0.00125\,(600)$$
$$= 2.5$$

<u>F_y</u>

$$F_y = -0.125\ (N.S.) - 0.002\ L + 1.35$$
$$= -0.125(2) - 0.002\ (100) + 1.35$$
$$= 0.9$$

<u>F_z</u>

$$F_z = -0.005\ (L) + 1.5$$
$$= -0.005\ (100) + 1.5$$
$$= 1.0$$

<u>F_t</u>

$$F_t = -0.075\ (N.S.) + 1.15$$
$$= -0.075\ (2) + 1.15$$
$$= 1.0$$

the weight of the bridge is computed as

w = 442.15 Kips

and the mass moment of inertia ($\overline{I}$) = 2273.63 Kip-in - sec^2

therefore, the induced seismic forces are computed as:

$$EQ_x = F_x \cdot A_x\ (\frac{W}{g})$$
$$= (2.5)\ (1.61\ g) \cdot \frac{W}{g}$$
$$= 4.025\ W$$
$$EQ_y = F_y \cdot A_y \cdot \frac{W}{g}$$
$$= (3.1)\ (0.9\ g)\ \frac{W}{g}$$
$$= 2.79\ W$$
$$EQ_z = F_z \cdot A_z \cdot \frac{W}{g}$$
$$= (1.0)\ (3.8\ g) \cdot \frac{W}{g}$$

$= 3.8\ W$

$EQ_t = F_t \cdot A_t \cdot \bar{I}$

$= (1.0)\ (1.5)\ (\bar{I})$

$= 1.5\ \bar{I}$

the equivalent induced uniform seismic forces q_x, q_y, q_z, q_t, are therefore:

$$q_x = \frac{EQ_x}{2 \times L} = \frac{4.025 \times 442.15}{2 \times 100 \times 12} = 0.7415 \text{ kips/in}$$

$$q_y = \frac{EQ_y}{2 \times L} = \frac{2.79 \times 442.15}{2 \times 100 \times 12} = 0.5140 \text{ kips/in}$$

$$q_z = \frac{EQ_z}{2 \times L} = \frac{3.8 \times 442.15}{2 \times 100 \times 12} = 0.7 \text{ kips/in}$$

$$q_t = \frac{EQ_t}{2 \times L} = \frac{1.5 \times 2273.63}{2 \times 100 \times 12} = 1.4210 \text{ kip-in/in}$$

Evaluation of the reactions of curved box girder bridge, using a straight girder gives:

i) x direction

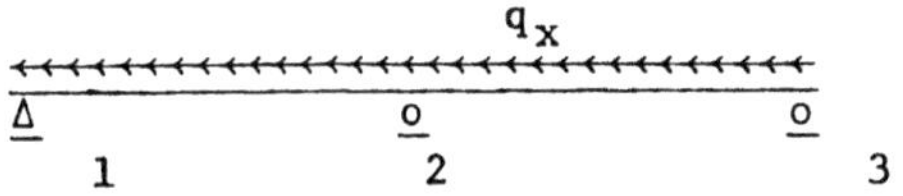

$Rx^1 = qx \cdot (2L) = EQx = 1779.65$ kips

$Rx^2 = 0$

$Rx^3 = 0$

ii) y direction

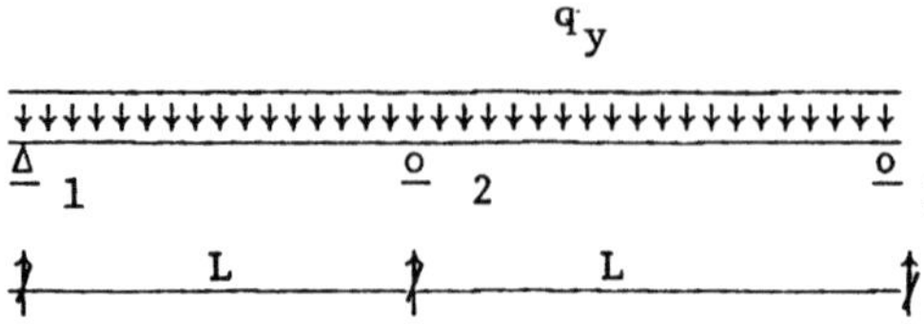

Ry^1 $= 0.375q_yL$ Ry^2 $= 1.25q_yL$ Ry^3 $= 0.375q_yL$

$$Ry^1 = 0.375q_yL$$
$$= 0.375\ (0.5140)\ (100 \times 12)$$
$$= 231.3 \text{ kips}$$

$$Ry^2 = 1.25q_yL$$
$$= 1.25\ (0.5140)\ (100 \times 12)$$
$$= 771 \text{ kips}$$

$$Ry^3 = Ry^1 = 231.3 \text{ kips}$$

iii) z direction

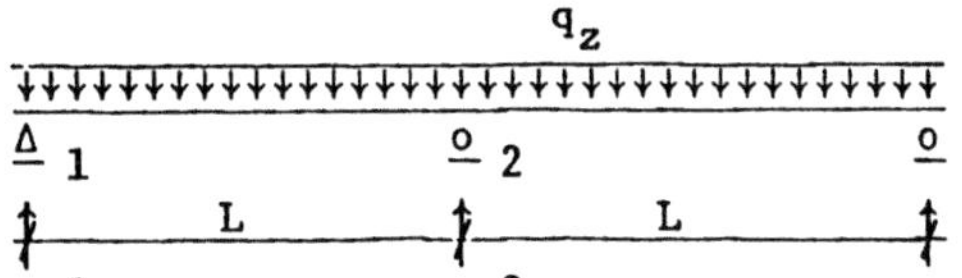

Rz^1 $= 0.375q_zL$ Rz^2 $= 1.25q_zL$ Rz^3 $= 0.375q_zL$

$$Rz^1 = .075q_zL$$
$$= 0.375\ (0.7)\ (100 \times 12)$$
$$= 315 \text{ kips}$$

$$Rz^2 = 1.25q_zL$$
$$= 1.25\ (0.7)\ (100 \times 12)$$

$= 1050$ kips

$Rz^3 = Rz^1 = 315$ kips

iv) Torsion

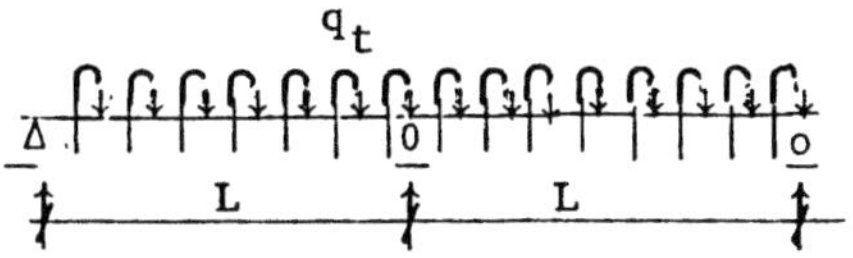

Rt^1 $= 0.375 q_t L$ Rt^2 $= 1.25 q_t L$ Rt^3 $= 0.375 q_t L$

$$Rt^1 = 0.375 q_t L$$

$$= 0.375\ (1.4210)\ (100 \times 12)$$

$$= 639.45 \text{ kip - in}$$

$$Rt^2 = 1.25 q_t L$$

$$= 1.25\ (1.4210)\ (100 \times 12)$$

$$= 2131.5 \text{ kip - in}$$

$$Rt^3 = Rt^1 = 639.45 \text{ kip - in}$$

In the interior support, we use $Ry \times d = T$, $d = 48''$.

$$Ry = \frac{T}{d} = \frac{2131.5}{48} = 44.41 \text{ kips}$$

comparing these results with the results obtained from the (SFS) computer program gives:

Support No. / Direction	1		2		3	
	Eqs	program	Eqs	program	Eqs	program
x direction Rx	1779.65	1298	0	223.9	0	0
y direction Ry	231.3	250	771	775.3	231.3	250
z direction Rz	315	417.5	1050	967.7	315	417.5
Torsion T	639.45	869.8	44.41*	55.5*	639.45	869.8

* Ry x d= T

which shows reasonable agreement.

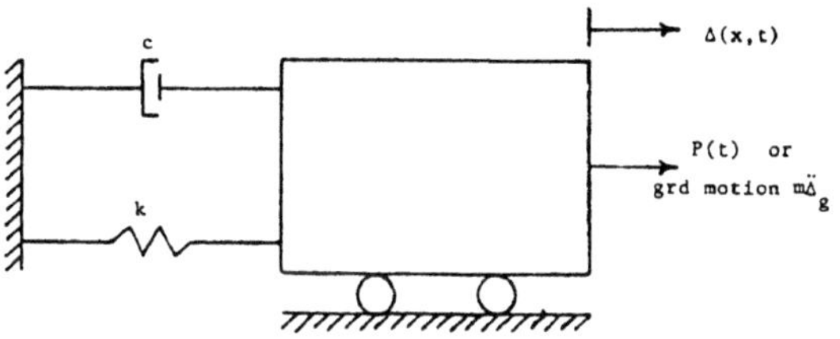

Fig. 1

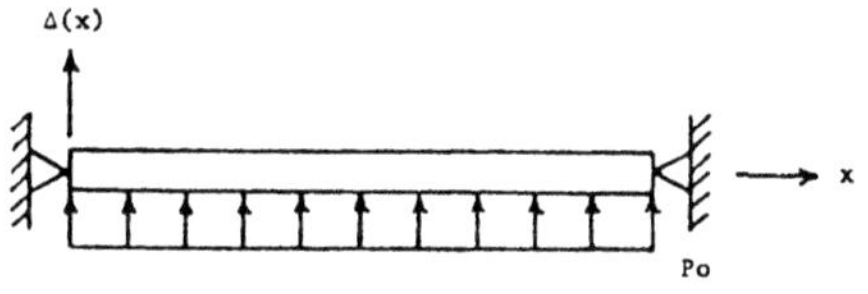

Fig. 2

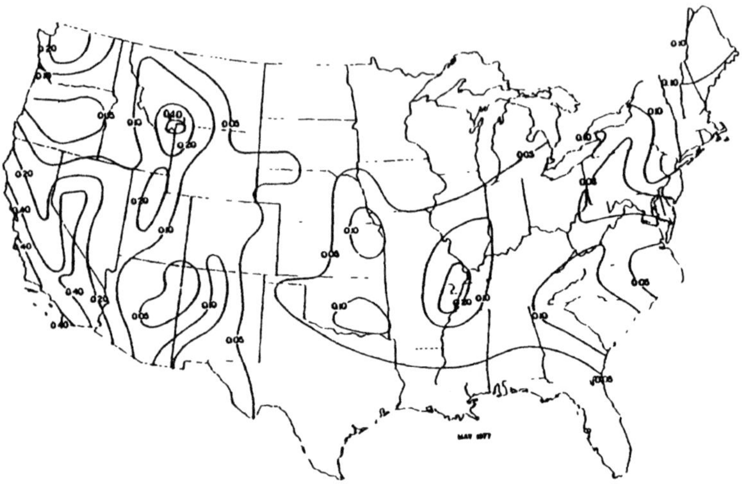

FIGURE 3. ACCELERATION COEFFICIENT – CONTINENTAL UNITED STATES
(An enlarged version of this map, including counties, is given in the back cover of the report)

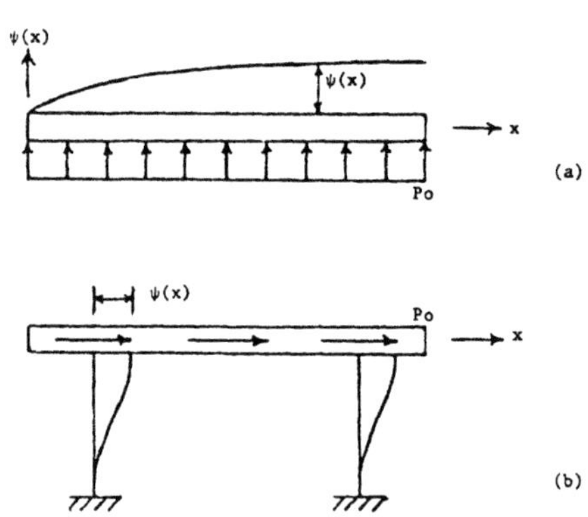

Fig. 4

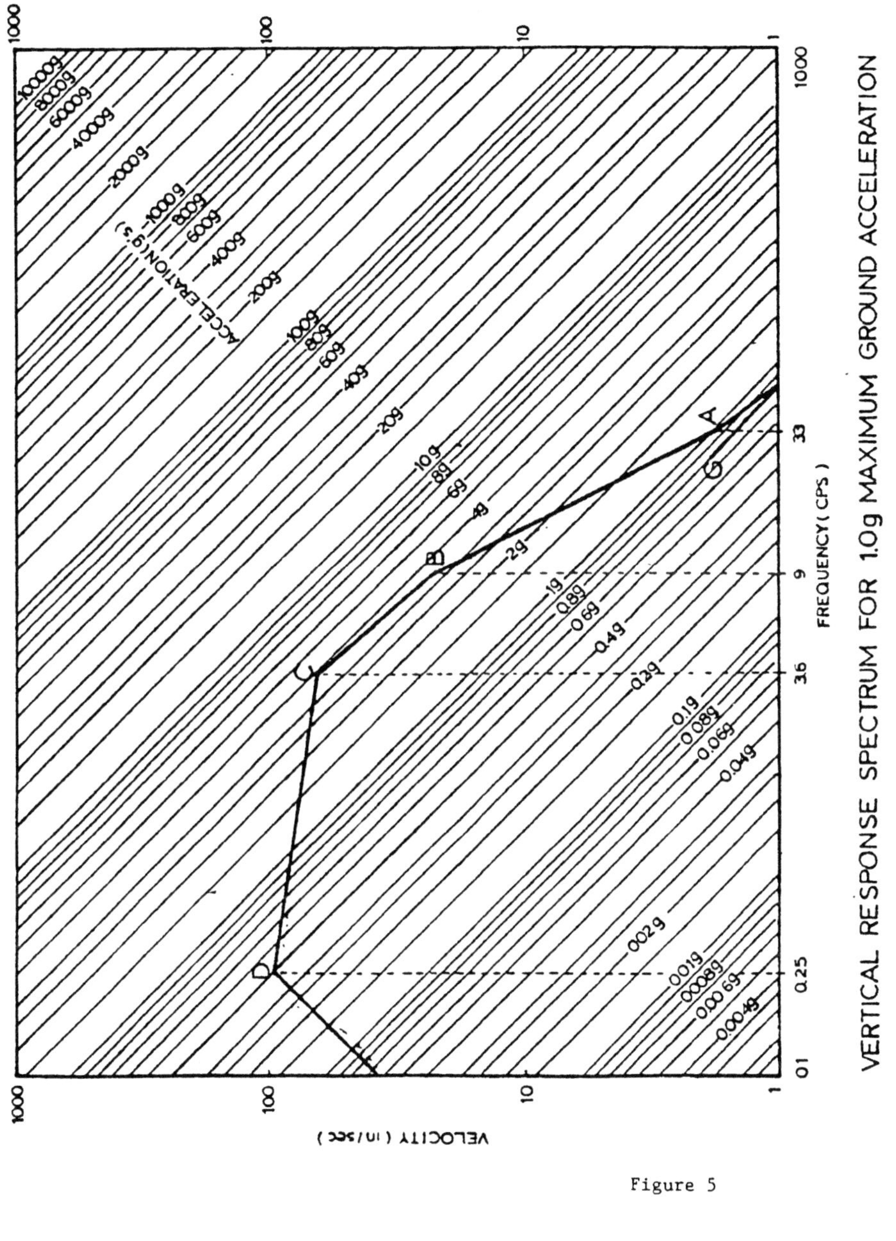
VERTICAL RESPONSE SPECTRUM FOR 1.0g MAXIMUM GROUND ACCELERATION
FREQUENCY (CPS)
VELOCITY (in/sec)
ACCELERATION (g's)
A
B
C
D
G

Figure 5

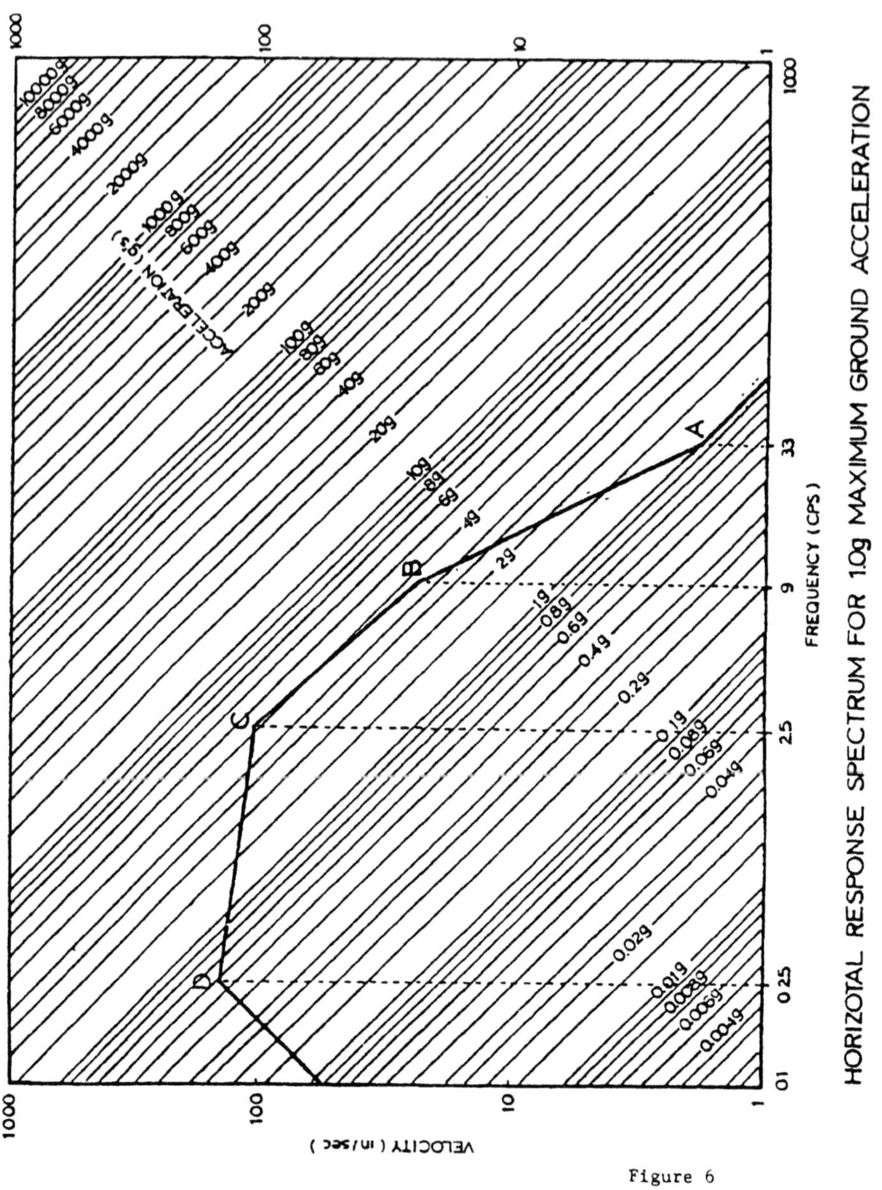

HORIZOTAL RESPONSE SPECTRUM FOR 1.0g MAXIMUM GROUND ACCELERATION
FREQUENCY (CPS)
VELOCITY (in/sec)
ACCELERATION (g's)
A
B
C
D

Figure 6

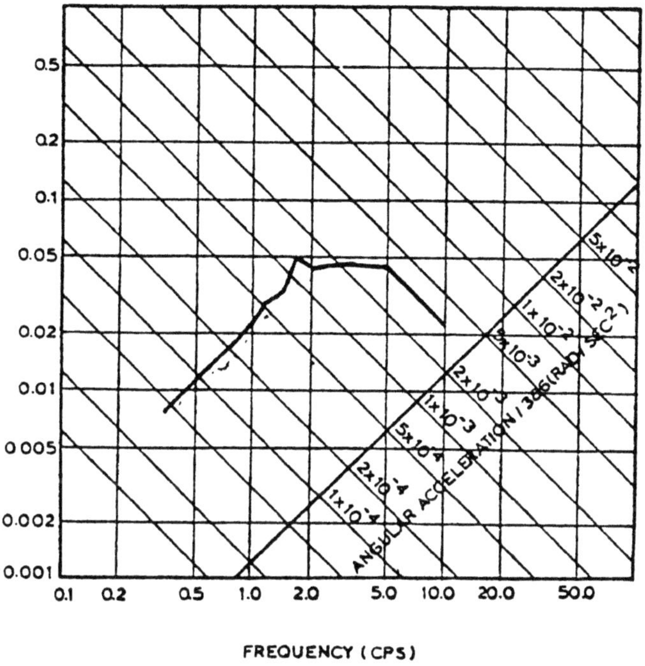

MEAN TORSIONAL SPECTRUM (AVERAGE SHEAR VELOCITY = 570 FT/S)

Figure 7

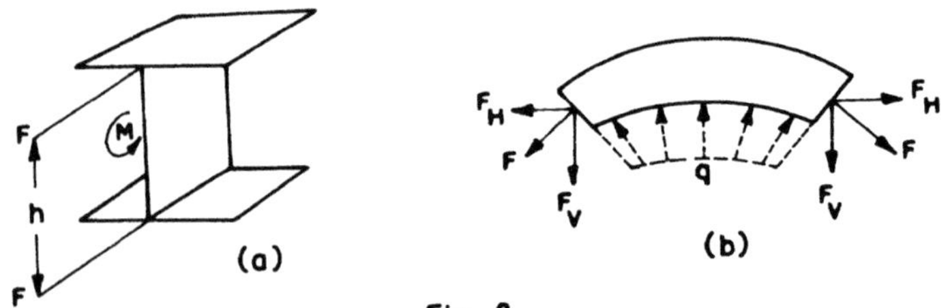

Fig. 8

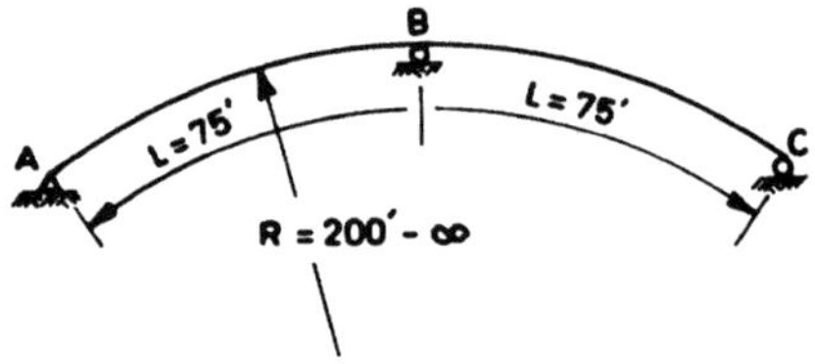

Fig. 9

REFERENCES

1. Miller, R. K., Felszeghy, S. F., "Engineering Features of the Santa Barbara Earthquake of Aug. 1978", BBRI, Berkeley, Calif., 1978.

2. "Eleventh Edition-AASHTO Bridge Design Specification", Washington, D. C.

3. Imbsen, R. A., Nutt, R. V., Pensien, J., "Evaluation of Analytical Procedure Used in Bridge Seismic Design Practice", Proceedings: Earthquake Resistance of Highway Bridges", ATC, Jan. 29/31, 1979.

4. Sharpe, R. L., Mayes, R. L., "Development of the Highway Bridge Seismic Design Criteria for the U. S.". Proceedings: "Earthquake Resistance of Highway Bridges", ATC, Palo Alto, Calif., Jan. 29.31, 1979.

5. "Seismic Design Guidelines for Highway Bridges:, ATC-6, Palo Alto, Calif. October, 1981.

6. Proceedings: "Earthquake Resistance of Highway Bridges", ATC, Palo Alto, Calif., Jan. 29/31, 1979.

7. Lin, I. C., "Equivalent Seismic Design of Curved Box Girder Bridges", M. S. Thesis, Civil Eng. Dept., University of Maryland, College Park, Md., April 1981.

8. Robinson, R. R. et al, "Structural Analysis and Retrofitting of Existing Highway Bridges Subjected to Strong Motion Seismic Loading", FHWA, May 1975.

9. Robinson, R. R., Longinow, A., Chu, K. H., "Seismic Retrofit Measures for Highway Bridges", Vol. 1, FHWA TS-79-216, Washington, D. C., April 1979.

10. Bathe, K, Wilson, E, Peterson, F., "SAPIV, A Structural Analysis Program for Static and Dynamic Response of Linear Systems", C. E. Dept., Univ. of Calif., Berkeley, Calif., June 1973.

11. Heins, C. P., "Bending and Torsional Design in Structural Members", Lexington Books, D. C. Heath Co., Mass., 1975.

12. Heins, C. P., "Box Girder Design, State of the Art". AISC Engr. Journal, Vol. 15, No. 4, 1978.

13. Heins, C. P., Sahin, M. A., "Natural Frequency of Curved Box Girder Bridges", ASCE Structural Journal, Vol. 105, No. ST12, Dec., 1979.

ANALYSIS AND DESIGN OF BRIDGES FOR EARTHQUAKE EFFECTS

Polat Gülkan

Faculty of Engineering, Hacettepe University, Ankara

ABSTRACT

The emergence of procedures related to increasingly more precise earthquake structural analysis has necessitated the implementation of similar procedures regarding the definition of the seismic input. It is evident that bridges, which form a significant group within lifeline structures, should also be given progressively greater attention during design and construction if earthquake induced loads are among the primary load conditions.

Following a review of damage to bridges during past earthquakes, an outline of methods on dynamic structural analysis is presented. A survey of code requirements in effect in several countries is presented next to emphasize the manner in which theoretical considerations are reflected in regulatory documents. This is followed by a thorough discussion of procedures for estimating earthquake ground motions. The report is concluded with the details of the earthquake analysis of a long multispan railway bridge planned to cross over a large dam reservoir.

1. INTRODUCTION

With the emergence of earthquake engineering as an independent branch of science from what was initially regarded as an esoteric area within structural engineering, the seismic design of bridges has also progressed from its infancy to a point where considerably more insight goes into the design of components than the mere application of an "equivalent" force at some convenient location. Much of this has been due to the impetus provided by the 1971 San Fernando event near Los Angeles which caused $6.5 million damage in the highway overcrossing near the epicenter. By comparison, the damage to bridges caused by all previous earthquakes in California had been minimal, and this had inevitably led to the comforting belief that the exacting requirements of superimposed gravity loads somehow led to structures with an adequate margin of safety against the effects produced by strong ground motion.

With particular reference to highway bridges, examination of damage patterns and reconstruction of the behavior scenarios suggested that the then standard requirements for earthquake design of bridges was inadequate and needed to be revised in conformity with the state of the knowledge (1). Specific conclusions derived from the San Fernando event were that in the epicentral region earthquake force levels greatly exceeded those specified by the design criteria, that vertical acceleration played a major role in some of the collapses, that the customary deficiencies in detailing connections may have led to some of the spectacular failures.

The establishment of vibratory design criteria for bridges is not a straightforward exercise, although code requirements frequently seem to be precisely that when viewed against the backdrop of the vast array of other standardized provisions. Particularly for "lifeline" class bridge structures considerably more thinking should ideally go into the process whereby earthquake effects are established during the design stage.

Advances in engineering seismology have made it possible to state what the general features of the earthquake ground motion at a particular site will contain. Establishing the seismic design parameters for a site requires consideration of a large body of geologic, geophysical and geotechnical information. Consideration of the site characteristics is frequently completed by information that characterizes the geologic area where possible future activity occurs, the earthquake source and wave propagation path and the local ground response.

It is not the purpose of this paper to advocate the implementation of detailed and sophisticated investigations for the purpose of determining the design earthquake criteria for every bridge structure. The majority of bridges can in fact be designed and detailed according to code requirements, and this process will result in perfectly acceptable structures. For the more important cases, however, the procedures described in the following should be considered, and code provisions should only be used for checking the results of the more detailed dynamic analyses or for "calibrating" them.

The rest of this paper is devoted to several considerations. First, observations based on the past seismic behavior of bridges and damage patterns will be described. This will be followed by a discussion of the code requirements in effect in several countries regarding bridge design for seismic effects. Considerable attention will be paid to the methodology for developing vibratory ground motion criteria at a specific site in a format suitable for use in analysis. In the last section, an example will be provided for a railway bridge currently under study and sample results will be presented.

2. BRIDGE BEHAVIOR DURING STRONG GROUND MOTIONS

By far the most extensive experience with the observed behavior of bridges during earthquakes has been accumulated in Japan because of the frequency of seismic activity in that country and the relatively longer period of bridge construction. To somewhat lesser degrees, damage to bridge superstructures observed during the Alaska earthquake of 1964, the Chilean earthquake of 1971 and the San Fernando earthquake of 1971 have enhanced the designer's understanding of typical deficiencies and modes of behavior. Data from European earthquakes, while not nonexistent, has been on a much smaller scale. A literature survey describing seismic effects on highway bridges has been compiled (2), and the following points have been cited as general conclusions :

1. Seismic damages, particularly to low bridges, are most commonly caused by foundation failures resulting from excessive ground deformation and loss of stability and bearing capacity of foundation soils. As a result, substructures often tilt, settle, slide, or even overturn leading to severe cracking or complete failure. These large displacements may cause relative shifting and damage the superstructure, and may include failures within the bearing supports.

2. Backfills exert large forces on abutments which can be in phase with the dynamic inertia forces developed in the superstructures. These forces in combination may cause severe failures, often of a brittle nature, in the substructures. It is common for wing walls to break loose from their abutments due to excessive backfill forces. Settlements

of backfills resulting from compaction is often observed.

3. Seismic damages due to vibration effects are less common than damages due to other effects; however they may occur in taller bridge structures which lack sufficient strength, stiffness, and strength. A commonly encountered form of failure is the insufficient length of bearing supports, hinges and other devices for preventing spans from falling off.

4. To minimize damages, bridge structures should be designed with proper recognition of the stability and bearing capacities of foundation soils, force-deformation and energy absorption characteristics of substructure, superstructure and linking elements, the dynamic nature of structural response, and the dynamic characteristics of all forces acting on the complete soil-structure system.

3. DYNAMIC ANALYSIS

In the dynamic response analysis of structural systems we seek the solution of the following set of equilibrium equations :

$$|M| \,.\, \{\ddot{x}\} + |C| \,.\, \{\dot{x}\} + |K| \,.\, \{x\} = \{R(t)\} \tag{1}$$

Here $\{R(t)\}$ can be a vector of arbitrary loads with prescribed variations in time or effective loads which result from ground motion, and $\{x\}$ is the vector of relative displacements of the structure with respect to the ground with primes denoting time differentiation. $|M|$, $|C|$, and $|K|$ are the standard expressions for mass, damping and stiffness matrices. In the case of ground motion, the right side of the equilibrium equations may be expressed as

$$\{R(t)\} = - |M| \,.\, \{e\} \; \ddot{x}_g(t) \tag{2}$$

Here, $\{e\}$ is a vector of influence coefficients relating those degrees of freedom that are directly affected by the motion of the ground; if a unit ground displacement causes a unit displacement in the n'th degree of freedom of the structure, the corresponding entry into the n'th row of $\{e\}$ is a one, otherwise it is zero. $\ddot{x}_g(t)$ is the acceleration vs. time function; in the majority of cases only one component of the ground acceleration record is employed, and the negative sign in Eq. (2) is ignored because of the arbitrary definition of the ground motion direction.

For the solution of Eq. (1) time domain or frequency domain approaches may be adopted. The time domain analysis can be carried out using mode superposition or step-by-step direct integration. In the frequency domain or response spectrum analysis it is first necessary to solve the corresponding eigenproblem.

3.1 Response History Analysis by Mode Superposition

In the mode superposition analysis, it is assumed that the structural response can be described adequately by a linear superposition of the p lowest modes where p is typically a fraction of the total number of degrees of freedom. Using the transformation

$$\{x\} = |\Phi|\{Y\} \tag{3}$$

where $|\Phi|$ are the lowest p eigenvectors commonly orthonormalized with respect to the mass matrix $|M|$, Eq. (1) may be written as

$$\{\ddot{Y}\} + \mathrm{diag}|2\omega_i\xi_i|.\{\dot{Y}\} + \mathrm{diag}|\omega_i^2|\{Y\} = |\Phi| \quad \{R(t)\} \tag{4}$$

In Eq. (4) it is assumed that the damping matrix $|C|$ also satisfies the modal orthogonality condition

$$\begin{aligned} \{\phi_i\}^T \; |C| \; \{\phi_j\} &= 0 & i \neq j \\ &= 2\omega_i\xi_i & i = j \end{aligned} \tag{5}$$

With this transformation Eq. (4) represents p uncoupled second order ordinary differential equations which may be solved by using any convenient step-by-step integration scheme (3). The same integration time step is used in the solution all components of Eq. (4) in order to simplify the calculation of stress components at specified times which are obtained from the nodal displacements.

3.2 Response History Analysis by Direct Integration

In direct integration the equations in (1) are integrated using a numerical step-by-step procedure, the term "direct" meaning that prior to the numerical integration, no transformation of the equations into a different form is performed (3). In essence, direct numerical integration is based on two ideas. First, rather than satisfying Eq. (1) at any arbitrary time t, it is aimed at satisfying it at discrete time points Δt apart. This implies that pseudo-static equilibrium, which includes inertia and damping effects, is sought only at discrete points within the interval of solution. Therefore, all solution techniques applicable to static analysis find broad application here. The second idea upon which direct integration is based is that implicit assumptions are made regarding the variation of displacements, velocities and accelerations within the time step Δt. It is the form of these assumptions that dictate the accuracy, stability and cost of the solution procedure. Package computer programs contain efficient built-in step-by-step direct integration algorithms (4).

3.3 Response Spectrum Analysis

In this type of analysis the ground acceleration vector in Eq. (2) is written as

$$\ddot{x}_g(t) = \ddot{x}_{gx}(t) + \ddot{x}_{gy}(t) + \ddot{x}_{gz}(t) \tag{6}$$

where $\ddot{x}_{gx}$, $\ddot{x}_{gy}$ and $\ddot{x}_{gz}$ are the ground accelerations in the three principal directions. The equation for the response in the r'th normal mode is then

$$\ddot{y}_r + 2\xi_r\omega_r\dot{y}_r + \omega_r^2 y_r = r_{rx} + r_{ry} + r_{rz} \tag{7}$$

where y_r is the r'th element in $\{Y\}$ and

$$\begin{aligned} r_{rx} &= -\{\phi_r\}^T \,|M|\, \ddot{x}_{gx} \\ r_{ry} &= -\{\phi_r\}^T \,|M|\, \ddot{x}_{gy} \\ r_{rz} &= -\{\phi_r\}^T \,|M|\, \ddot{x}_{gz} \end{aligned} \tag{8}$$

Using the definition of spectral displacement, the maximum absolute modal displacements of the system subjected to, say, an acceleration in the x direction are

$$\{x_{rx}^{(max)}\} = \{\phi_r\} \mid \{\phi_r\}^T \,|M|\, \{I_x\} \mid S_{dx}(\omega_r,\xi_r) \tag{9}$$

where $S_{dx}(\omega_r,\xi_r)$ is the spectral displacement in the x-dinection corresponding to the frequency ω_r and the fraction of damping ξ_r and $\{I_x\}$ is a null vector except that those rows corresponding to the x direction translational degrees of freedom are equal to one. The concept of the response spectrum and its consideration for a given seismic environment will be given further attention.

Similarly for the responses due to ground acceleration in the y- and z-directions

$$\{x_{ry}^{(max)}\} = \{\phi_r\} \mid \{\phi_r\}^T \,|M|\, \{I_y\} \mid S_{dy}(\omega_r,\xi_r) \tag{9a}$$

$$\{x_{rz}^{(max)}\} = \{\phi_r\} \mid \{\phi_r\}^T \,|M|\, \{I_z\} \mid S_{dz}(\omega_r,\xi_r) \tag{9b}$$

The total maximum response in the r'th mode is then assumed to be

$$\{x_r^{max}\} = \{x_{rx}^{max}\} + \{x_{ry}^{max}\} + \{x_{rz}^{max}\} \tag{10}$$

The total response in terms of displacement and stress resultants may be calculated as the square root of the sum of the squares of the modal maximum responses expressed by Eq. (10) provided the modes are well separated in the frequency space so that no strong coupling exists between them (5).

4. SURVEY OF CODE REQUIREMENTS

The primary function of code requirements in relation to earthquake analysis is to express the dynamic effects of the ground motion as a set of "equivalent" static forces which will produce a "similar" influence on the structure. Given the almost unlimited variations of the physical parameters that govern the behavior of any structure, it is evident that gross approximations are inherent in such approaches.

As a whole, current codes for the earthquake analysis of bridge structures depend wholly or mostly on the equivalent lateral force method. As the name implies, this method replaces the dynamic forces exerted by the movement of the ground on a structure by constant horizontal static forces. The substitute force P can be obtained from

$$P = \alpha W$$

$$= \alpha . gM \qquad (11)$$

where W or M stand for the weight or mass of a structure or a structural component and α or $\alpha .g$ represent the seismic coefficient or substitute acceleration, respectively.

In most cases α can be expressed as the product of a series subfactors :

$$\alpha - a_o \cdot \nu_1 \cdot \nu_2 \cdot \ldots \nu_n \qquad (12)$$

In this equation,

a_o = the basic coefficient

ν_1 = factor related to regional seismicity (i.e. zone factor)

ν_2 = factor related to the importance of the structure, which ideally contains its vulnerability

ν_3 = factor that considers the vibratory characteristics of the structure, most typically its fundamental period

ν_4 = factor related to structural type (ductility)

ν_5 = soil factor .

For superstructures with uniform distribution of mass over the height of the structure most norms portray the distribution of the substitute forces with consideration of the fundamental mode of vibration. For bridges with significant concentration of mass along the pier height this will not be the case.

In the following, short descriptions will be provided for codes or specifications in effect in the following countries : Switzerland (6), United States (7), Japan (8), New Zealand (9). Attention will also be paid to the California Seismic Design Criteria (10).

Switzerland

The basic seismic design coefficient is taken as

$$\alpha g = \frac{b_o}{\sqrt[3]{T}} \quad \text{with} \quad T \geq 0.5s \qquad (13)$$

In this equation $b_o = g/20$ for "higher" seismic hazard areas which are defined as having an annual probability of experiencing a ground motion with intensity equal to or larger than I = 7 (MSK) and $b_o = g/50$ for all other areas. T is the computed fundamental period of the bridge pier, and a seismic hazard map is given together with the specifications.

United States

Here the lateral force is specified by

$$\alpha = F\ C \qquad (14)$$

where

F = framing coefficient, equal to 1.0 for single piers

C = combined response factor which is graphically specified as being dependent on fundamental period T, on underlying soil properties and on the earthquake zone (for Zone I, A = 0.09g, Zone II, A = 0.22g and Zone III, A = 0.50g).

New Zealand

Similar to the U.S. requirements the basic lateral force is specified by

$$\alpha = C\ F \qquad (15)$$

where, for the direction being considered,

C = basic seismic coefficient. Value of C for fully ductile structures shown in Fig.1. Composite values CF for "non-

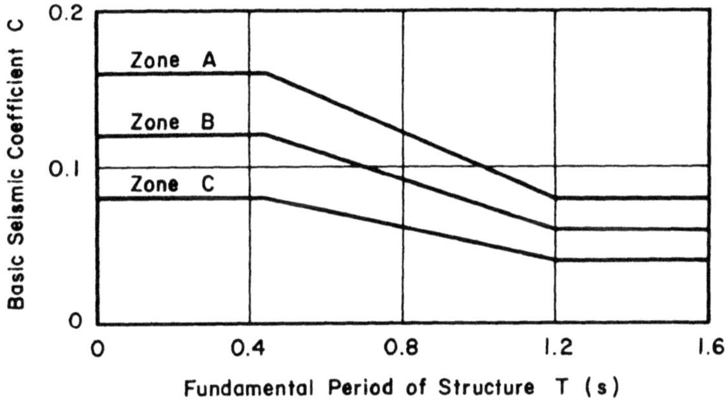

Figure 1. Variation of the Basic Seismic Coefficient C in the New Zealand Code

ductile" structures are given in Table 1.

F = importance factor, as in Table 2.

Table 1. Values of CF for "Non-Ductile" Structures, NZMWD Design Brief

Category	Zone A	Zone B	Zone C
1	0.24	0.18	0.12
2	0.20	0.15	0.10
3	0.17	0.13	0.09

Note : Categories are defined in Table 2

Table 2. Importance Factor, NZMWD Design Brief

Category	Description	Minimum Value of F
1	Bridges carrying more than 2500 vehicles per day and all bridges under or over motorways or railways	1.0
2	Bridges carrying between 250 and 2500 vehicles per day	0.85
3	Bridges carrying less than 250 vehicles per day	0.7

Note : Choice of category in Table 2 should be based on the average number of vehicles per day current at the time of design.

California

These criteria specify the following value for the lateral force coefficient

$$\alpha = \frac{ARS}{Z} \tag{16}$$

In Eq. (16)

A = peak rock acceleration, determined from seismological studies of fault activity and attenuation data gathered from past events

R = acceleration spectra in rock, based on actual earthquake data recorded on rock

S = soil amplification factor, based on analytical studies and actual recorded data

Z = ductility/risk reduction factor based on observed damage and assumed ductility.

The product A.R.S gives an elastic response spectrum curve for the site that would result from a maximum credible event on the closest fault. Division by the factor Z then gives a design force for the portion of the structure under consideration. The factor Z is component oriented, thus the design force depends not only on seismicity and site conditions but on the actual structural component being designed.

5. PROCEDURES FOR ESTIMATING EARTHQUAKE GROUND MOTIONS

Determination of the seismic design parameters for a particular site requires consideration of a large body of geologic, geophysical, seismological and geotechnical information. Basically, seven steps are followed in specifying the characteristics of the ground motion required for earthquake resistant design (Fig.2).

- Determination of the seismicity of the geographic location where the application is planned,
- Identification of the seismotectonic features,
- Estimation of the regional seismic attenuation,
- Estimation of the vibratory ground motion parameters (for example peak acceleration or intensity) at the site,
- Definition of the ground motion response spectra for the site,
- Determination of the local amplification effects and modification of the design response spectra for the site as necessary, and
- Estimation of the uncertainty in the ground motion design values.

5.1 Determination of Seismicity

The objective of this step is to use the past seismicity record to define earthquake recurrence relations applicable to the region of interest, to provide a basis for correlating earthquake epicenters and tectonic structure, and to define seismic source zones. From the seismicity record, one attempts to establish such basic information as :

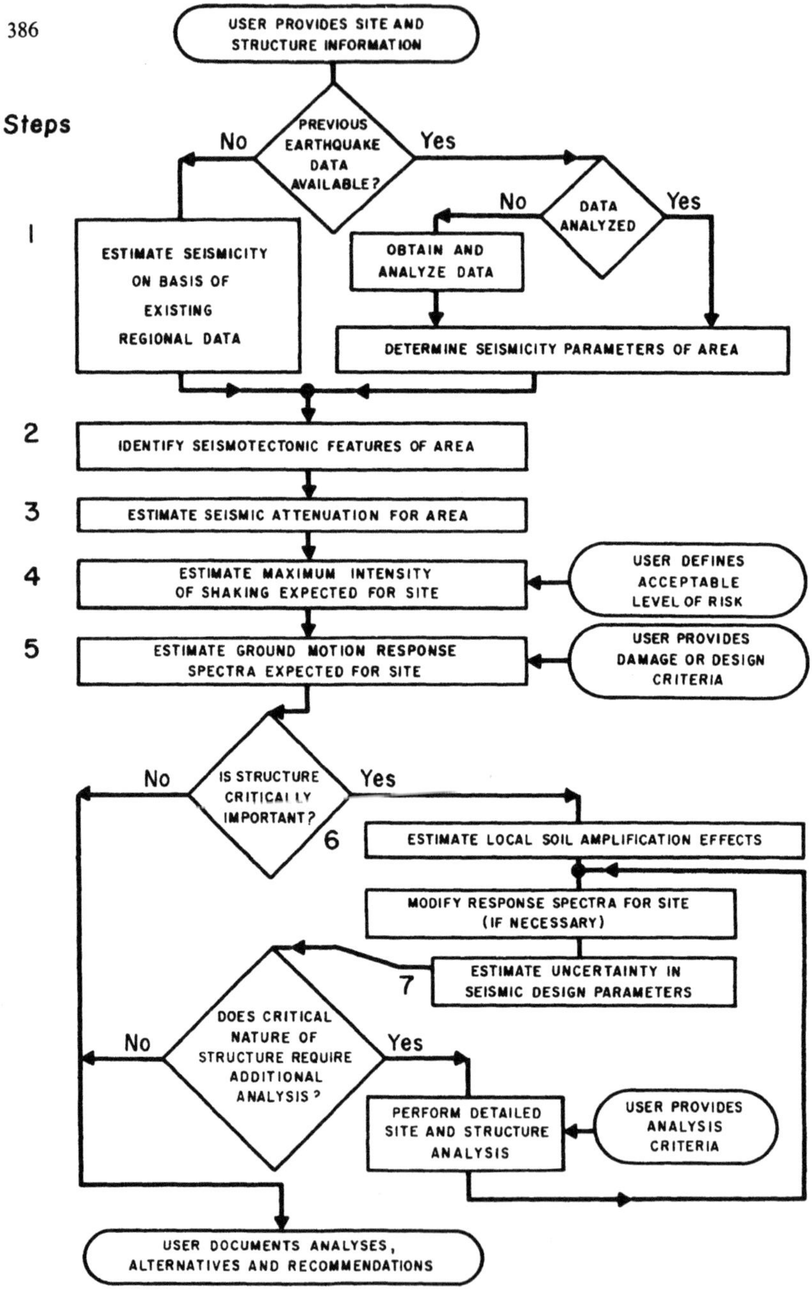

Figure 2. Steps in Estimating Ground Motion for Earthquake Resistant Design

- Date of occurrence
- Epicentral intensity, and where possible, magnitude
- Epicenter and hypocenter locations
- Epicentral maps showing the position of all reported earthquakes within an 80 km radius of the site, and, ideally, all earthquakes with intensity larger than V (or magnitude larger than about 4.2)
- Correlation of epicenter location with tectonic features
- Frequency of occurrence for earthquakes in terms of magnitudes or intensities.

This task is often undertaken within the format of an earthquake catalog compilation. Catalogs of felt or recorded earthquakes do not in general cover a long enough time interval to permit meaningful extrapolations into the future except on a board regional scale (13).

Earthquake recurrence law is of the inverse logarithmic form

$$\log N(M) = a - bM \qquad (17)$$

where N(M) is the number of earthquakes occurring within a region during a given time period with magnitude equal to or larger than M, and a and b are regression constants. The seismic activity index b is commonly in the range 0.8 to 1.0 for most parts of the world.

5.2 Identification of the Seismotoctonic Features

The location and characteristics of faults in the region surrounding the site are essential information along with historical seismicity for estimating the upper bound magnitude and locations of potential earthquakes. A fault is a fracture or fracture zone along which displacement of the two sides relative to one another has occurred. A capable fault is defined as one that may undergo future surface displacement. A fault is classified as capable if (i) it has had Quaternary or more recent movement, or (ii) macroseismic activity has been associated with it, or (iii) it has a demonstrated structural relation to a known capable fault such that movement of one may trigger a movement of the other during the lifetime of the project under consideration. An analysis of faulting and tectonic activity in a region can be complex, requiring data from : (i) geologic data on faults and the ages of rock formation they displace, (ii) seismological data on the areal distribution of seismicity in relation to known faults and tectonic structures, and (iii) historical accounts which provide evidence that a fault has ruptured at the surface.

One rationale for classifying fault activity uses the available geologic, seismological and historical data to specify the fault activity descriptions. Faults are considered to be significant if they satisfy one of the following conditions : (i) faults crossing the site vicinity are capable of fracturing during the lifetime of the proposed structure, (ii) faults near the site are recognized as being capable of generating large damaging earthquakes. Current criteria indicate that a fault should be considered capable of permanent surface displacement if movement occurred at least once during the last 35 000 years or more than once during the last 500 000 years. Empirical relationships relating fault length and earthquake magnitude have been derived from observations .

5.3 Determination of Regional Attenuation

An important consideration in specifying the nature of the ground motion at a given site is related to the knowledge of how ground motion is attenuated with distance away from causative fault. As seismic waves propagate though the rock formations in the crust anisotropy and inhomogeneity of the earth, geometric damping, absorption, scattering, reflection, refraction and wave interference play complex roles in modifying and reshaping them. Strong motion records are the best guides in deriving attenuation relations because statistical information can be cited together with such information but an adequate number of strong motion records are not available for most regions. In most cases isoseismal maps from past earthquakes are currently the best basis for deriving an intensity attenuation relation. An isoseismal map contains the maximum ground motion intensity reported at various locations. These contours may overestimate the ground motion level at a site of interest because they are usually tainted with non-quantified local soil amplification effects. Data from a number of earthquakes which occurred in Turkey has been anlyzed to give the following relation for intensity attenuation :

$$I(R) = 7.56 + 1.07M - 2.04\,\ell n(R + 30) \tag{18}$$

In Eq. (18) R is the epicentral distance (with no azimuthal bias) and M is the magnitude. The equations has a standard deviation of 0.7 in intensity units (14).

The preferred procedure for defining the attenuation function for a region is based on processing recorded strong motion data so that a family of empirical acceleration-distance curves can be derived. Such a family of mean curves proposed for "rock" sites with data obtained from Western United States events is shown in Fig. 3.

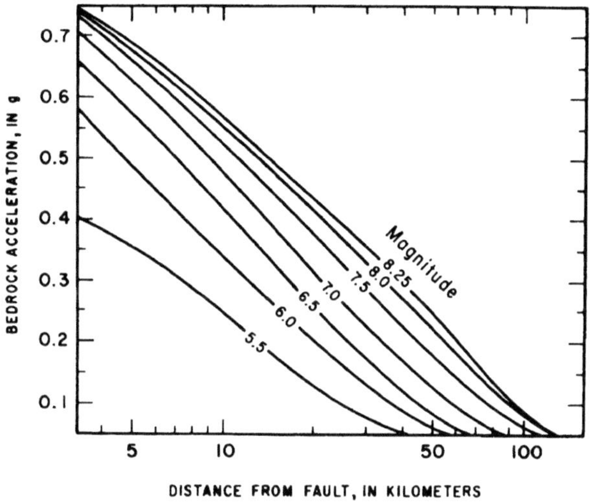

Figure 3. Mean Values of Peak Acceleration vs. Distance for Sites on Rock

5.4 Estimation of the Vibratory Ground Motion Characteristics

After determining the location and magnitude of the design earthquake the characteristics of the ground shaking at the site can be established with the aid of a proper attenuation function. Current procedures for this step are empirical in nature.

The characteristics of ground motion produced by an earthquake are dependent on the source mechanism, epicentral distance and the geometry and physical properties of the geologic structures though which the waves pass as they propagate from the source to the site. Seismograms recorded in the "near field" are particularly complex and site properties are overshadowed by source properties in defining the characteristics of the ground shaking.

Ground motion can be specified in different ways. The commonly used formats include the time records, spectra or peak ground motion parameters. These will be described in the following.

5.4.1 Ground Motion Records. A collection of ground motion seismograms constitutes a key element in seismic design procedures because they contain significant information related to the following items

- Magnitude
- Fault dimensions and configuration
- Seismic energy released
- Average displacement across the fault
- Fault orientation, depth and sense of slip
- Rupture velocity
- Elastic constants associated with the fault area
- Stress drop across the fault
- Complexity of energy release.

5.4.2 Peak Ground Motion Parameters. In spite of its limitations current practice in the design of earthquake resistant structures employs the peak ground acceleration as a measure of the severity of the ground motion. This may be true for stiff structures but for intermediate and low frequency systems peak ground velocity and displacement are more applicable.

It is commonly assumed that the maximum ground accelerations in the two horizontal directions are equal and that the maximum vertical acceleration is two-thirds of this value. In statistical terms, this assumption is true for about 80 percent of accelerograms in current use. There is a growing body of information which appears to contradict the data contained in Fig. 3.

Ground motion can also be characterized by velocity and displacement histories and their peak values. The peak values seem to have more stable upper bounds than acceleration.

5.4.3 Duration. The duration of strong ground shaking has been recognized as being of paramount importance regarding damage potential. There are numerous examples of accelerograms with peak accelerations equal to 0.5g or higher with no associated structural damage. The 1940 Imperial Valley earthquake has been regarded as a model of a destructive earthquake primarily because of its long duration. The dominant parameters in the destructiveness of a given ground shaking have been shown to be (1) the amplitude, (2) the duration, and (3) dominant frequencies at which the shaking occurs (16). General trends regarding duration are that as the overall characterization of the recording site ranges from "hard" to "soft" the average duration will increase, that larger magnitude events are of longer duration and that increasing epicentral distances are accompanied by increasing durations.

5.4.4 Earthquake Spectra. In engineering seismology Fourier techniques are the most commonly used form for the analysis of seismograms. A Fourier spectrum decomposes the ground motion record into its harmonic components in tne frequency domain, and is uniquely related to the original seismogram in terms of amplitude and phase relationships.

For structural applications the response spectrum is preferred because structural systems respond to ground motion within fairly well defined frequency limits. The motion of simple viscously damped elastic systems to ground excitations is governed by Eq. (1). The response spectrum technique is a method for determining the maximum amplitudes of response for structural systems from the maximum responses of an ensemble of simple oscillators (which are essentially narrow band filters) when they are excited by a given ground motion. There are various response spectra such as pseudo absolute acceleration (PSAA), pseudo relative velocity, absolute acceleration (AA), relative velocity (RV) and relative displacement (RD). The following relationships exist between relative displacement and the pseudo quantities :

$$PSRV = \omega.(RD) \qquad (19)$$

$$PSAA = \omega\,(PSRV) = \omega^2(RD) \qquad (20)$$

5.4.5 Intensity. It has been observed that intensity is a more stable quantity to consider in attenuation relationships because factors which affect this parameter are smoothed and averaged for a given region. The current accepted procedure in earthquake resistant design is to establish the expected ground motion intensity and then to convert this information to maximum ground acceleration so that the design response spectrum can be scaled. Western United States earthquake data (17) yield the following empirical relations between Modified Mercalli intensity I_{MM} and peak horizontal and vertical ground motion parameters.

$$\log A_h = -0.014 + 0.30\, I_{MM} \quad (\mathrm{cm/s^2}) \tag{21a}$$

$$\log A_v = -0.18 + 0.30\, I_{MM} \quad (\mathrm{cm/s^2}) \tag{21b}$$

$$\log V_h = -0.63 + 0.25\, I_{MM} \quad (\mathrm{cm/s}) \tag{21c}$$

$$\log V_v = -1.10 + 0.28\, I_{MM} \quad (\mathrm{cm/s}) \tag{21d}$$

$$\log D_h = -0.53 + 0.19\, I_{MM} \quad (\mathrm{cm}) \tag{21e}$$

$$\log D_v = -1.13 + 0.24\, I_{MM} \quad (\mathrm{cm}) \tag{21f}$$

In these equations A, V and D represent the peak acceleration, velocity and displacements, respectively while the subscripts h and v denote the horizontal or vertical directions. These average trends and the spread in the data expressed as ±1 standard deviation are shown in Fig. 4.

5.4.6 Probabilistic Estimates of Effective Ground Motion Parameters. With the help of probability theory coupled to information on regional geology, seismic history and attenuation functions one can make estimates of ground motion parameters for given return periods (13). Probabilistic estimates are not without their own particular pitfalls, however, because our knowledge of the seismic past, which is essentially extrapolated hundreds or even thousands of years into the future, is frequently too short and incomplete.

5.5. Definition of Design Response Spectra

After the characteristics of ground shaking at a site have been established design response spectra can be constructed. Procedures for developing vibratory ground motion parameters in terms of the design spectra may be considered within two categories: site independent and site dependent. Both approaches were developed primarily as a response to stringent nuclear plant siting requirements.

There is considerable controversy regarding the definition of seismic input and site response. This is caused by the debate about whether available geologic, geophysical, seismological and geotechnical data are adequate to specify the seismic input and site response in any precise manner. Controversy also centers on whether judgments about conservatism in the seismic designs are

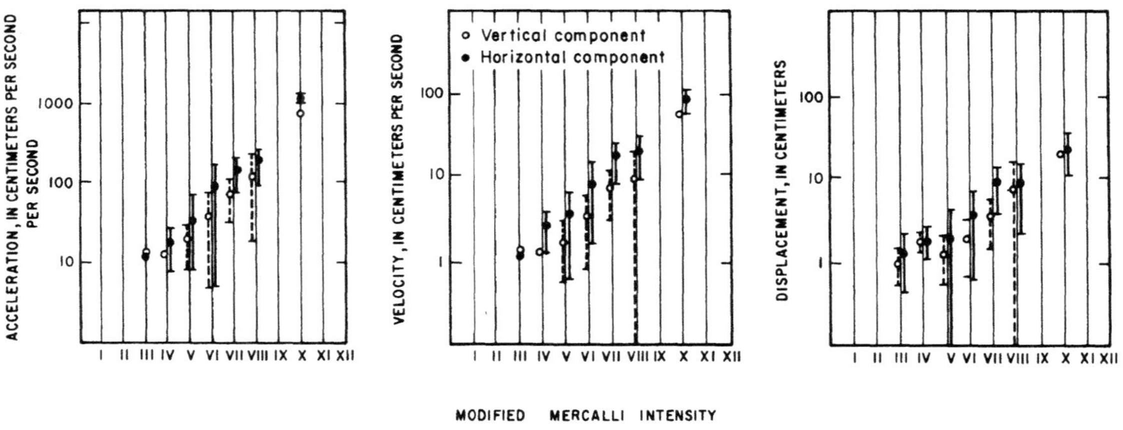

Figure 4. Correlation of the Modified Mercalli Intensity Scale with Peak Ground Motion Parameters

justifiable in view of uncertainties in the data. Information for evaluating the ground response in the site vicinity is obtained from a combination of borings, trenches, laboratory measurements and geophysical investigations (18). The purpose is to (1) determine the classification, lateral distribution, layering, geologic structure and physical properties of materials underlying the site, (2) obtain samples and cores for testing, (3) establish the elevation of underground water level.

Vibratory response of geologic materials is governed by the following parameters : shear wave velocity, density, shear modulus, damping, Poisson's ratio, bulk modulus, and shear strength. These physical properties are based on data from laboratory and field tests.

5.5.1 Site-Independent Spectra. Response spectra which do not explicitly take into account the characteristics of the construction site fall into this group. These spectra have well defined standard shapes, and because they are based on accelerograms recorded on a wide range of geologic and seismologic conditions, are considered representative of general conditions when local site properties are not expected to play a significant role in shaping the response.

The earliest attempt at obtaining site independent spectra was by Housner (19). His spectrum shapes were simply the average of two components each from four large earthquakes available at that time. Later, armed with a larger collection of ground motion records Newmark and Hall (20) proposed a standard procedure for drawing design response spectra. The technique is based on the observation that the spectral ordinates are proportional to the peak values of ground motion over well defined frequency ranges. Construction of the response spectrum is accomplished by multiplying the maximum ground acceleration, velocity and displacement values by amplification factors dependent on damping. In the absence of data to the contrary, the maximum values of ground motion parameters may be assumed to be proportional to each other, and for a 1 g peak acceleration, peak velocity and displacement values are to be taken as 122 cm/s and 91 cm, respectively. In statistical terms, the amplification factors shown in Table 3 are mean plus one standard deviation numbers in a lognormal distribution, so that they may be exceeded in only 16 percent of the time.

Independent studies performed by different groups were consolidated into a single document (21) as the official position of the U.S. Atomic Energy Commision regarding response spectra for the design of nuclear plants in the United States. To use the R.G. 1.60 spectrum shapes maximum horizontal ground acceleration and displacement must first established. The maximum displacement is assumed to be proportional to the maximum acceleration regardless of the soil conditions and is set as 91 cm for a 1 g acceleration.

Table 3. Aplification Factors in the Newmark-Hall Procedure (20)

Percent Damping	Amplification Factor		
	Acceleration (PSAA)	Velocity (PSRV)	Displacement (RD)
0	6.4	4.0	2.5
1	5.2	3.2	2.0
2	4.3	2.8	2.0
5	2.6	1.9	1.8
10	1.5	1.3	1.2
20	1.2	1.1	1.0

The response spectra are established by a series of straight lines but the control points have definite frequencies for all horizontal spectra. Amplification factors are a function of the percentage of critical damping, as listed in Table 4. A horizontal response spectrum drawn for a 1 g maximum acceleration earthquake is shown in Fig. 5. For the vertical direction design response spectra are drawn in much the same way as for the horizontal direction.

Table 4. Amplification Factors in R.G. 1.60 for the Horizontal Direction

Percentage of Critical Damping	Amplification Factors for Control Points			
	Acceleration			Displacement
	A(33 Hz)	B(9 Hz)	C(2.5 Hz)	D(0.25 Hz)
0.5	1.0	4.96	5.95	3.20
2.0	1.0	3.54	4.25	2.50
5.0	1.0	2.61	3.13	2.05
7.0	1.0	2.27	2.72	1.88
10.0	1.0	1.90	2.28	1.70

Note that the factors given for the construction of the vertical spectra are to be applied to the horizontal components of acceleration and displacement.

5.5.2 Site Dependent Spectra. The data base used for determining the site independent spectrum shapes can be reexamined with regard

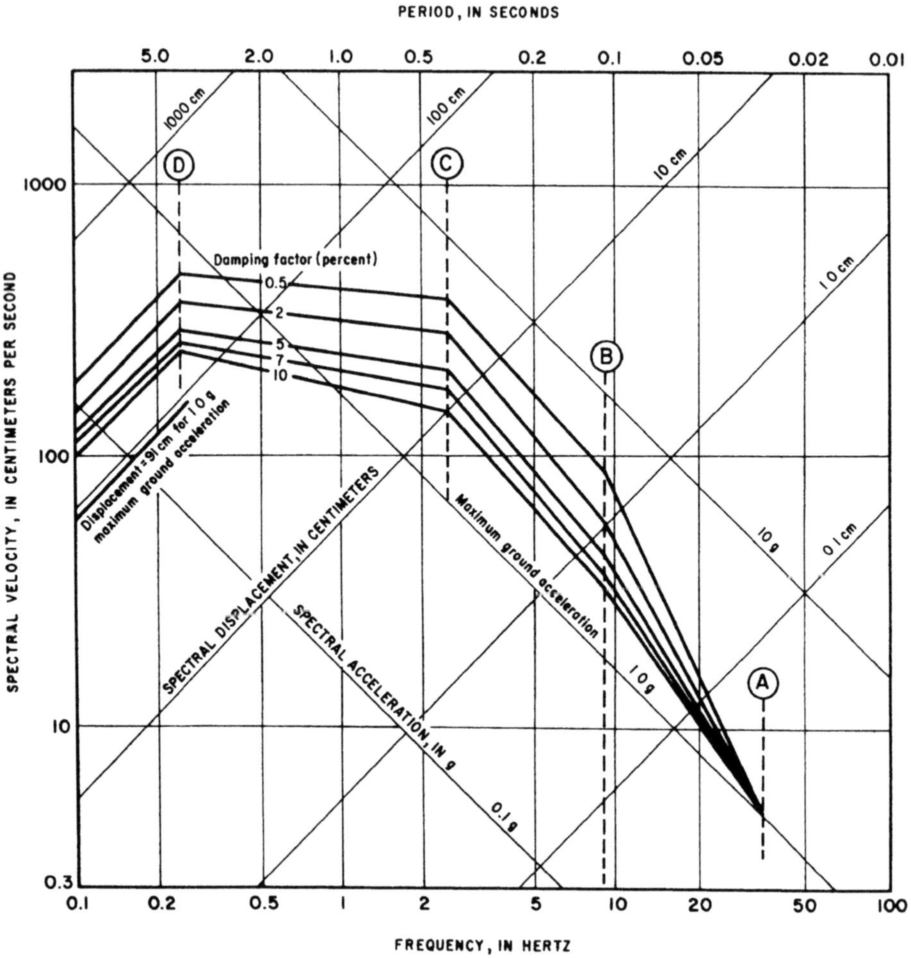

Figure 5. Horizontal Design Spectrum Scaled to 1 g

to representative samples of a variety of site conditions. In some cases it is possible to select a family of response spectra that were derived from seismograms for which soil conditions were similar to those of the proposed site. The basic idea in the site matched response spectrum concept is to categorize the geological conditions into one of the broad classes for which a standard shape definition is available.

A comprehensive study of site dependent spectra was performed by Seed et al. (22). Ensemble average and mean plus standard deviation spectra were developed from 104 records classified as "soft-medium", "deep cohensionless", "stiff", and "rock". Comparison of the ensemble average and mean plus one standard deviation spectra shown in Fig. 6 indicate trends that can be attributed to local soil conditions. In the low frequency range, spectra from soft to medium soil sites and deep cohensionless soil sites have larger amplitudes than the spectra from rock and stiff soil sites. For frequencies above about 6 Hz spectra from stiff or rock sites exhibit amplitudes larger than the soft to medium or the deep cohensionless soil sites. Both trends are clearly visible in Fig. 6 in which the R.G. 1.60 spectra are also presented for comparison purposes.

5.6 Determination of Local Ground Response

From the earliest times of scientific observation in relation to seismic events it has been noted that otherwise similar structures situated on different soil formations tend to experience different degrees of damage. For example, structures founded on unconsolidated soils are frequently more severely damaged by ground shaking, and damage distribution has been positively related to site characteristics. Buildings of a certain class or type with a certain natural period are often damaged from ground shaking when they are located on geologic formations with similar periods. Mismatch of structure and soil periods has the beneficial effect of attenuating the effects of shaking.

Local ground response is complicated because earth materials behave nonlinearly, even at low strain levels. In the infinitesimal strain ranges below the order about 10^{-5} the deformations exhibited by most soils are nearly elastic and recoverable. Over the intermediate ranges of strain from 10^{-4} to 10^{-2}, the behavior of soils is elasto-plastic and produces irrecoverable permanent deformation. When even larger strains are imposed, failure generally occurs (23).

Site amplification modeling is especially required for strong ground motions. Comprehensive parametric curves cannot be developed because of the increasing number of parameters that need to be considered and gaps of knowledge pertaining to natural materials. Some general curves that have been published permit estimates to be

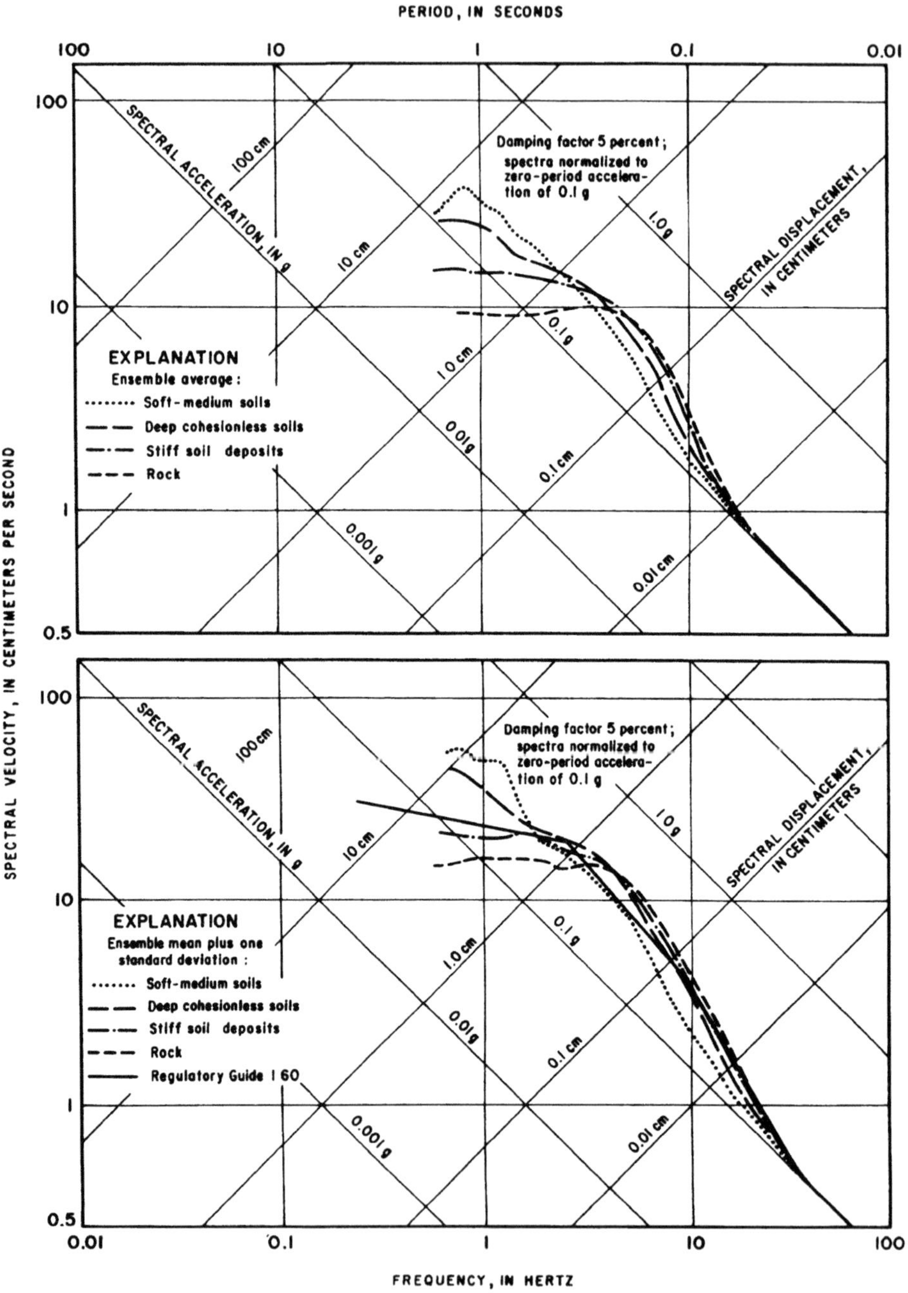

Figure 6. Site Dependent Mean and Mean Plus One Standard Deviation Response Spectra

made.

When generalized parametric curves are judged to be insufficiently precise, it might be desirable to utilize finite element or similar procedures (24, 25). The degree of accuracy with which the physical characteristics of the earth materials at a site are known should be evaluated critically before embarking upon a costly site response analysis investigation.

6. THE RAILWAY BRIDGE OVER THE KARAKAYA DAM RESERVOIR : A CASE STUDY

With the impoundment of water behind the Karakaya dam which is currently under construction, a part of the railway linking the cities of Malatya and Elazığ in Southeastern Turkey will be inundated. Economic studies comparing the various possible alternatives have shown that moving the bridge to point downstream from the dam site or beyond the upper reaches of the lake were not feasible so that a new route with an approximately 2.1 km long crossing over the reservoir has been planned. In the absence of an accepted seismic design code for bridges, it was decided to develop project specific criteria for the bridge. In the following these will be outlined in a format similar to that described in Section 5, and some sample results based on a response spectrum approach as described in Section 3 will be presented.

6.1 Description of the Structure

Cost constructions dictated the form and type of the bridge. The guiding principle agreed upon at the outset was that it should be possible to erect the structure within a minimum time and complete all the structural work prior to the completion of the dam proper itself. To minimize foreign currency requirements for construction it was also considered appropriate to impose the constraint that it should be possible to produce the necessary parts and materials domestically. Factors regarding easy installation, maintainability and previous experience with long span railway bridges led to the adoption of a 29-span simple truss type bridge supported on a succession of piers which vary in height from approximately 30 m to 70 m. Figure 7 illustrates the overall view of the planned structure and in Fig. 8 a typical reinforced concrete pier elevation and cross section are shown. Each pier supports one fixed and one roller support, and each span is 70 m long. It is obvious that under normal operational conditions water will seep into the closed center of the box section in Fig. 8, so portholes will be left in the walls of the piers to facilitate this action.

The selected truss type is conventional, and represents what the supplier declared could be built with existing technology rather

Figure 7. The Railway Bridge at the Karakaya Reservoir : An Artist's Conception of the Overall View

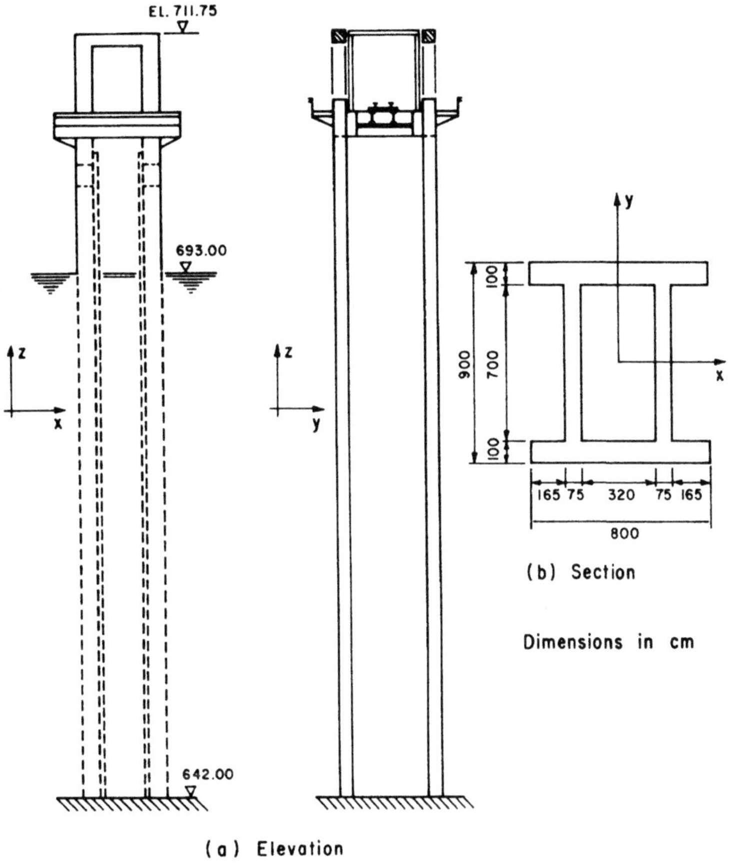

Figure 8. Typical Pier Elevation and Cross Section

than the end product of a broad optimization effort aimed at reducing the number of foundation excavations in difficult conditions and a more economical use of materials.

6.2 Determination of the Seismic Input

Current thinking in relation to meeting seismic requirements for the construction of critical facilities is based on the two-level approach inspired by the approach to nuclear design. The lower intensity earthquake motion is defined as corresponding to the event which may reasonably be expected to occur once during the lifetime of the facility. The more intense second level corresponds to the strongest earthquake ground motion which could possibly be recorded at the same site. While the two-level philosophy has found widespread acceptance among engineers the exact quantification of these events has been rife with controversy. The former event controls the elastic limit of behavior while the latter governs strength requirements. Studies aimed at establishing these levels were conducted in several steps.

6.2.1 Regional Seismotectonics and Regionalization. Seismotectonic features of Anatolia (Asia Minor) may be considered within five major groups : (1) North Anatolian fault zone, (2) Aegean - Marmara-Thrace region, (3) Central Anatolia, (4) Eastern Anatolia fault structures, and (5) Western Tauroids.

The left-lateral strike slip Eastern Anatolia fault extends in the northeast-southwest direction, and can be considered within two main parts. The first is approximately 120 km long and follows the Murat river valley while the second forms the 65 km long Göynük Valley. These two main extensions are separated by a 15 km long discontinuity section, and have been formed as a result of inter-plate compressive forces causing reverse faulting in southwestern Turkey. This fault zone extends southward as far as the Dead Sea rift valley. The major seismotectonic features of the area are shown in Fig. 9.

The effort aimed at establishing characteristics of the ground motion at the site is based on an expression of the general seismo-tectonic features in the form of a regionalization. To this end, seismic events of the last 15 years within a 300 km radius of the site were reviewed and their relationship to the faulting was investigated. This study resulted in the definition of the major sources indicated in Fig. 9.

6.2.2 Earthquake Hazard Estimation. In the two widely used approaches towards establishing some index of the ground motion such as peak acceleration or intensity a deterministic or a probabilistic methodology may be implemented. Calibration of the final results is

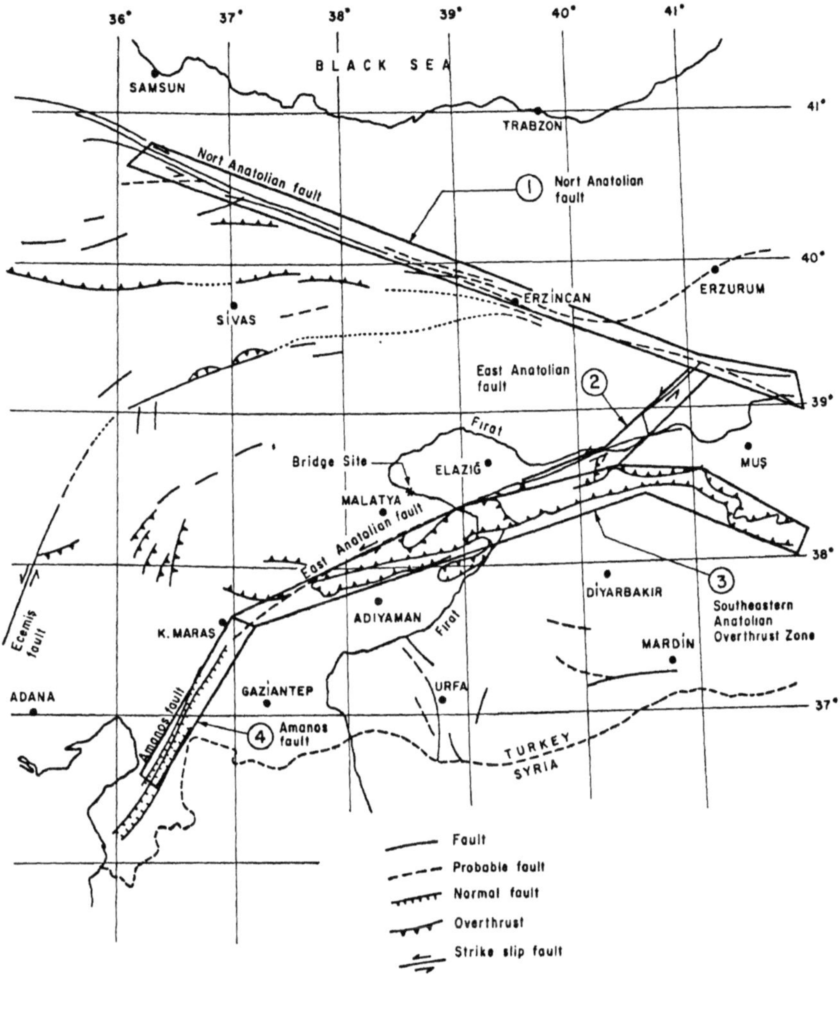

Figure 9. Seismotectonic Features of Eastern Turkey and Source Regionalization

done by ensuring that both will yield similar results. In hazard estimates the most important relationship which is utilized is the attenuation expression which indicates the variation in the ground motion index as a function of distance. Given the paucity of recorded strong ground motion, and the more stable variation of intensity over distance, it was judged more appropriate to employ Eq. (18) and to calibrate the design response spectrum through the peak ground acceleration derived from Eq. (21).

For the bridge under study a maximum ground acceleration value of 0.20 g was calculated for the upper limit event, and this value was corroborated by both probabilistic and deterministic procedures. For the lower limit event 0.13 g was computed. Consideration of the general tectonic features led to the decision that either component of the Taft record of 1952 scaled with respect to the appropriate acceleration could be employed in amplification studies for the local geotechnical structure as well as the bridge piers and their interaction with the fluid and foundation media.

6.2.3 Assessment of Local Amplification Effects. An extensive series of borings and geophysical investigations were made along the bridge axis. The results of these were carefully evaluated in defining four major soil profiles for which amplification studies were performed. The ultimate aim of soil amplification studies is to determine the ground motion at some control point (typically the free surface) given the input motion, typically defined at the bedrock level or to determine the response spectrum at the free surface given the response spectrum for the motion at the bedrock level and physical properties of the overlying layers of soil.

A truly nonlinear model was used in ascertaining the modifications in the input response spectrum to the four major soil profiles. Compared to equivalent linearization techniques, this type of modeling results in spectra in which one can have a greater degree of confidence, especially at higher frequencies. All analyses were performed in the time domain and the N21E component of the 1952 earthquake recorded at Taft was used as the input motion without any modification.

Viewed in relation to the customary differences in spectrum ordinates for a given ground motion which are smoothed for design purposes, the conclusion was reached that for two of the rock or stiff soil profiles the input motion reached the surface without substantial modifications, whereas for the other profiles through cohensionless layers the spectral acceleration ordinates were amplified by 50 percent. On this basis two design spectra were obtained. These are illustrated in Fig. 10 : the smooth curve denoted as "rock" corresponds to the stiff profiles, and the curve labeled as "deep alluvium" refers to the latter profiles. For

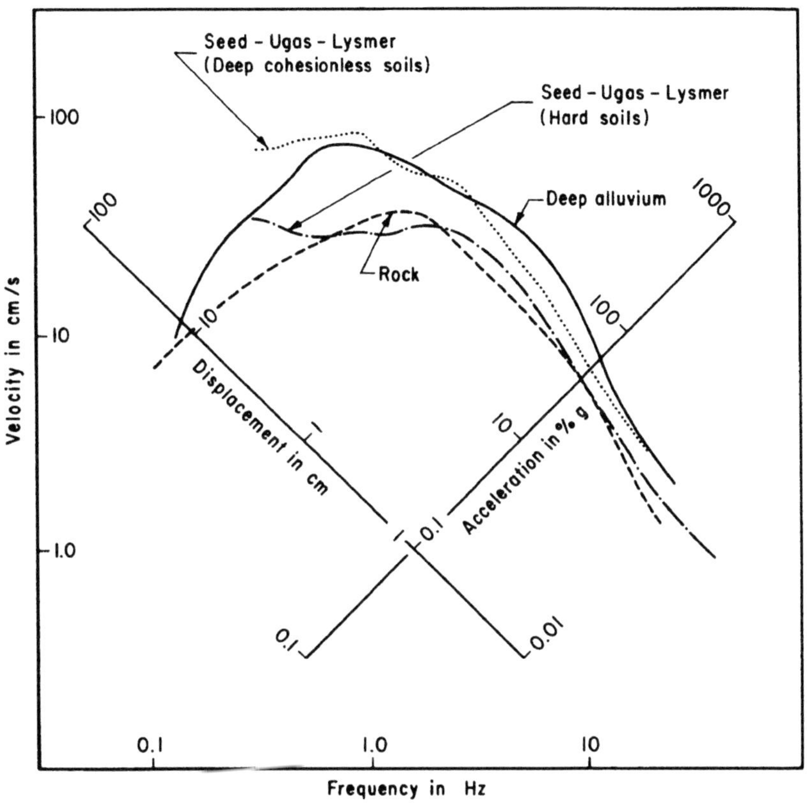

Figure 10. Design Response Spectra

comparison purposes two design spectra for "hard soils" and "deep cohensionless" soils reported in (23) scaled to the appropriate maximum acceleration are shown in Fig. 10. It is noted that the general dynamic features of the soil columns are reflected in these curves.

6.3 Analysis of Dynamic Response

For long lifeline structures such as continuous bridges it may be desirable to investigate the effect of finite wave propagation velocity on dynamic response. The relative motion between successive piers may be accentuated when out-of-phase ground displacements are input from the foundations. The same effect may be observed for two adjacent piers with greatly dissimilar heights. For the bridge under consideration this will not be critical in the longitudinal direction because of the discontinuous manner in which the trusses are connected to the piers. In the transverse direction the vibratory response will be primarily controlled by horizontally polarized shear waves and Love waves. However, there is negligible inertial coupling between the piers in this direction because the weight of the truss structure is only a fraction of the weight of a given pier-typically of the order of 10 percent or less. This judgment made it possible to calculate the dynamic response of each pier in either direction on an individual basis because it is not dependent upon the behavior of adjoining piers.

The direct use in design of the spectra shown in Fig. 10 in conjunction with the response spectrum method would lead to unacceptably large design forces because no modifications have been made on them to account for the effects of cracking and other nonlinear actions. Most design codes based on the equivalent lateral force concept contain requirements which recognize the reduction in force levels resulting from cracking and yield. Calculations aimed at the determination of the design forces on a succession of piers with heights varying from 30 m to 70 m were performed for both principal directions and for two different stages of construction corresponding to empty and full reservoir.

The principal special problem which arises in the evaluation of system property matrices for the pier-superstructure system results from the inertial effects of water in the reservoir surrounding the piers. As piers respond to earthquake ground motion, they cause corresponding motions in the water in contact with the pier surfaces, and the resulting changes of water pressure acting on the piers in turn affect their dynamic response. This phenomenon is called fluid-structure interaction.

In the case of gravity dams this hydrodynamic effect is considered by assuming that the pressure changes at the face of the dam result from the action of a certain volume of water attached to the surface (26). For flexible structures interaction effects are more complex but they are again approximated by an extension of the added mass concept.

Earthquake induced vibrations of submerged structures can be studied satisfactorily by assuming that wave action is negligible and the velocity of the structure relative to the surrounding water is low. Under these conditions the analysis can be performed by adding to the mass of the structure the mass of a certain volume of water which gives a new virtual mass but otherwise treating the structure as if it stood in vacuo. Classical normal modes do not exist for the piers when hydrodynamic effects are included, but their displacements in terms of their modes of vibration in vacuo may still be used to approximate the actual displacements.

Analysis of the response of structure interacting with fluids is best carried out in the frequency domain. The response to unit harmonic ground acceleration.

$$\ddot{x}_g(t) = \mathrm{Re}\{e^{i\omega t}\} \tag{22}$$

may be expressed by

$$Y_n(t) = \mathrm{Re}\{\bar{Y}_n(\omega)e^{i\omega t}\} \tag{23}$$

where $\bar{Y}_n(\omega)$ represents the complex frequency response of the n'th generalized coordinate. It can then be shown that the governing equations can be expressed as

$$S(\omega)\ \bar{Y}(\omega) = L(\omega) \tag{24}$$

In Eq. (24) the frequency dependent coefficient matrix $S(\omega)$ relates the generalized displacement vector $\bar{Y}(\omega)$ to the corresponding generalized loads $L(\omega)$ which include the effects of hydrodynamic pressure. Unlike the classical mode-superposition analysis outlined in Section 3.1, the matrix $S(\omega)$ is not diagonal because the mode shapes $|\phi_i|$ are not the normal modes of the pier-water system.

Under a set of reasonable assumptions the motion of the water in the reservoir is governed by the wave equation

$$\frac{\partial^2 P}{\partial x^2} + \frac{\partial^2 P}{\partial y^2} + \frac{\partial^2 P}{\partial z^2} = \frac{1}{c^2}\frac{\partial^2 P}{\partial t^2} \tag{25}$$

in which $P(x, y, z, t)$ is the hydrodynamic pressure and c is the velocity of sound in water.

Evaluation of Eq. (24) and (25) with the appropriate boundary conditions and for a range of values of ω to permit synthesizing earthquake motions into Fourier components requires considerable computational effort. Therefore, the alternative of solving the equations of motion for a combined pier-reservoir system using finite elements in which the water is represented by plane strain quadrilateral elements with appropriately modified elasticity matrices was adopted. In later stages the results obtained from this approach were checked against a more refined model in which the piers were modeled as axisymmetric shells surrounded by eight noded bricks of water. In both types of analyses the body of water inside the piers was taken simply as a non-interacting mass, and the mode superposition technique was employed. The results obtained from these studies corroborated one another : in general it was observed that the inclusion of hydrodynamic effects tended to increase the elastic forces by about 40 percent as compared with the case when these are not considered.

The effect of soil structure interaction was also investigated for the piers which were reduced to single degree systems vibrating in their wirst generalized modes, and tne pile cap foundations were modeled as rigid rectangular masses with one translational and one rocking degree of freedom. The elastic constants of the springs connecting the foundation to the surrounding soil were based on a range of values representative of soil type and geophysical investigation results. The simplified model (27) showed that soil-structure interaction was insignificant. The slight reduction in base shear and overturning moment was of the order of 5 percent, and the top displacements increased by a maximum of 15 percent.

6.4 Derivation of Equivalent Lateral Forces

The computation outlined in the previous section were performed in both principal directions for the range of piers considered. Also included in the studies were the spectra shown in Fig. 10 and the response spectrum of the original Taft record. The average of these computations for spectra adjusted to 0.2 g zero period spectral acceleration is indicated by the jagged lines shown in Fig. 11 drawn for the two principal directions. The ordinate denotes the ratio of the maximum shear force at the ground level to the total weight of the pier in vacuo. In these analyses rather than explicitly considering the water surrounding the piers, the mass density of the submerged part of the pier was modified in order to reduce computation time.

Tall flexible and statically determinate structures such as the piers considered for the Karakaya Bridge are sufficiently long period from the dynamic point of view, and the modification

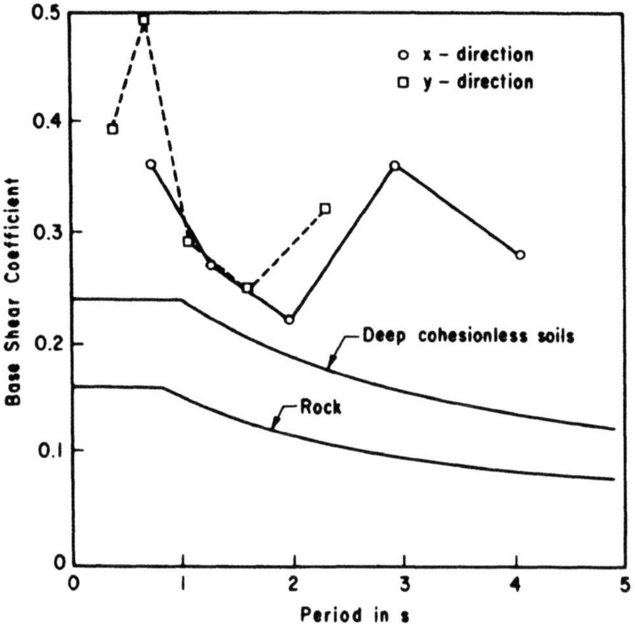

Figure 11. Equivalent Static Forces

of the elastic response spectrum can be done by simply dividing the elastic ordinates by the postulated ductility factor (28). The design response spectra shown as smooth curves in Fig. 11 are the attenuated replicas of the computed elastic forces. It is to be noted that these curves contain the hydrodynamic effects, therefore they were applied to the piers as if they stood in air. The lateral force curves corresponding to the case when these effects are not considered would be obtained by further dividing the ordinates by a factor representative of the hydrodynamic influences.

6.5 Sample Results

The design elastic forces for which the cross sections shown in Fig. 8 were designed were derived from the spectra shown in Fig. 11. It is of interest to note that good agreement was obtained between the "enhanced" design spectrum applied to the structure assumed to be standing in air and the reduced spectrum excluding hydrodynamic forces applied to the piers with appropriately modified mass distribution derived from the added mass concept to account for the submerged portions. Although eight of the lowest modest were considered the fundamental mode was by far the most dominant, and results based upon it would have been quite satisfactory.

A sample of typical results is given in Fig. 12 for a 70 m tall pier in terms of displacements, shear forces and bending moments in the direction transverse to the bridge axis. Reinforcement requirements were basically determined on the basis of working stress procedures although the ultimate strength theory was used in checking the strength against critical modes of failure. Contrary to most design codes the computations were performed for the case when half of the design train load was effective at the same time as the design earthquake. This explains the finite value of the shear force at the top of the pier in Fig. 12; it stems from the mass of the truss and the train lumped at the top of the pier. Because of these additively conservative assumptions and because of the rather strong ground motion considered for design substantial overturning moments were obtained for which the pile foundations need to be designed.

7. CONCLUSIONS

Lifeline structures such as the long multispan railway bridge described in Section 6 require careful analysis regarding the seismic design aspects. Whether as a consequence of structural complexity, unusual ground conditions or high cost or risk potential similar bridges will merit special studies. It is judged appropriate that dynamic analyses to obtain maximum earthquake forces and displacements or ductility demand should be carried out for bridges

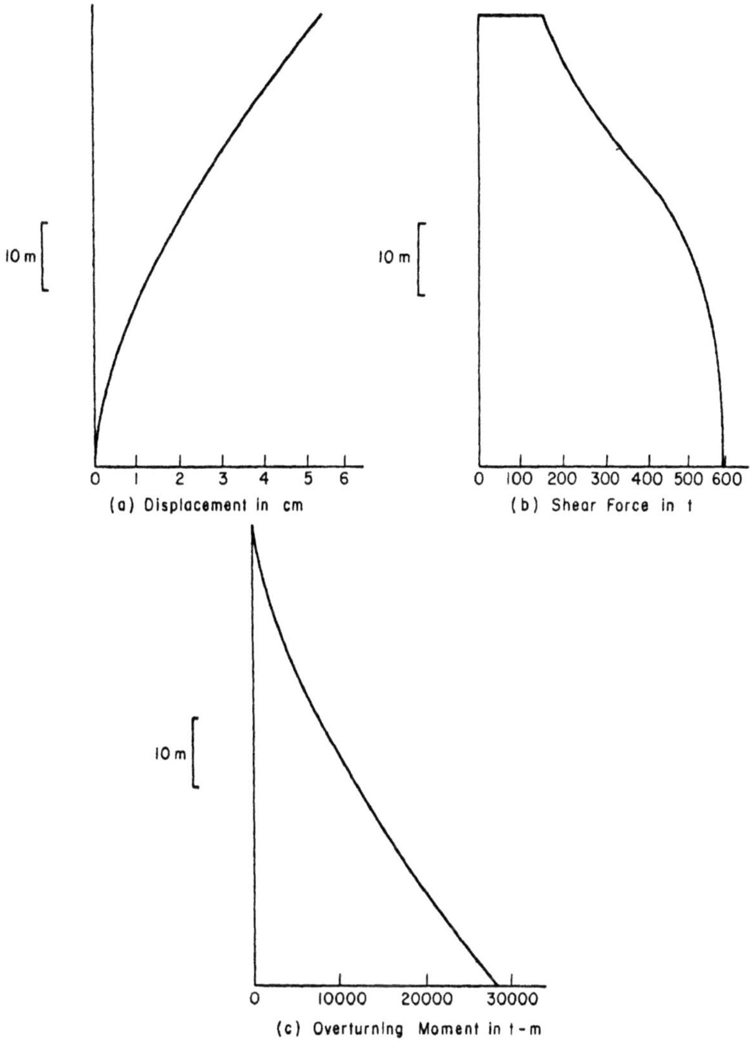

Figure 12. Computed Results for a Typical Tall Pier

all into one or more of the following categories :

- Major bridge structures considered to be of particular importance
- Mass of any pier, including any allowance for hydrodynamic effects greater than 20 percent of the mass of the superstructure estimated to contribute to the inertia on the pier
- Bridges with superstructure joints that are designed to allow significant relative movement between the separated spans
- Structures in which the lateral load resistance is provided by structural systems other than conventional piers and abutment structures
- Suspension, cable stayed or long span arch bridges.

Establishing the design criteria for a bridge structure and verification of their adequacy requires the mobilization of a significant range of activities within earthquake engineering.

8. REFERENCES

1. Jennings, P.C., (Editor), "Engineering Features of the San Fernando Earthquake of February 9, 1971," California Institute of Technology Report No. EERL 71-02, June, 1971, Pasadena, California.

2. Iwasaki, T., J. Penzien, and R. W. Clough, "Literature Survey-Seismic Effects on Highway Bridges," Earthquake Engineering Research Center Report No. EERC 72-11, University of California, Berkeley, November, 1972.

3. Bathe, K.J., and E.L. Wilson, Numerical Methods in Finite Element Analysis, Prentice Hall, 1976.

4. Bathe, K.J., E.L. Wilson, and F.E. Peterson, "SAP : A Structural Analysis Profram for Static and Dynamic Response of Linear Systems," Earthquake Engineering Research Center Report No. 73-11, University of California, Berkeley, June, 1973, revised April, 1974.

5. Wilson, E.L., A. der Kiureghian, and E.P. Bayo, "A Replacement for the SRSS Method in Seismic Analysis," International Journal on Earthquake Engineering and Structural Dynamics, Vol.9, pp. 187-194, 1981.

6. "Norm für die Belastungsannahmen und die Überwachung der Buaten," SIA 162, Art. 22, Schweizerischen Ingenieur und Architekten Verein, Zürich, 1970.

7. American Association of State Highway and Transportation Officials, "Standard Specifications for Highway Bridges," Twelfth Edition, Washington, 1977.

8. Bridge and Structural Committee, Japan Society of Civil Engineers, "Earthquake Resistant Design of Bridges," in Earthquake Resistant Design for Civil Engineering Structures, Earth Structures and Foundation in Japan, Tokyo, 1980.

9. New Zealand Ministry of Works and Development, "Highway Bridge Design Brief," NZMWD Publication CDP 701/D, Wellington, New Zealand, 1978.

10. Gates, James H., "Factors Considered in the Development of the California Seismic Design Criteria for Bridges," Workshop Proceedings on the Research Needs of Seismic Problems Related to Bridges, Applied Technology Council, Palo Alto, California, 1979.

11. Hays, Walter H., "Procedures for Estimating Earthquake Ground Motions," U.S. Geological Survey Professional Paper No. 1114, Government Printing Office, Washington, D.C., 1980.

12. Werner, S.D., "Procedures for Developing Vibratory Ground Motion Criteria at Nuclear Plant Sites," Nuclear Engineering and Design, Vol. 36, pp. 410-441, 1976.

13. Gülkan, P., and M.S. Yücemen, "Seismic Risk Analysis for Nuclear Power Plants," Journal of Pure and Applied Sciences, Middle East Technical University, Ankara, April, 1977.

14. Gülkan, P., M. Öner, M. Erdik, and M.S. Yücemen, "A Methodology for the Assessment of Ground Motion Parameters for Nuclear Power Plants," Proceedings of the Seventh European Conference on Earthquake Engineering, Athens, September, 1982.

15. Schnabel, P.B., and H.B. Seed, "Accelerations in Rock for Earthquakes in the Western United States," Bulletin of the Seismological Society of America, Vol. 62, pp. 501-516, 1973.

16. Trifunac, M.D., and A.G. Brady, "A Study on the Duration of Strong Motion," Bulletin of the Seismological Society of America, Vol. 65, pp. 581-626, 1975.

17. Trifunac, M.D., and A.G. Brady, "On the Correlation of Seismic Intensity Scales with the Peaks of Recorded Ground Motion," Bulletin of the Seismological Society of America, Vol. 65, pp. 139-162, 1975.

18. Waldron, H.H., R.P. Miller, and S. D. Werner, "Geotechnical Investigations at Nuclear Power Plant Sites," Nuclear Engineering and Design, Vol. 36, pp. 397-406, 1976.

19. Housner, G.W., "Behavior of Structures During Earthquakes," Journal of the Engineering Mechanics Division, ASCE, Vol. 85, pp. 109-129, 1959.

20. Newmark, N.M., and W.J. Hall, "Seismic Design Criteria for Nuclear Reactor Facilities," Proceedings of the Fourth World Conference on Earthquake Engineering, Santiago, Chile, Vol. B, pp. 37-50, 1969.

21. U.S. Atomic Energy Commission, "Design Response Spectra for Seismic Design of Nuclear Power Plants (Revision)," Regulatory Guide 1.60, 1973.

22. Seed, H.B., C. Ugas, and J. Lysmer, "Site Dependent Spectra for Earthquake Resistant Design," Bulletin of the Seismological Society of America, Vol. 66, pp. 221-244, 1976.

23. Ishihara, K., "Strength of Cohesive Soils under Transient and Cyclic Loading Conditions," in State-of-the-Art in Earthquake Engineering 1981, edited by O. Ergünay and M. Erdik, Turkish National Committee on Earthquake Engineering, Ankara, 1981.

24. Idriss, I.M., J. Lysmer, R. Hwang, and H.B. Seed, "QUAD-4 : A Computer Program for Evaluating the Seismic Response of Soil Structures by Variable Damping Finite Element Procedures," Earthquake Engineering Research Center Report No. EERC 73-16, University of California, Berkeley, 1973.

25. Joyner, W.B., and A.T.F. Chen, "Calculation of Nonlinear Ground Response in Earthquakes," Bulletin of the Seismological Society of America, Vol. 65, pp. 1315-1336, 1975.

26. Clough, R.W., and A.K. Chopra, "Earthquake Response Analysis of Concrete Dams," in Structural and Geotechnical Mechanics, edited by W.J. Hall, Prentice Hall, 1977.

27. Veletsos, A.S., "Dynamic of Structure-Foundation Systems," in Structural and Geotechnical Mechanics, edited by W.J. Hall, Prentice Hall, 1977.

28. Chopra, A.K., and N.M. Newmark, "Analysis," in Design of Earthquake Resistant Structures, edited by E. Rosenblueth, Wiley, 1980.

WIND LOADS ON BRIDGE STRUCTURES

E. ZELLER

Associate Professor
Department of Civil Engineering - Institut Universitaire
de Technologie - Université STRASBOURG-III (FRANCE)

Introduction

In the history of construction, men have always had to deal with the effects of wind. Any time a structure or construction was unable to resist it, the wind always made a ruin of it. Each time, man was given a lesson by the confrontation of his structural concepts with reality.

Thus for ages, the knowledge of wind effects has remained inseparable from pratical experience. A structure was considered stable if it could resist wind ; thus, in many languages, transversal stability was expressed in a similar fashion... "wind-bracing", "contreventement", "controvento"...

Certainly from the end of the last century, in dealing with certain bold projects (the Eiffel Tower, skyscrapers, the first long span bridges) research had to be done. However there remained a world of difference between knowledge of structural wind effects and the otherwise advanced science of aerodynamics (aviation).

And so, in 1940, when the Tacoma bridge collapsed, other similar bridges had already met the same fate. But in the case of the Tacoma bridge, this occurred during daylight hours. The resonance effect was observable for a long time, and was even recorded on film, as is well known. This served as a starting point in the scientific investigation of the aeroelastic phenomenon.

Nowadays, in constructions or structures whose structural characteristics (dimensions, sections, mass and stiffness distribution, damping...) stay in a certain defined range, national wind codes, although sometimes presenting differences, bring ample security to classic constructions.

In the setting of the European Convention for Constructional Steelwork, a great amount of work has been achieved by some twenty experts from ten nations . This group has just brought together the main strong points of various wind codes, the results of wind-tunnel work and in situ measurement, such as constructive experiments. Because the author has had the privilege of participating in the Technical Committee on Wind Effects since 1968, the analytical approach in the beginning of his report, chapter I, reflects the results of this work-group.

Nowadays, aeroelasticity has made notable strides, but in direct consequence of the bridge sections being made of sharp edges, the position and intensity of shedding vortices have not been revealed by a faithful mathematical model.

That is why in great bridges with low frequencies, wind tunnel study on complete or partial models remains the tool number one for guaranteeing the required serviceability and security in design.

When the construction of a bridge has been decided, the theoretical analysis in relation to the parameters (length, structure type, load, material...) leads inevitably to a solution in which the two following aspects are sufficient to determine the behaviour in the wind :

Point 1 - mechanical characteristics of the bridge : rigidity, frequency, damping...

Point 2 - a geometrical section determining the wind flow conditions

When the corresponding aerolastic model leads us to a structure whose response remains inferior to allowed limit values in all cases of possible excitation, behaviour will be acceptable.

In the opposite case, when excitation conditions (different wind directions, speed, wind turbulence) produce flexural or torsional instability of the bridge, the solution must be modified by changing the frequencies or the section, until the modified model yields a satisfying aerodynamic response.

The author will show the influence that the two above aspects can have on actual bridges in chapter II of this report.

During the construction of a cable stayed bridge "Pont des Meules" Mr. J. WIANECKY completed a wind tunnel study on a partial aeroelastic model in the Laboratories of CEBTP at St Rémy les Chevreuse near Paris. The tests render all wind conditions and made it possible, by independent adjustment of vertical and torsional frequencies and with different damping values to foresee the bridge response and to avoid critical velocity and flutter. The systematic exploration of all parameters with an interesting measuring technique also recorded as a film, represents for the bridge designer a pedagogical tool of the highest order.

On the other hand, it may sometimes act to improve the dynamical behaviour of an already existing bridge. Mr. R.L. Wardlaw had to solve this problem on the ong's Creek cable stayed bridge (chapter III). This time with mechanical characteristics already imposed, the section for the existing bridge had to be modified to reduce oscillations. The investigation of several sections in the wind tunnel made possible a spectacular improvement in bridge's behaviour by changing its profile. The amplitude of the vibrations could be reduced by a factor of more than three to one.

The author wishes to record his gratitude to

- D. Sfintesco - Chairman Technical Committee T 12 on Wind Effects, European Convention for Constructional Steelwork, E.C.C.S.

- H. Van Koten - Director of Dynamic of Structures and Aerodynamic Department, Institute TNO for Building Materials and Building Structure, Delft, The Netherlands

- R.L. Wardlaw - Deputy Head, Low Speed Aerodynamics Laboratory National Research Council of Canada, Ottawa.

- J. Wianecky - Chief Engineer, Division of Aerodynamique du Bâtiment et des Ouvrages d'art, CEBTP, Paris.

for sharing some of their practical knowledge gained in their long and distinguished research careers.

CONTENTS

I. Calculation of wind effects on bridges

1. Wind characteristics, spectrum

- nature of the wind : air movement to balance the atmospheric pressure differences. The wind velocity, as well as the wind direction depend on great number of factors : latitude, Coriolis acceleration, relief, roughness of the ground, etc. The variation in time can only be approximated in statistical terms. For a given point in space corresponding direction and velocity of wind variations are shown in fig. 1.

- statistical distribution : in most countries, statistical data have now been gathered -- in some instances, for almost 100 years. Generally, only the maximum of the wind velocity, obtained each day, week, or month, is recorded. One of the greatest difficulties for an accurate definition of wind velocity for the design derives from the turbulent character of air flow.

- The fluctuation of the wind velocity can be considered as a random process. It can be written in a Fourier series.

Wind velocity is a function of the frequency n of its fluctuations as follows :

$$V = \int_0^{\infty} a(n) \sin (2\pi nt + \theta(n)) \, dn$$

where

$$a(n) = \sqrt{\left[\int_0^T V \cos (2\pi nt) \, dt\right]^2 + \left[\int_0^T V \sin (2\pi nt) \, dt\right]^2}$$

$$\theta(n) = \text{Arc tg} \left[\frac{\int_0^T V \cos (2\pi nt) \, dt}{\int_0^T V \sin (2\pi nt) \, dt}\right] \qquad \text{phase angle}$$

The variance of the wind velocity over a period T long enough in order to have $(\frac{1}{T}\int V \, dt)^2$ negligible with respect to $\frac{1}{T}\int_0^T V^2 \, dt$ can be written :

$$\sigma_V^2 = \frac{1}{T}\int_0^T V^2 \, dt = \int_0^{\infty} \frac{a^2(n)}{2} \, dn = \int_0^{\infty} S_V(n) \, dn = \int_0^{\infty} nS_V(n) \, d(\log_e n)$$

$S_V(n)$ is the spectrum or the power spectral density of the wind velocity at the frequency n

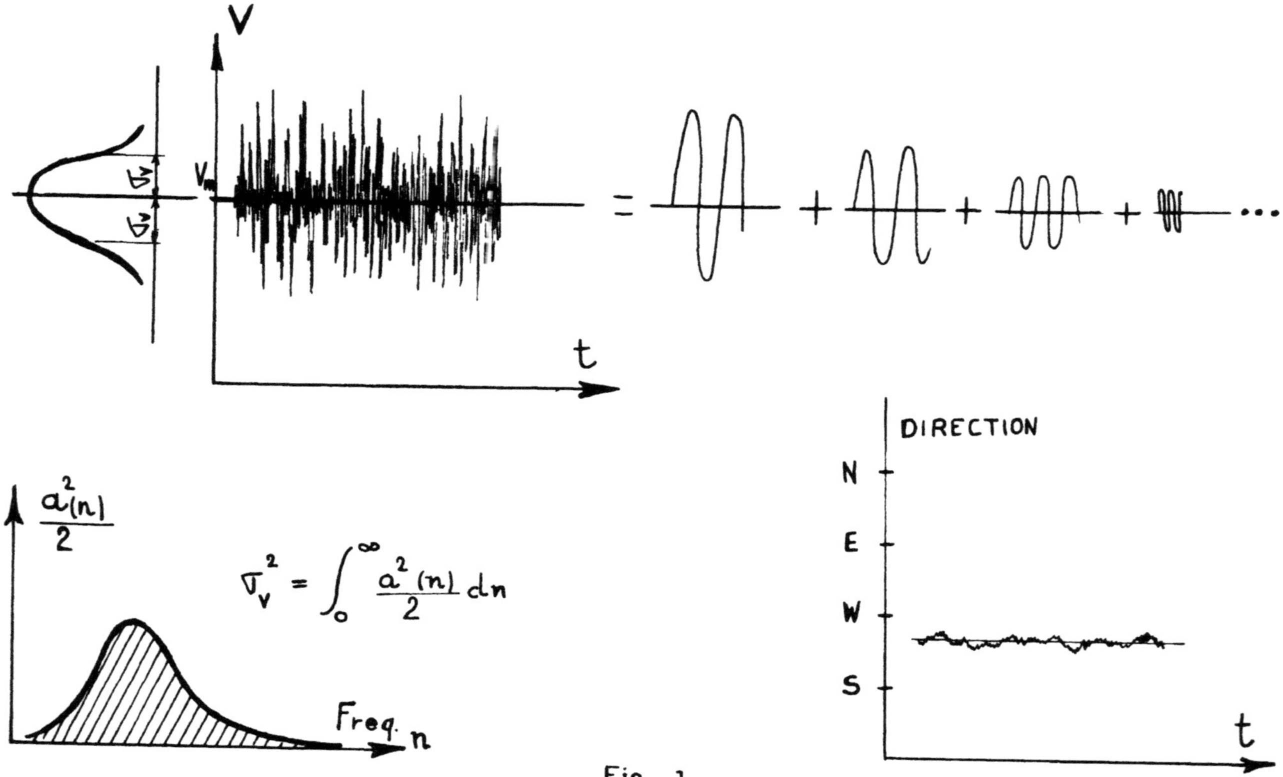
V
V_m
σ_v
σ_v
t
=
+
+
+
...
$\frac{a^2(n)}{2}$
$\sigma_v^2 = \int_0^\infty \frac{a^2(n)}{2} dn$
Freq. n
DIRECTION
N
E
W
S
t

Fig. 1

The analysis of the spectrum of the wind velocity shows that the wind energy presents some maximum values :

- the first peak at a period of 4 days corresponds to the transit time of fully developed weather systems (depressions),
- the second peak at a 1/2 day period corresponds to diurnal heating and cooling,
- the third peak at a period of 1 minute is due to the atmospheric turbulence.

The wind spectrum measured by DAVENPORT and VAN DEN HOVEN is nowadays unanimously accepted.

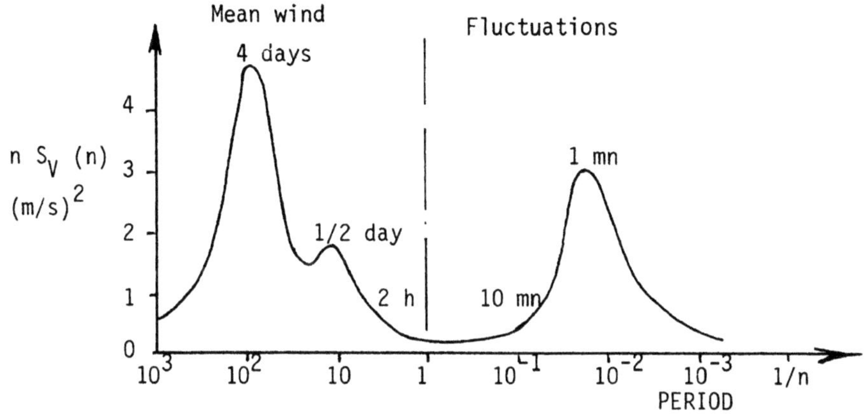

Fig. 2
Wind spectrum measured by van den Hoven

In the aerodynamicsof structures the velocity fluctuation of turbulent wind need to choose a base of time T of the order of minutes and then the expression of wind speed becomes :
$V = V_m + V_f$

where

V_m is the mean wind speed defined over a period of 10 minutes :

$$V_m = \frac{1}{T}\int_0^T V\, dt$$

V_f is the fluctuation of the wind speed about its mean value V_m due to the turbulence of the atmosphere.

2. Design wind speed

V_m at high Z during t years $V_{m,t,z}$ measured over a long time (years) show that the distribution of the extreme values of the mean wind speed can be approximated in terms of a low of "FISCHER-TIPET type I", statistical distribution :

- the mean speed at the height z occuring on an average once during t years that is of return period t years :

$$V_{m,t,z} = U - \frac{1}{a} \log_e \left[- \log_e \left(1 - \frac{1}{t}\right) \right]$$

The parameters U and a define the statistical distribution of the speed V_m , and depend on the considered country.

By considering the probability P that the mean wind speed occurs at the height Z during N years, this can also be written :

$$V_{m,t,z} = U - \frac{1}{a} \log_e \left[- \log_e (1 - P)^{1/N} \right]$$

For the calculation of the mean speed $V_{m,t,z}$ of return period t one starts from the mean speed $V_{m,50,z}$ of return period t = 50 years and applies the relationship $V_{m,t,z} = k_t V_{m,50,z}$ with k_t

$$k_t = \frac{1}{1,507} \left[1 - 0,13 \log_e \left(- \log_e \left(1 - \frac{1}{t}\right)\right)\right]$$

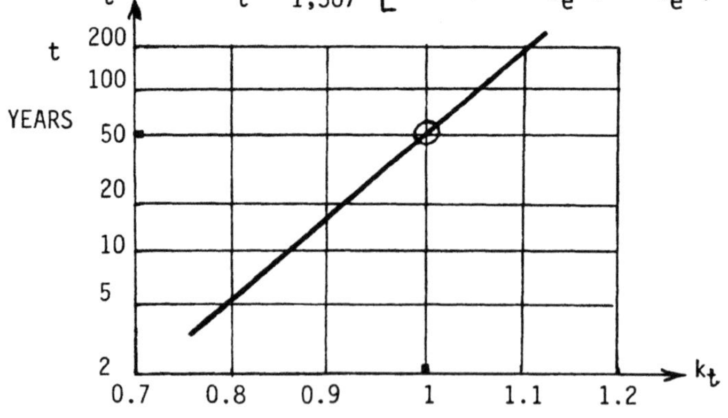

Fig. 3

Wind direction :

As a general rule, the design speed $V_{m,t,z}$ is determined taking no account of direction. But in all cases the wind is quasi-parallel to the ground surface.

Turbulence intensity :

The statistical analysis of the anemometer records shows that the standard deviation σ_{V_f} of the speed fluctuation is proportional to the mean velocity $V_{m,t,z}$:

$$\sigma_{V_f} = V_{m,t,z} \, I \qquad I = \frac{\sigma_{V_f}}{V_{m,t,z}}$$

I is the intensity of the turbulence and is a function of terrain roughness and of the height above ground, but independent of the long-term variations of wind speed.

The spectrum of the speed fluctuation can also be defined as a function of the mean speed $V_{m,t,10}$, of the variance $\sigma^2_{V_f}$ and of the frequency n by the following empirical formula :

$$\frac{n \, S_{V_f}(n)}{\sigma^2_{V_f}} = \frac{0,6 \, x}{(2 + x^2)^{5/6}}$$

where

$S_{V_f}(n)$ = power spectral density of wind at frequency n

$$x = \frac{1800 \, n}{V_{m,t,10}}$$

$V_{m,t,10}$ as defined hereafter.

Influence with height

The mean wind speed $V_{m,t,z}$ increases with height according to some empirical laws up to the height of gradient z_G where this increase shows a point of inflexion. The wind speed at this point is called the gradient speed V_G ; it is independent on the terrain roughness.

The gradient height z_G defines the thickness of the boundary layer;it varies from 250 to 600 m according to the terrain roughness.

Below gradient height, the mean wind speed can be defined by following relationship

$$\frac{V_{m,t,z}}{V^{II}_{m,t,10}} = k \log_e \frac{z}{z_0}$$

where

$V^{II}_{m,t,10}$ basic speed, i.e. mean speed of return period t at 10 m above the ground in zone II

k, z_0, b and α parameters depending on the terrain roughness, values of which are given in the table below.

Bridges are mostly constructed near the ground surface but there are exceptions in which the height over the level 0 can be important.

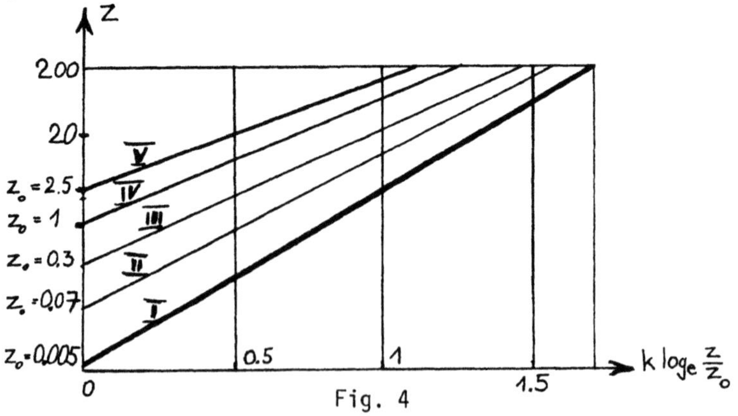

Fig. 4

<u>Remark</u> :

$V_{m,t,z}$ does not take into account some phenomena such as tornadoes, local wind, local topographic effect (escarpments, hills, valleys).

The vertical gradient of wind velocity depends on the terrain roughness of the five conventional roughness classes, defined as:

I sea or lake shore with more than 5 km of fetch over the water

II flat terrain with some isolated obstacles (trees, buildings,...)

III rural areas with low buildings, trees, hedges

IV urban, industrial or forest areas

V centres of large cities

Roughness class		I	II	III	IV	V
Logarithmic law	z_0 (m)	0,005	0,05	0,3	1	2,5
	k	0,16	0,19	0,23	0,26	0,29

The turbulence intensity I is a function of height of the air above the ground and of roughness. I is given by the following relation

$$I = \frac{K}{k \log_e \frac{z}{z_0}}$$ in which the parameter K

K is dependent on the terrain roughness class and has the following values :

	I	II	III	IV	V
K	0,16	0,19	0,22	0,23	0,23

Spatial correlation of the speed fluctuation

The determination of the wind pressure acting on a bridge can not only be based on the mean wind speed. In fact the wind speed V_f at a given moment varies from one point to another in a vertical as well as in a horizontal direction.

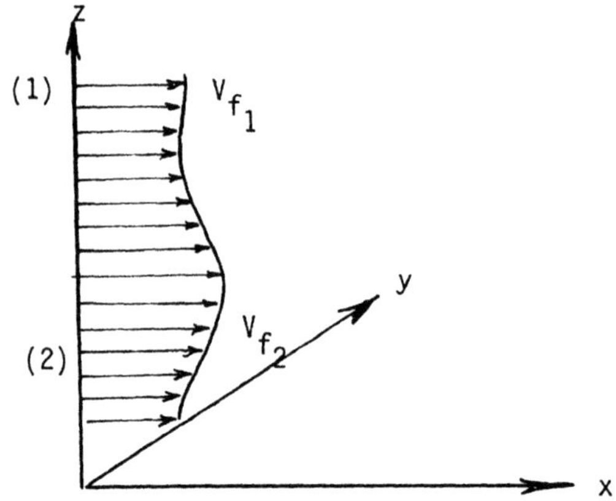

Fig. 5

This variation can be described by a correlation function ρ expressing the simultaneity of occurrence of gusts at two points (1 and 2) :

$$\rho = \frac{\frac{1}{T}\int_o^T V_{f_1} V_{f_2} \, dt}{\sigma_{V_{f_1}} \sigma_{V_{f_2}}}$$

It is given by the following empirical relationships :

$$\rho_z = e^{-\frac{\Delta_z}{L_z}}$$ in the vertical direction (application to bridge towers)

$$\rho_y = e^{-\frac{\Delta_y}{L_y}}$$ in the horizontal direction (application to bridge spans)

L_z and L_y being the scale parameters of the correlation and

Δ_z and Δ_y being the distances which separate the two particular points.

The parameters L_z and L_y are defined by the empirical relationship

$$L_z = \sqrt{37\ z} \qquad (z \text{ in m})$$

$$L_y = 42\ (\frac{z}{20})^{0,25}$$

The passage from the wind V speed to the dynamic pressure q is linked by the relation

$$q = \frac{1}{2} \rho\ V^2$$

with q expressed in N/m^2, V in m/s, ρ air density (at 15°C and 1 bar) = 1,225 kg/m^3.

This permits the equivalence of the statistical wind speed with the statistical wind pressure $w = w_m + w_f$ with

w_m = mean pressure over 10 minutes

w_f = pressure fluctuation about the mean value w_m.

3. Wind effect on bridges

A bridge is a flexible structure and the dynamic action must be taken into account. The developed pressure on a bridge shows fluctuations which depend:

- on the incident wind characteristics which affect the free stream incident flow (turbulence, gust).

- on the air flow around the bridge : turbulent shear layer between the air flow and the value, vortices, turbulence...

The resultant pressure fluctuation produces oscillations and vibrations of the bridge - The mechanical response of the bridge shows vibrational modes in bending and torsion :

the resultant movement can modify the action (lift, drag, torsion) of the air flow and the bridge can so be dynamically exited.

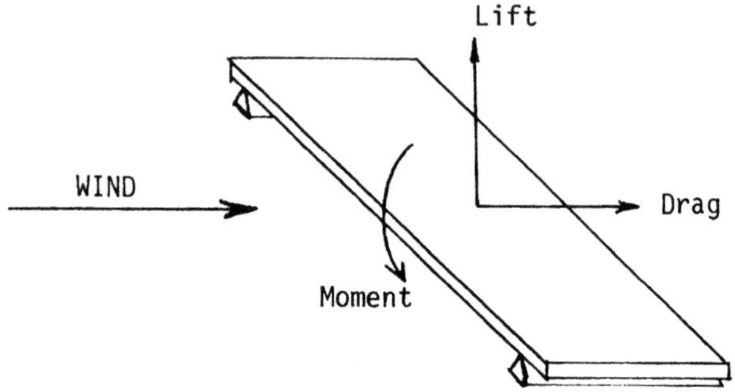

Fig. 6

When a slender structure oscillates in bending perpendicular to the wind, a periodic variation of pressure can be produced and an amplification of vertical oscillation can lead to aerodynamic instability if critical wind speed is reached.

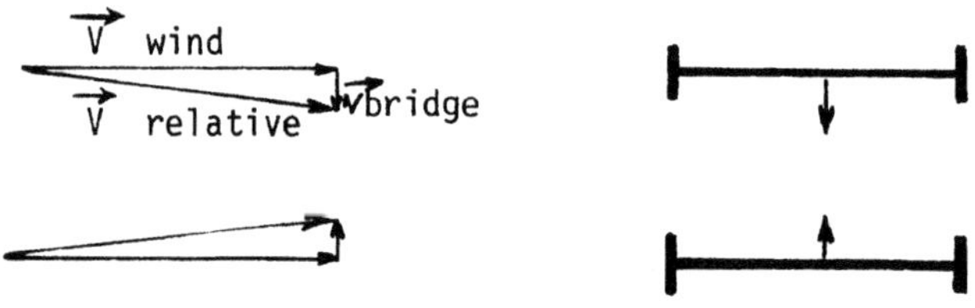

Fig. 7

The torsion of the bridge section modifies the angle of incidence ; the resultant pressure variation can produce instability in torsion.

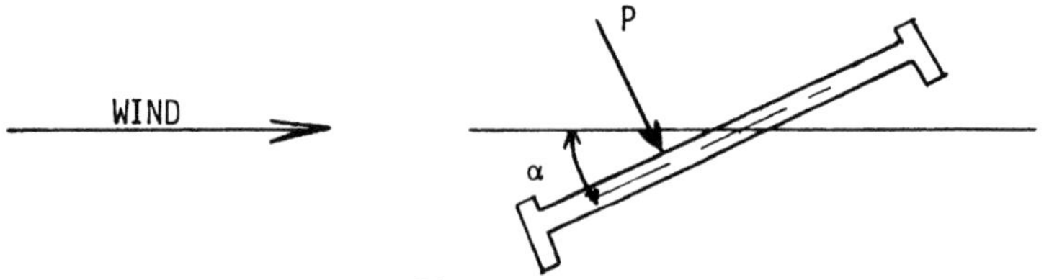

Fig. 8

In addition, when the natural torsional frequency approaches natural bending frequency, the coupling of both movements can produce instability.

The theoretical amplitude of the flutter is not easily obtained because different nondimensional aerodynamic coefficients have to be evaluated experimentally over the corresponding reduced frequency range, and the damping value which can only be estimated, has a great importance.

Scruton proposed a well-known formula for obtaining the bending amplitude at the resonance, based on partial model study in wind tunnel

$$Y_{max} = \frac{\pi}{\delta_K} C_L \frac{\rho V^2}{2} S$$

$C_L \frac{\rho V^2}{2} S$ is non stationary lift action in with C_L is the experimental lift coefficient

δ damping and K vertical rigidity.

By contrast, it will be easier to determine the critical wind speed of the different instability types in order to avoid them.

4. Aerodynamic instability

Schematically let us consider the interaction wind flow-bridge response as follows : Fig. 9.

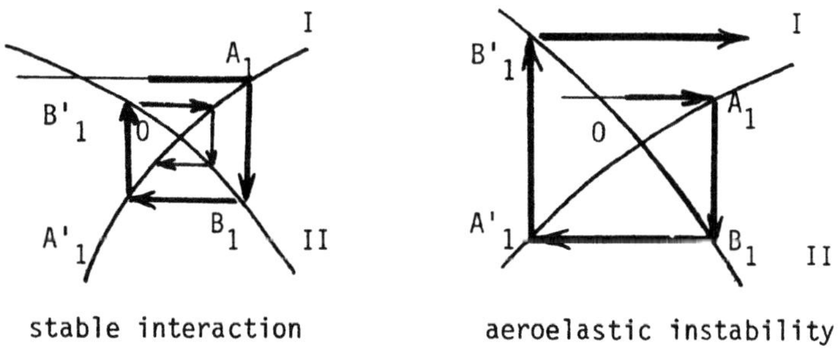

Fig. 9

The wind flow around the bridge section produces the aerodynamic action on the bridge deck. The different possibilities A_i (position, rotation...) are located on curve I. Curve II represents the resultant mechanical response B_i of the bridge (vertical or horizontal bending, torsion) corresponding to the above mentioned action. From an initial equilibrium point 0, a small perturbation can displace the bridge slightly to point A_1. The corresponding modified wind flow reciprocally induces a new mechanical response B_1 which leads the system at position A_1'. The feedback system involved can follow two extreme possibilities :

1° the energy input in the bridge deck at each cycle is dissipated quickly enough so that the amplitude will be limited or diminished: the equilibrium is stable ;

2° the energy input can not be dissipated by the internal damping ; the dynamic amplitude with the corresponding stress in the material grows, and the aerodynamic instability can lead the bridge to destruction.

Instabilities can arise with cross-sections that are aerodynamically unstable. Hovewer the general features of vibration do not basically differ in turbulent wind or in smooth wind ; the transition from turbulence generated buffeting to fluttering can be continuous but in any case the instability occurs only if a critical wind speed is reached. The determination of the critical wind velocity is of highest importance for the bridge designer.

Oscillation in pure first mode bending :

This must be taken into consideration for a rectangular or polygonal structure if

$$V_c \leqslant V_{design}$$

for a constant cross section

$$V_c = \frac{1,63\ \delta\ n_1\ m}{\sigma \times b}$$

with n_1 = first mode bending frequency

m = mass per unit length (kg/m)

b = width of the structure (m) perpendicular to the wind direction

σ = coef. of aerodynamic stability

Example

σ =

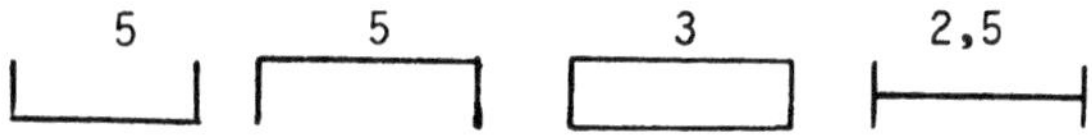

δ = logarithmic decrement $= \frac{2\pi\ \xi}{\sqrt{1-\xi}} \sim 2\pi\xi$ if ξ small

(ξ = % of critical damping)

the damping $\delta = \delta_s + \delta_a + \delta_d$

. structural damping δ_s

depends on material, type of construction, connection, effect of the prestressed concrete elements, bearings...

mostly $0.01 < \delta_s < 0.06$

. aerodynamical damping δ_a

is generally negligible except for cable or very slender elements or bridge decks

$$\delta_a = \frac{Q_w}{n_1\ mV_{design}}$$

Q_w = total force produced by the design wind

m = oscillating mass (kg)

. logarithmic decrement δ_d due to possible additional damping devices.

- oscillation in pure torsion :

the critical wind speed for pure torsion can be approximated with the same calculation as V_c for coupled oscillations bending -torsion.

- oscillation in bending torsion :

the critical wind speed V_c is given by the formula :

$$V_c = 3{,}14\ \eta\ n_{1\,flex}\ d\left[1 + \left(\frac{n_{1\ tors}}{n_{1\ flex}} \cdot 0{,}5\right)\sqrt{\frac{1{,}5\ \sqrt{mI}}{d^3}}\right]$$

η : the value is function of the section and is given in the figure n° 10

d : along wind dimension of the section

m : mass per unit length 1

I : torsional stiffness factor per unit length

η is a empirical coefficient equal to the ratio between the critical speed for the bridge section under consideration and the critical speed for a flat plate

$$\left(\eta = \frac{V_c^{\overline{\mathrm{TT}}}}{V_c^{-}}\right)$$

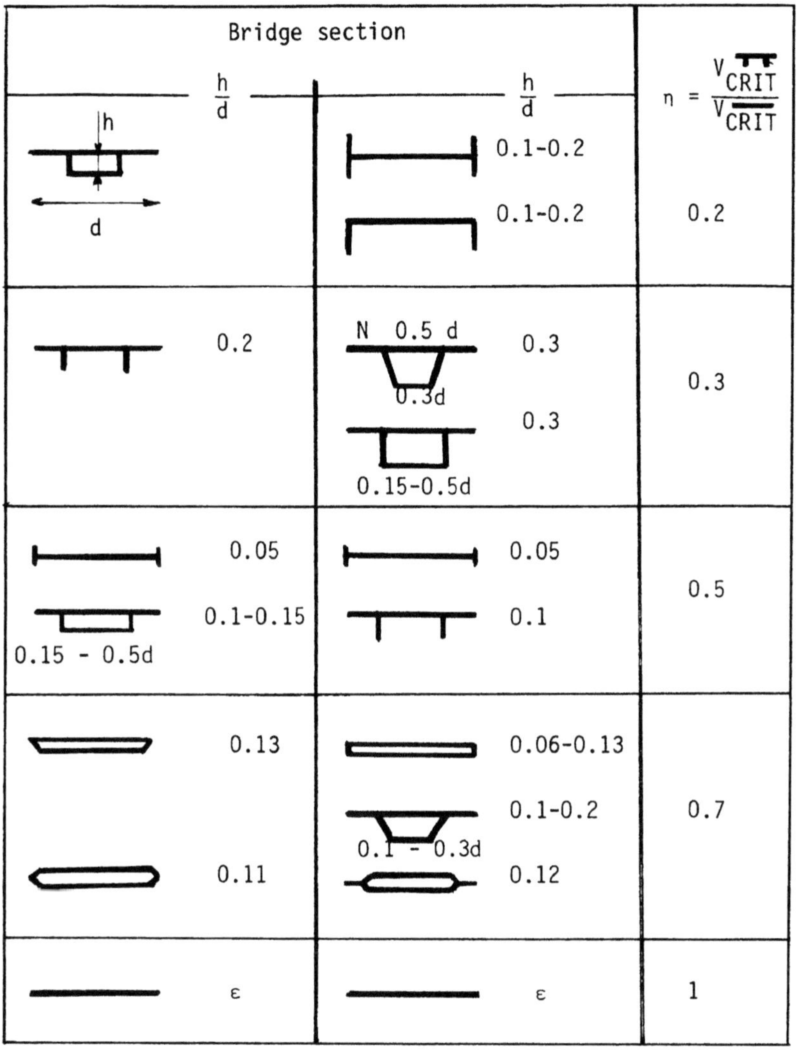

Bridge section
h/d
h/d
η = V CRIT / V CRIT
h
d
0.1-0.2
0.1-0.2
0.2
0.2
N 0.5 d
0.3
0.3d
0.3
0.15-0.5d
0.3
0.05
0.1-0.15
0.15 - 0.5d
0.05
0.1
0.5
0.13
0.06-0.13
0.1-0.2
0.1 - 0.3d
0.11
0.12
0.7
ε
ε
1

Fig. 10

Remark 1 : aerodynamic instabilities should be avoided. This is not always possible in the first stage of design. If $V_c \leqslant V_{design}$ it is necessary to check by a more detailed calculation of the critical speed ; or to check the behaviour from an aeroelastic model in a windtunnel.

Remark 2 : In case of instability for given bending and torsional frequencies the damping is the mechanical characteristic number one which can limit the amplitude.

Remark 3 : This formulation of V_c is approximate but for the designer this formulation allows him to define simply structures which are certainly not sensitive to tne instability phenomenon.

II. Wind tunnel approach :

1. Dynamic simulation on sectional model

Wind tunnel tests can be made on full rigid models which reproduce entire bridges in a given geometrical scale. Now, more and more, sectional models mounted on springs allow the complete simulation of the dynamic behaviour. The vertical frequency and the torsional frequency can be chosen separately and all conditions of wind excitation can be systematically explored in order to establish that the mechanical response will not exceed fixed limits of acceptability (amplitude, stress...). But the extrapolation from the wind tunnel response to the actual bridge is only assured if certain conditions are respected.

- all scaling parameters must be observed : Froude number $\frac{V^2}{dg}$ (V = wind speed, d bridge width , g gravity acceleration) Reynolds number, etc...

- the characteristics of the actual wind flow which act on the bridge must be introduced in the wind tunnel : properties of the wind layer, turbulence intensity, scale of the turbulence relative to the size of the bridge span, angle of incidence of the wind (horizontal, vertical)

- the buffeting response must be checked, the energy of turbulence can be important compared with the mass of the bridge deck ; the buffeting can affect the stress level and the number of cycles

- Flutter and vortices : for turbulent as well as for uniform flow, critical wind speed should be above the expected wind speed in the future site.

The experimental determination of the aerodynamic forces and moments must be completed with a full dynamic simulation. The energy input in the successive cycles increases with the amplitude of vibration (Fig. 11). As the steady state deflection is a function of the mechanical damping, a small variation of the introduced value can modify considerably the stored mechanical energy and the resultant amplitude and, furthermore, the critical velocity. This is why the wind tunnel studies have to be made with different values of the damping. The cable stayed bridge "Pont des Meules" was studied with a partial aeroelastic model ; a special electrical device allowed precise control of bending and torsion damping (Fig. 12).

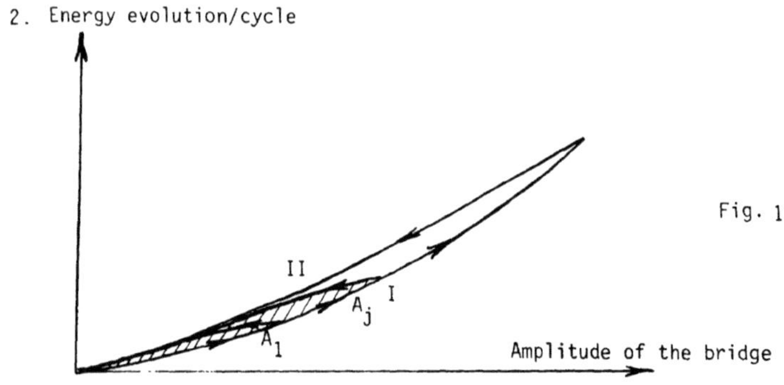

Fig. 1i

- curve I : elastoplastic energy of deformation in the bridge
- (hatched area): successive energy input due Σ(wind effect - damping effect)
- curve II : total energy stored in the bridge after a cycle
  A_1 , A_j amplitude corresponding to the successive incremental energy input

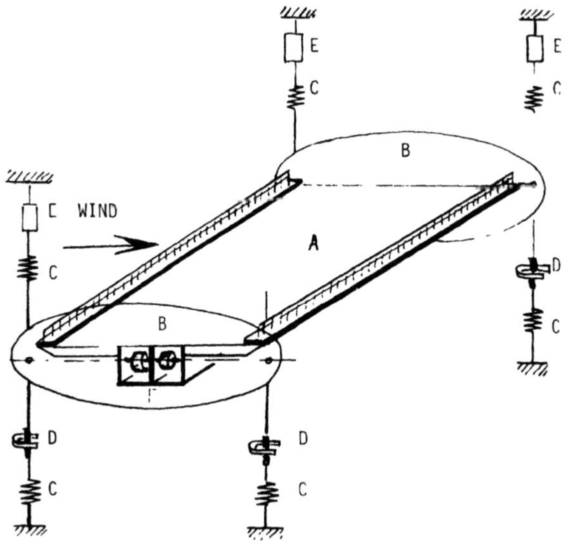

Fig. 12

A partial aeroelastic model (130 x 65 cm)
B end plate of the model
C spring for suspension (torsion, translation)
D electrical damping device with control
E strain gage
F mechanical device for torsional inertia control

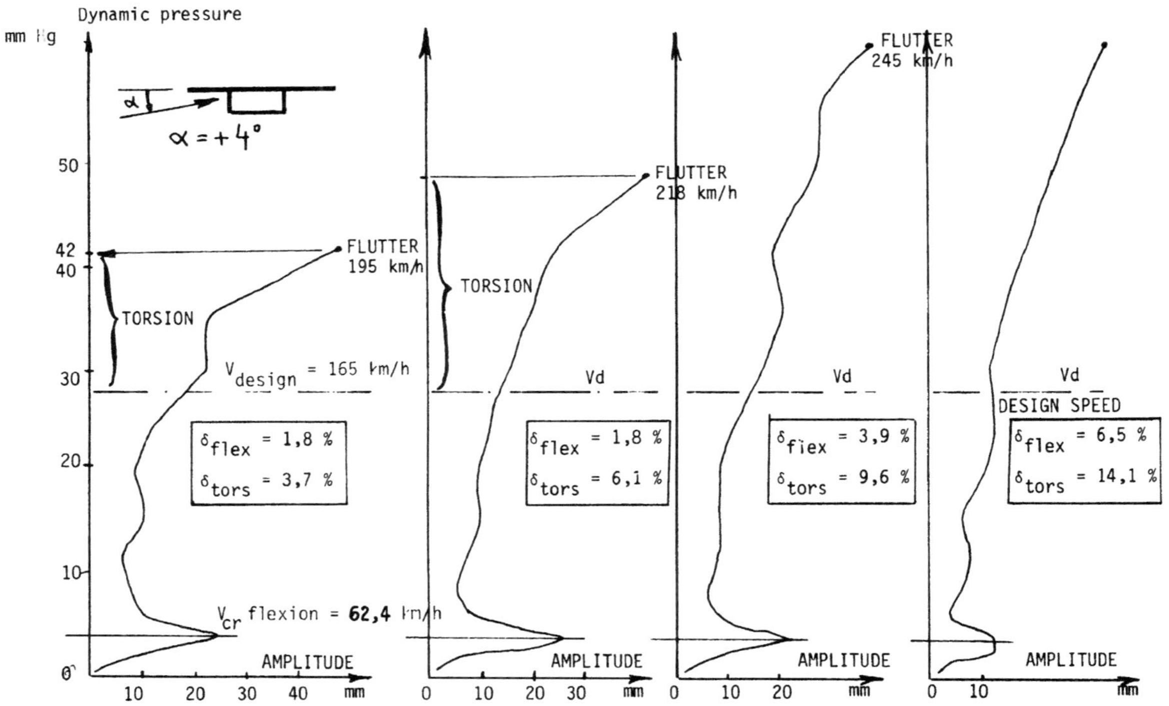

Fig. 13 - Damping and Amplitude of the aeroelastic model (for steel solution)

3. Damping effect of amplitude, example

Increased values of damping in flexion and torsion reduce for a given wind speed the response amplitude, and the critical velocity for flutter increases markedly over the design speed (Fig. 13).

If the bridge deck behaviour has been tested and if the dynamic behaviour is acceptable, some elements must also be checked:

- example : the vibration of the cables or the stays.

When the cable stayed bridge "Pont des Meules" (100 km from Paris) was under construction, a storm struck in Oct. 1976. As the wind speed reached 15 m/sec blowing with a incidence of 20-30° with the longitudinal bridge axis, the vibration amplitude of a 100 m cable was between ± 30 cm for the first mode; after several hours, the storm abated ; with 15 m/sec the cable reached again the critical resonance for 30 minutes.

A tunnel experiment showed that the 165 mm diameter cable became a sufficient damping if the surrounding duct was not injected and for the same critical wind speed the amplitude remained insignificant.

- the behaviour of the mast must also be checked.

III. Improvements of dynamic behaviour of existing bridges

1. Aerodynamic shape modification

The bridge design needs a good choice of mechanical and aerodynamic data in order to minimize the risk of dynamic excitation and instabilities. Howewer in some cases, in spite of a rational selecting of the technical data some initial conditions may change ; and the aerodynamic stability of a bridge deck can become unacceptable. Solutions must be found to reduce dynamic response of the existing bridge.

Aerodynamic shape modification :

A small modification of the section can be the best solution for improving the aerodynamic stability. For example the Long's Creek bridge

- characteristics : cable stayed orthotropic girder bridge
  main span of 217 m,
  structural steel weight 4 270 kg/m

- vibration in vertical flexion has been produced after opening the bridge at a velocity of ~ 40-50 km/hr
  frequency in vertical bending $N_z = 0.59$ HZ
  vertical acceleration 0.3 g unacceptable
  amplitude 5-10 cm

As the handrail was blocked by snow the amplitude reached an excessive value : ~ 20 cm.

Mr R.L. Wardlaw had to find a solution to improve the bridge response without making structural modification. The dimensionless frequency range was between 0.11 and 0.13 ($\frac{N_z \cdot h}{V}$, h girder depth = 2.5 m), and the expected Strouhal number : 0.13 ; the large amplitude was only produced in the speed range 40-50 km/hr : excitation by vortex shedding. A corrective modification of the bridge section had to be found in wind tunnel study but, for safety reasons, a temporary solution had to be applied in order to reduce the amplitude.

2. Additional damping device

As was shown in chapter II (for the Pont des Meules) an increased damping value reduced the amplitude. The measured damping log. decrement by a vertical impulse method reached 0.065, was not sufficient. An interesting and simple additional damping system has been used. A load consisting of rocks was put in a box (2 x 2 x 0.5 m) and suspended in the water 30 m below the bridge. Fig. 14.

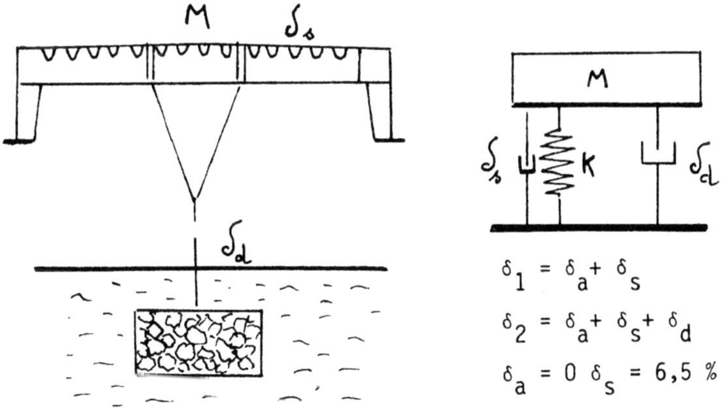

$\delta_1 = \delta_a + \delta_s$

$\delta_2 = \delta_a + \delta_s + \delta_d$

$\delta_a = 0 \quad \delta_s = 6{,}5\ \%$

Fig. 14

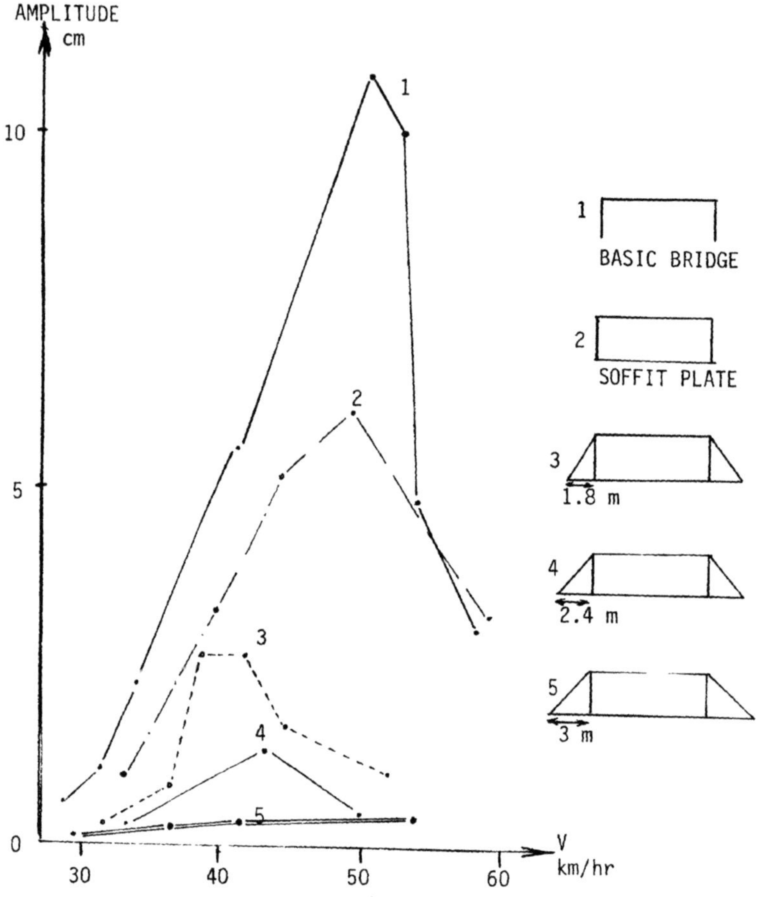
AMPLITUDE
cm
10
5
0
30
40
50
60
V
km/hr
1
2
3
4
5
1
BASIC BRIDGE
2
SOFFIT PLATE
3
1.8 m
4
2.4 m
5
3 m

Fig. 15

The initial damping δ_1 became increased with the temporary damping device δ_d . With δ_2 the pseudo periodical movement had now an amplitude $\simeq$ 3 cm .

Definitive solution :

The investigation on a model in a wind tunnel gave clearly the same mechanical behaviour as observed on the bridge. After exploring different shapes for the bridge as well as the use of perfored girders (permeability modifies wind flow favorably) the definitive solution consisted in a simple external modification of the bridge deck : adding soffit plating. The effectiveness of different dimensions of fairing is illustrated in Figure 15.

Importance of small details :

Furthermore, such insignificant details for a bridge designer, such as handrails, can considerably modify the amplitude of the response when blocked by snow. The sectional model response of the unmodified Long's Creek bridge allowed to reproduce this clearly. Fig. 16.

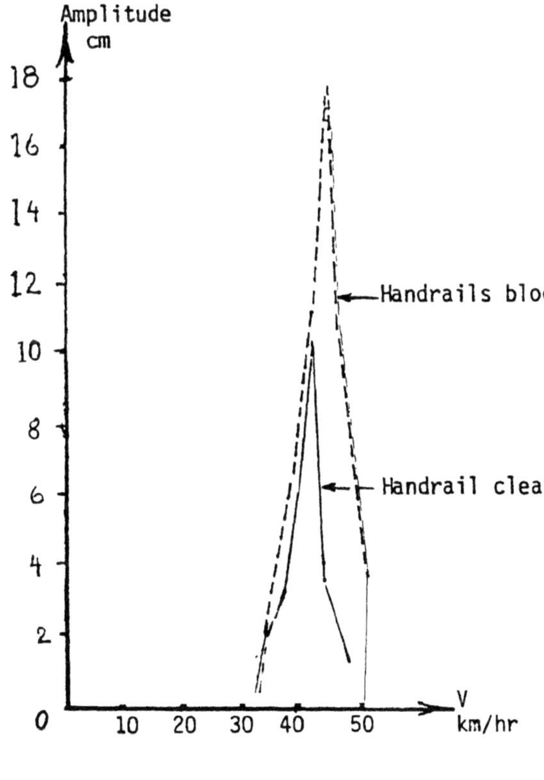

Fig. 16

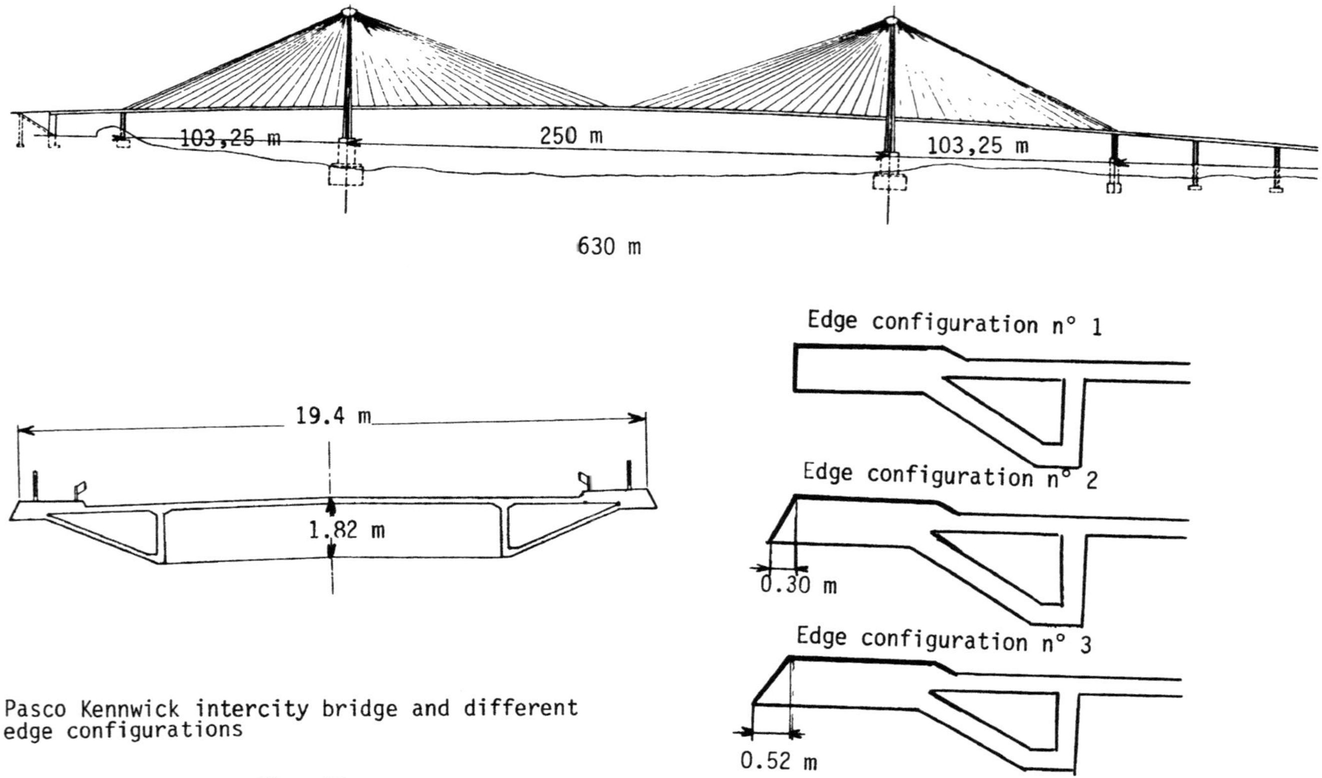

Pasco Kennwick intercity bridge and different edge configurations

Fig. 17

3. Example : bridge section and dynamic response

It is useful to note how a small change on the external bridge section can modify the wind flow and consequently the vortex shedding and flutter velocities. For example different edge configurations showed in Fig. 17 have been examined by Mr R.L. Wardlaw for the Pasco-Kennewick intercity bridge. In spite of small damping values used in the test $\delta_{flex} = 1{,}9\ \%$ and $\delta_{tors} = 0{,}8\ \%$, and with the largest incidence angles ($\alpha = \pm\ 8°$) and highest wind to be expected in the site, a satisfactory response could be obtained with configuration n° 3. Figure 18 gives the critical wind speeds (test without groundboard), with the maximum amplitude in bending and torsion, and the vortex shedding range. The torsion values in the table 18 are given by the vertical displacement at bridge edge.

CONF. NUMBER	WIND ANGLE	VORTEX SHEDDING				FLUTTER VELOCITY, MPH	
		VELOCITY, MPH		MAXIMUM AMPLITUDE, FT		VELOCITY INCREASING	VELOCITY DECREASING
	α	LOWER LIMIT	UPPER LIMIT	TORSION*	BENDING		
1	+2					179	171
	+4					135	122
	+6	29.0	43.7		0.409	100	70
	+8	25.2	41.7	0.101	0.469	77	60
	-6					157	148
	-8					122	106
2	+4						147
	+6	26.4	37.3	0.261			110
	+8	24.5	51.1	0.139	1.24		80.2
	-6					175	152
	-8	26.3	44.3		0.438	111	102
3	+4					155	153
	+6	26.3	45.7		0.335	132	126
	+8	26.3	41.1		0.468	99	88
	-6					134	125
	-8					124	117

Fig. 18

IV. Security for Wind Bridge Design

1. Design wind speed and structural safety

Meteorological wind data : velocity over 10 minutes, turbulence intensity, return period can be properly used to determine the design wind speed. The dynamic response is distributed around the mean pressure response ; the range of the response data is a function of the turbulence spectrum, the aerodynamic admittance, the mechanical admittance of the structure and the joint acceptance. Aeroelastic instabilities of the bridge section, in vertical bending as in torsional mode, must be checked with all actual possibilities of incidence angle, velocities and turbulence intensity. The resultant amplitude in all cases must stay below fixed limits. Fig. 19.

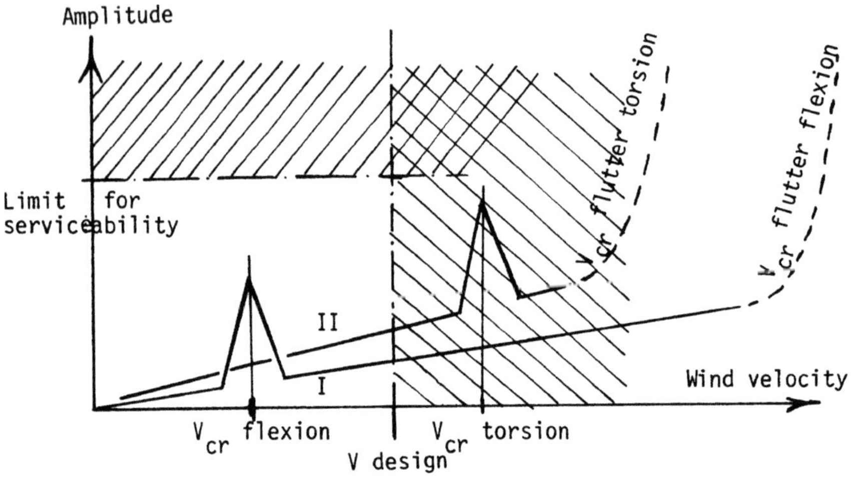

Fig. 19

For the bridge "des Meules", curve I shows schematically the solution of flexural and curve II that of torsional movement with the wind speed. In fact, each response type is a combination of response I and II. In order to improve structural safety, the range V_{cr} flexion V_{cr} torsion has to be large ; bending and torsion or coupled flutter have to be situated largely over the design speed.

Without structural modification, the two examples in chapter III illustrate how with only small changes in external shape, the dynamical response of a bridge section can be optimized in order to diminish the amplitude.

2. Damping, critical value

In the successive dynamic oscillation, the wind energy input tends to increase the elastical energy, therefore the amplitude. In bending, the bridge model in wind tunnel (chapter II) showed that the amplitude decreases with increase of damping, the intersection with the horizontal axis gives directly the critical damping value, Fig. 20.

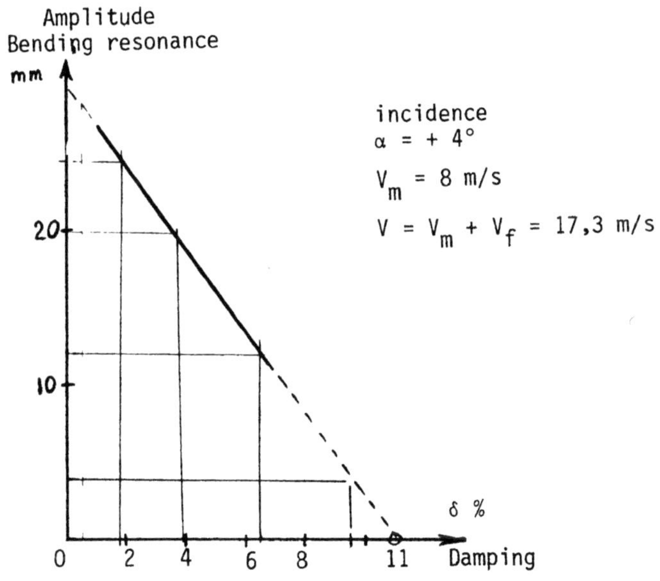

Fig. 20

The benefit obtained by a high degree of damping is important for safety : decreasing of the maximum stress level and reduction of the highstress cycles (fatigue). In mechanical engineering, dampers are frequently used, but in civil engineering, only a small number of solutions have been proposed for bridges.

3. Aerodynamic stability of a bridge under erection

It must also be underlined that wind sensitive bridges are sometimes more vulnerable during the construction stages, which must be studied with special attention. The progressive steps of the bridge under erection must be checked, the vibration modes, the unfinished shape can contribute to diminish the critical velocity : The different stages must be studied with special attention so that method and technological devices allow a safe erection of the bridge.

4. Security criteria of bridge design :

Wind action on bridges is a stochastic process, changing with time. Mostly, only the peak values of the response under different wind conditions are considered so that, generally, in the different wind codes the aerodynamic behaviour of bridges, in spite of statistical wind data, is, in fact, treated in a deterministic way.

Each load (dead and live load, wind...) which acts on a bridge can be defined by a density of probability of the load effect ; the structural resistance can be defined with the density function so that the probability of failure is obtained by convolution. At present time, design criteria for bridges, subjected to different loads avoid the full statistical treatment ; simplified design rules, load combination, the so called load factors and characteristics loads are used.

For important bridges a better approach for security is now developed and based on a more complex but improved analytical process :

The probability for a bridge to present a failure under combined actions of time variable loadings is the probability for the given period to exceed, at minimum, one of the different limit conditions. For a structural system the whole limit conditions together define the safe domain. The design life time in this involved formulation is nothing more than the duration which needs the corresponding vector process to make the first excursion from this domain.

REFERENCES

CHAPTER I

E.C.C.S. - Recommendations for the calculation of wind effects on buildings and structures, 1978.

SCRUTON C., NEWBERRY C. - On the estimation of wind loads for Buildings and structural design, 1963.

SFINTESCO D., ZELLER E. - Prise en compte des effects du vent sur les constructions, 1975.

VAN KOTEN - Windbelasting, TNO, 1974.

WYATT T.A. - A note on the significance of Cross-Modal Correlations on the Evaluation of quasi-static response to wind, 1974.

CHAPTER II

DAVENPORT A.G. - The Use of Taut-Strip Models in the prediction of the response of long span bridges to turbulent Wind, 1972.

KLOPPEL K. - Modellversuche im Windkanal zur Bemessung von Brücken gegen die Gefahr Winderregter Schwingungen, 1967.

WIANECKY J. - Wind excited vibration of cable stayed bridge, 1980.

CHAPTER III

BORGES J.F., CASTANHETA M., BORGES A.R.J. - Design criteria for wind loads on statistical Bases, Proceedings Wind effects on Buildings and Structures, Tokyo, 1971.

WARDLAW R.L. - Approaches to the suppression of wind induced vibrations on structures, 1979.
Laboratory report on the aerodynamic stability of the Pasco-Kennewick intercity bridge, 1974.

LIST OF PARTICIPANTS

Mr. K. Akıllılar (Turkey)

Dr. N. Akkaş (Turkey)

Mr. H. Belkaya (Turkey)

Mr. V. Caby (Belgium)

Dr. J. Church (U.K.)

Mr. C. Debuysscher (Belgium)

Mr. E. Erk'akan (Turkey)

Dr. H.R. Evans (U.K.)

Mr. T. Friedrich (Germany)

Prof.Dr. P. Gülkan (Turkey)

Dr. H. Günerman (Turkey)

Prof.Dr. C.P. Heins (U.S.A.)

Dr. E. Karaesmen (Turkey)

Prof.Dr. I. Kaya (Turkey)

Prof.Dr. M. Kıcıman (Turkey)

Mr. C. Kozanoğlu (Turkey)

Mr. N. Kök (Turkey)

Prof.Dr. C.N. Köstem (U.S.A.)

Dr. A.R.G. Lamas (Portugal)

Prof.Dr. F. Leonhardt (W. Germany)

Mr. S. Loseth (Norway)

Prof.Dr. Y. Mengi (Turkey)

Prof.Dr. C. Menn (Switzerland)

Dr. A.P. Parker (U.K.)

Prof. A.C. Scordelis (U.S.A.)

Dr. İ.İ. Sungur (Turkey)

Dr. T. Tankut (Turkey)

Dr. C. Trezos (Greece)

Prof.Dr. O. Ural (U.S.A.)

Dr. P. Waldron (U.K.)

Prof.Dr. S.T. Wasti (Turkey)

Dr. A. Yargıçoğlu (Turkey)

Prof.Dr. B.T. Yen (U.S.A.)

Prof.Dr. V. Yerlici (Turkey)

Dr. Ç. Yılmaz (Turkey)

Lieut. R. Yılmaz (Turkey)

Prof.Dr. E. Zeller (France)

Dr. H. Sucuoğlu (Turkey)

GPSR Compliance

The European Union's (EU) General Product Safety Regulation (GPSR) is a set of rules that requires consumer products to be safe and our obligations to ensure this.

If you have any concerns about our products, you can contact us on ProductSafety@springernature.com

In case Publisher is established outside the EU, the EU authorized representative is:

Springer Nature Customer Service Center GmbH
Europaplatz 3
69115 Heidelberg, Germany

Batch number: 10815974

Printed by Printforce, the Netherlands